# Mittag-Leffler Euler Difference Techniques

Tianwei Zhang · Yongkun Li · Jianwen Zhou

# Mittag-Leffler Euler Difference Techniques

Springer

Tianwei Zhang
Department of Mathematics
Yunnan University
Kunming, Yunnan, China

Yongkun Li
Department of Mathematics
Yunnan University
Kunming, Yunnan, China

Jianwen Zhou
Department of Mathematics
Yunnan University
Kunming, Yunnan, China

ISBN 978-981-96-6227-2    ISBN 978-981-96-6228-9 (eBook)
https://doi.org/10.1007/978-981-96-6228-9



This Springer imprint is published by the registered company Springer Nature Singapore Pte Ltd.
The registered company address is: 152 Beach Road, #21-01/04 Gateway East, Singapore 189721, Singapore

# Preface

Over the years, different numerical methods have been proposed to solve differential equations. At the same time, different types of differential equations have been generated in areas from mathematics to computer science, engineering and biology. The simplest numerical methods were proposed by Euler in the 18th century. They involved replacing the derivatives of differential equations by difference quotients. In 1883, Bashforth and Adams proposed the use of linear multi-step differences to discuss difference equations, which extended Euler's differences by making sufficient use of the previous multi-step information of the approximate solutions. Around 1900, the use of the weighted average of the slopes at several points as an approximation of the average slope by Runge and Kutta introduced another important type of numerical integrator. In addition, there are other useful numerical methods for studying differential equations. These include Runge-Kutta differences, Taylor series methods, and exponential Euler differences.

The main stimulus for this book is the exponential Euler differences for semi-linear differential equations. Exponential Euler difference methods have recently gained significant attention in research, particularly in the context of variation-of-constants methods. These methods include exponential Euler integrators, exponential Runge-Kutta methods, multi-step exponential integrators, exponential general linear methods, and exponential Rosenbrock-type methods. These numerical methods exhibit perfect linear stability and can be executed efficiently. Adapting exponential Euler integrators to semi-linear ordinary differential equations yields the corresponding exponential Euler differences. Numerous existing literature discusses various dynamical behaviours for a variety of exponential Euler difference models based on this technique. These behaviours include periodicity, almost periodicity, stability, randomness, and control, etc.

The question that arises is whether the exponential Euler differences for semi-linear differential equations of integer order can be applied to the study of semi-linear problems of differential equations of fractional order. This book aims to show that the concept of exponential Euler differences can indeed be applied to studying semi-linear problems involving fractional-order differential equations. The book refers to these differences as Mittag-Leffler Euler differences. These methods are also applicable to the study of other discrete-time models in science and engineering.

The main contributions of this book are summarised below.

(1) The Mittag-Leffler Euler difference is a new numerical method introduced using the variation-of-constants formula for fractional-order differential equations. This method expands on the approach of exponential Euler differences.
(2) In reference to fractional-order differential equations, the most commonly used and effective difference methods are the fractional Euler differences found in existing literature. Compared to other techniques, Mittag-Leffler Euler differences are better suited to depict and retain the dynamic characteristics of corresponding fractional-order differential equations. They also have high implementation efficiency in practical applications.
(3) The book proposes difference schemes and defines various difference operators, including the Euler, fractional Euler, exponential Euler, and Mittag-Leffler Euler difference operators.
(4) Through the use of Mittag-Leffler Euler difference operators, we introduce the Mittag-Leffler Euler difference equations (MLEDEs). This introduces a new avenue for the research of discrete-time fractional-order models.
(5) MLEDEs have variation-of-constants formulas that use continuous Mittag-Leffler functions. These formulas are easier to study and calculate in practical applications. The variation-of-constants formula is important in discussing ordinary differential or difference equations, especially in determining the existence and uniqueness of various solutions. Several variation-of-constants formulas for fractional-order difference models have been proposed, involving discrete Mittag-Leffler functions. However, many properties of discrete Mittag-Leffler functions still require further research. Additionally, some properties of continuous Mittag-Leffler functions may not be applicable to discrete Mittag-Leffler functions. The application of the variation-of-constants formula involving discrete Mittag-Leffler functions is more difficult than that of MLEDEs due to these reasons.
(6) The structure of the exponential Euler differences is designed for differential equations with the Caputo-Fabrizio operator, which falls under the implicit Euler difference category. Subsequently, the fractional PECE algorithm is suggested to solve them. This approach offers a new method to examine discrete-time models for differential equations with Caputo-Fabrizio operator.
(7) This book introduces a new research area in the theories of numerical computations and difference equations. From a numerical computations theory perspective, the exponential and Mittag-Leffler Euler differences presented in this book can be refined into more precise numerical methods, such as exponential Runge-Kutta, exponential multi-step, and exponential Rosenbrock-type differences. The theory of difference equations suggests that it is important to consider the massive dynamical behaviours of various exponential Euler difference models in fields such as science, engineering, biology, economy, and industry. These

behaviours include periodicity, almost periodicity, almost automorphy, stability, randomness, chaos, and control, etc.

The book mainly consists of Chaps. 1–6, which provide frameworks for the theories and applications of exponential and Mittag-Leffler Euler differences to different types of semi-linear (fractional-order) differential equations. Chapter 1 introduces the research background of various numerical methods used to solve differential equations, the Riemann-Liouville and Caputo fractional calculus, and Caputo-Fabrizio operator, etc. Chapter 2 establishes the Delta Mittag-Leffler Euler differences for semi-linear fractional-order differential equations in the sense of Caputo. Chapter 3 discusses the dynamics of the Mittag-Leffler Euler difference model corresponding to a fuzzy fractional-order Cohen-Grossberg neural network. In contrast to Chap. 2, Chap. 4 introduces the Nabla Mittag-Leffler Euler differences to the semi-linear fractional-order differential equations in the sense of Caputo. Additionally, Chap. 5 focuses on the exponential Euler differences to a type of semi-linear differential equations with Caputo-Fabrizio operator. Chapter 6 presents the exponential Euler differences of a fuzzy bidirectional associative memory neural networks with Caputo-Fabrizio operator. This chapter investigates some important dynamical behaviours of the proposed difference model.

This research monograph is intended for undergraduate and graduate students and researchers. It requires elementary knowledge of numerical computations, difference equations, fractional calculus, nonlinear analysis, and control theory, etc.

Kunming, China
February 2024

Tianwei Zhang
Yongkun Li
Jianwen Zhou

# Acknowledgements

This work was supported by the National Natural Science Foundation of China under Grant Nos. 62463032, 12261098, 11861072 and 11961078. Next, this book is derived in part from our recent publications in

1. Applied Mathematics Letters, 2022, copyright © by Elsevier Ltd., available online: https://doi.org/10.1016/j.aml.2021.107709.
2. Results in Physics, 2022, copyright © by the authors, available online: https://doi.org/10.1016/j.rinp.2022.105482.
3. Knowledge-Based Systems, 2022, copyright © by Elsevier Ltd., available online: https://doi.org/10.1016/j.knosys.2022.108675.
4. International Journal of Systems Science, 2022, copyright © by Taylor & Francis, available online: https://doi.org/10.1080/00207721.2022.2051093.

# Contents

# Symbols and Acronyms

## Symbols

| | |
|---|---|
| $\mathbb{R}^n$ | Space of $n$-dimensional real vectors; $\mathbb{R}_0 := [0, +\infty)$ |
| $\mathbb{C}^{m\times n}$ | Space of $m \times n$ complex matrices |
| $\mathbb{Z}$ | Field of integral numbers |
| $\mathbb{N}$ | Field of positive integral numbers |
| $\mathbb{N}_0$ | Field of nonnegative integral numbers |
| $C^m(D, \mathbb{R}^n)$ | Space of all $m$-order continuously differentiable functions from $D$ to $\mathbb{R}^n$ |
| $L^p(D, \mathbb{R}^n)$ | Space of all $p$-integrable functions from $D$ to $\mathbb{R}^n$ |
| $AC^m(D, \mathbb{R}^n)$ | Space of all absolutely continuous functions with derivative of order $m-1$ from $D$ to $\mathbb{R}^n$ |
| $\mathfrak{R}(z)$ | Real part of $z \in \mathbb{C}$ |
| $\arg(z)$ | The principal argument of $z \in \mathbb{C}$ |
| $\mathcal{L}(\cdot)$ | Laplace transform |
| $\mathcal{F}(\cdot)$ | Fourier transform |
| $\mathrm{diag}(a_1, \ldots, a_n)$ | Diagonal matrix with the main diagonal elements $(a_1, \ldots, a_n)$ |
| $\mathbb{I}_n$ | $n$-order identical matrix |
| $\mathbb{O}_{m\times n}$ | Zero matrix in $\mathbb{R}^{m\times n}$; $\mathbb{O}_n := \mathrm{diag}(0, \ldots, 0) \in \mathbb{R}^{n\times n}$ |
| $\mathbf{1}_n$ | A vector in $\mathbb{R}^n$, whose each component equals to 1 |
| $\mathbf{0}_n$ | Zero element in $\mathbb{R}^n$ |
| $I_J$ | Intersection of sets $I$ and $J$ in $\mathbb{R}$ |
| $\|\cdot\|$ | $\|\|A\|\| := \max\limits_{1\le i\le m, 1\le j\le n} \|a_{ij}\|$ for $A = (a_{ij})_{m\times n} \in \mathbb{R}^{m\times n}$ |
| $\|\cdot\|_{\mathrm{r}}$ | $\|\|A\|\|_{\mathrm{r}} := \max\limits_{1\le i\le m} \sum\limits_{j=1}^{n} \|a_{ij}\|$ for $A = (a_{ij})_{m\times n} \in \mathbb{R}^{m\times n}$ |
| $[\cdot]_{\mathrm{abs}}$ | $[A]_{\mathrm{abs}} := (\|a_{ij}\|)_{m\times n}$ for $A = (a_{ij})_{m\times n} \in \mathbb{R}^{m\times n}$ |
| $\|A\|_{\det}$ | Determinant of $A = (a_{ij})_{m\times n} \in \mathbb{R}^{m\times n}$; $[A]_{\mathrm{dia}} := (a_{11}, \ldots, a_{nn})^T$ and $\langle A\rangle_i := a_{ii},\ i = 1, 2, \ldots, n$ |

## Acronyms

| | |
|---|---|
| (E)ED(s) | (Exponential) Euler difference(s) |
| (E)EDEs | (Exponential) Euler difference equations |
| (E)EDO | (Exponential) Euler difference operator |
| (P)MLED(s) | (Piecewise) Mittag-Leffler Euler difference(s) |
| (P)MLEDEs | (Piecewise) Mittag-Leffler Euler difference equations |
| (P)MLEDO | (Piecewise) Mittag-Leffler Euler difference operator |
| (P)FED(s) | (Piecewise) Fractional Euler difference(s) |
| (P)FEDEs | (Piecewise) Fractional Euler difference equations |
| (P)FEDO | (Piecewise) Fractional Euler difference operator |
| BAMNNs | Bidirectional associative memory neural networks |
| CF(OEs) | Caputo-Fabrizio (operator equations) |
| CGNNs | Cohen-Grossberg neural networks |
| FODCEs | Fractional-order difference equations |
| FODEs | Fractional-order differential equations |
| PECE | Predict-Evaluate-Correct-Evaluate |

# Chapter 1
# Introduction

This chapter introduces the research background of various numerical methods used to solve differential equations, including Euler difference, Runge–Kutta difference, exponential Euler difference, etc. Furthermore, this chapter introduces the concepts and properties of Riemann–Liouville and Caputo fractional calculus, Caputo–Fabrizio operator, etc.

## 1.1 Euler Difference

In 18th century, Euler proposed the simplest numerical methods (i.e., Euler differences) to produce difference equations by replacing the derivatives of differential equations with the difference quotients.

In engineering applications, it usually solves the differential equations

$$\begin{cases} \dfrac{\mathrm{d}x(t)}{\mathrm{d}t} = f(t, x(t)), \\ x(0) = \xi_0 \in \mathbb{R}, \end{cases} \tag{1.1}$$

where $f$ is continuously smooth and satisfies the Lipschitz condition, $\xi_0 \in \mathbb{R}$ denotes the arbitrary initial value. The existence and uniqueness of the solution x to the Cauchy problem (1.1) can be theoretically proved; see book [1].

The Euler difference seeks approximate values $x_1, x_2, \ldots, x_k, x_{k+1}, \ldots$ of solution $x$ at a series of discrete nodes $0 = t_0 < t_1 < t_2 < \cdots < t_k < t_{k+1} < \cdots$. In this case, the distance between two adjacent nodes is called step length, that is, $h = t_{k+1} - t_k$ for all $k = 0, 1, 2, \ldots$. Unless otherwise stated in this chapter, $h$ is always assumed to be constant and the node is $x_k = x(t_k)$, $t_k = kh$ for all



T. Zhang et al., *Mittag-Leffler Euler Difference Techniques*,
https://doi.org/10.1007/978-981-96-6228-9_1

$k = 0, 1, 2, \ldots$. In Eq. (1.1), replacing $\frac{dx(t)}{dt}$ and $f(t, x(t))$ by $\frac{x_{k+1}-x_k}{t_{k+1}-t_k}$ and $f(t_k, x_k)$, respectively, we get the following relationship

$$\frac{x_{k+1} - x_k}{t_{k+1} - t_k} = f(t_k, x_k), \quad k = 0, 1, 2, \ldots.$$

Transforming the above equation, it obtains the Euler difference formula

$$x_{k+1} = x_k + hf(t_k, x_k), \quad k = 0, 1, 2, \ldots.$$

The derivative of Eq. (1.1) at point $x_{k+1}$ is depicted by

$$\left.\frac{dx(t)}{dt}\right|_{t=t_{k+1}} = f(t_{k+1}, x_{k+1}), \quad k = 0, 1, 2, \ldots. \tag{1.2}$$

With backward difference quotient $\frac{x_{k+1}-x_k}{h}$ instead of the derivative $\left.\frac{dx(t)}{dt}\right|_{t=t_{k+1}}$, the discretization equation of Eq. (1.2) is given by

$$\frac{x_{k+1} - x_k}{h} = f(t_{k+1}, x_{k+1}), \quad k = 0, 1, 2, \ldots.$$

Then the implicit Euler difference formula is depicted by

$$x_{k+1} = x_k + hf(t_{k+1}, x_{k+1}), \quad k = 0, 1, 2, \ldots. \tag{1.3}$$

In order to solving implicit difference form (1.3), the simplified predicted and corrected approaches are proposed by

$$\begin{cases} \tilde{x}_{k+1} = x_k + hf(t_k, x_k), \\ x_{k+1} = x_k + f(t_{k+1}, \tilde{x}_{k+1}), \\ x_0 = \xi_0 \in \mathbb{R}, \end{cases}$$

where $\tilde{x}_{k+1}$ is called the predicted value, $k = 0, 1, 2, \ldots$.

Currently, the Euler difference method is a crucial tool for studying applied discrete-time models in the fields of science and engineering. For more information, please refer to the literature [2–14]. In addition to Euler differences, there are other precise numerical methods that can be used to discuss differential equations. These include linear multi-step differences, Runge–Kutta differences, Taylor series methods, and multi-value multi-stage differences. For further details, please consult book [15].

## 1.2 Exponential Euler Difference

Recently, there has been significant research on a variety of exponential Euler difference methods, such as exponential Euler integrators [16–21], exponential Runge–Kutta methods [22], multi-step exponential integrators [23], exponential general linear methods [24], and exponential Rosenbrock-type methods [25], due to the rapid development of exponential Euler differences in light of the variation-of-constants formula. These numerical methods exhibit perfect linear stability and can be implemented effectively.

Concretely, consider semi-linear differential equations as described below

$$\begin{cases} \dfrac{\mathrm{d}X(t)}{\mathrm{d}t} = AX(t) + F(t, X(t)), \quad \forall t \in (0, +\infty), \\ X(0) = \xi \in \mathbb{R}^n. \end{cases} \tag{1.4}$$

Here, $X = (x_1, \ldots, x_n)^T \in \mathbb{R}^n$, $A$ is a nonsingular $n$-order real matrix, $F : [0, +\infty) \times \mathbb{R}^n \to \mathbb{R}^n$. Let $h > 0$ be a given time step length. Adapting exponential Euler integrators in [19–21] to Eq. (1.4) yields the exponential Euler difference equations as noted below

$$\begin{cases} X(kh + h) = e^{Ah}X(kh) + (e^{Ah} - \mathbb{I}_n)A^{-1}F(kh, X(kh)), \quad k \in \mathbb{N}_0, \\ X(0) = \xi \in \mathbb{R}^n. \end{cases} \tag{1.5}$$

Based on Eq. (1.5), literature [19, 21, 22, 26–32] were devoted to discussing the various dynamical behaviours for a variety of exponential Euler difference models, e.g., periodicity [28], almost periodicity [26, 27, 29], almost automorphy [31, 32], stability [21, 26, 27, 31, 32], stochastic disturbances [20, 21, 27, 30], etc.

For instance, Huang et al. [29] discussed the following exponential Euler difference cellular neural networks

$$\begin{cases} x_i(k+1) = e^{-a_i(k)}x_i(k) \\ \qquad + \dfrac{1 - e^{-a_i(k)}}{a_i(k)}\left[\displaystyle\sum_{j=1}^{n} b_{ij}(k)f_j(x_j(k)) + I_i(k)\right], \quad k \in \mathbb{N}_0, \\ x_i(0) = \xi_{i0} \in \mathbb{R}, \end{cases} \tag{1.6}$$

where $i = 1, 2, \ldots, n$. In [29], sufficient conditions were obtained for the existence of a unique stable almost periodic sequence solution of Eq. (1.6).

In 2014, by using the exponential Euler difference technique, Huang et al. [26] obtained the following discrete-time models for a class of general neural networks

$$
\begin{cases}
x_i(k+1) = e^{-a_i(k)} x_i(k) + \dfrac{1 - e^{-a_i(k)}}{a_i(k)} \\
\qquad \times \left[ \displaystyle\sum_{l=1}^{m} \sum_{j=1}^{n} b_{ijl}(k) f_j(x_j(k - \tau_{ijl})) + I_i(k) \right], \quad k \in \mathbb{N}_0, \\
x_i(0) = \xi_{i0} \in \mathbb{R},
\end{cases}
\tag{1.7}
$$

where $i = 1, 2, \ldots, n$. The authors [26] derived the existence of locally exponentially convergent $2^N$ almost periodic sequence solutions of Eq. (1.7).

Paper [30] considered the following fuzzy exponential Euler difference cellular neural networks with stochastic perturbations

$$
\begin{cases}
x_i(k+1) = e^{-a_i(k)} x_i(k) + \dfrac{1 - e^{-a_i(k)}}{a_i(k)} \left[ \displaystyle\sum_{j=1}^{n} b_{ij}(k) f_j(x_j(k)) \right. \\
\qquad + \displaystyle\sum_{j=1}^{n} c_{ij}(k) g_j(x_j(k - \tau_{ij}(k))) + \bigwedge_{j=1}^{n} \alpha_{ij} g_j(x_j(k - \tau_{ij}(k))) \\
\qquad + \displaystyle\bigwedge_{j=1}^{n} T_{ij}\mu_j + \bigvee_{j=1}^{n} \beta_{ij} g_j(x_j(k - \tau_{ij}(k))) + \bigvee_{j=1}^{n} S_{ij}\mu_j \\
\qquad \left. + \displaystyle\sum_{j=1}^{n} d_{ij}(k) \sigma_j(x_j(k - \eta_{ij}(k))) \Delta w_j(k) + I_i(k) \right], \quad k \in \mathbb{N}_0, \\
x_i(0) = \xi_{i0} \in \mathbb{R},
\end{cases}
\tag{1.8}
$$

where $\Delta w_j(k) := w_j(k+1) - w_j(k)$, $k \in \mathbb{N}_0$, $w_j$ is the standard Brownian motion defined on a complete probability space, $i, j = 1, 2, \ldots, n$. The existence of at least one $p$th mean almost periodic sequence solution and $p$th moment global exponential stability of Eq. (1.8) had been studied. In addition to the works mentioned above, the exponential Euler difference technique has been used to discuss other applied models, such as prey-predator models [33–36].

## 1.3 Fractional Euler Difference

Consider the following nonlinear fractional-order differential equation

$$
\begin{cases}
{}_0^C D_t^{\alpha} x(t) = f(t, x(t)), \quad \forall t \geq 0,\ 0 < \alpha < 1, \\
x(0) = \xi_0 \in \mathbb{R},
\end{cases}
\tag{1.9}
$$

whose equivalent form is the second kind of Volterra integral equation

$$x(t) = \xi_0 + \frac{1}{\Gamma(\alpha)} \int_0^t \frac{f(s, x(s))}{(t-s)^{1-\alpha}} \mathrm{d}s, \quad \forall t \geq 0. \tag{1.10}$$

In Eq. (1.9), $f$ is continuously smooth and satisfies the Lipschitz condition.

Using R0-algorithm (see [37]) to Eq. (1.10) derives the numerical solution of Eq. (1.9) in the form of

$$x(kh+h) = \xi_0 + h^\alpha \sum_{j=0}^{k} a_{j,k+1} f(jh, x(jh)), \quad x(0) = \xi_0 \in \mathbb{R}, \tag{1.11}$$

where $h > 0$ is a given time step length,

$$a_{j,k} = \frac{(k-j)^\alpha - (k-j-1)^\alpha}{\Gamma(1+\alpha)}, \quad j = 0, 1, \ldots, k, k \in \mathbb{N}_0.$$

Formula (1.11) is called the fractional Euler difference or fractional Adams–Bashforth difference.

Using R2-algorithm (see [37]) to Eq. (1.10) leads to

$$x(kh+h) = \xi_0 + h^\alpha \sum_{j=0}^{k+1} b_{j,k+1} f(jh, x(jh)), \quad x(0) = \xi_0 \in \mathbb{R}, \tag{1.12}$$

where

$$b_{j,k} = \frac{1}{\Gamma(2+\alpha)} \begin{cases} (1+\alpha)k^\alpha - k^{1+\alpha} + (k-1)^{1+\alpha}, & j = 0; \\ (k-j+1)^{1+\alpha} - 2(k-j)^{1+\alpha} + (k-j-1)^{1+\alpha}, & 1 \leq j \leq k-1; \\ 1, & j = k \end{cases}$$

for $j = 0, 1, \ldots, k+1, k \in \mathbb{N}_0$. The difference scheme (1.12) is called the Adams-Moulton difference.

Since both sides of Eq. (1.12) include the unknown variables $x(kh+h)$ and due to the non-linearity of $f$, it is often difficult to derive $x(kh+h)$. To achieve a better approximate solution, let $\tilde{x}(\cdot)$ be the predicted value, which can be obtained by the fractional Euler difference Eq. (1.11) in the shape of

$$\begin{cases} \tilde{x}(kh+h) = \xi_0 + h^\alpha \sum_{j=0}^{k} a_{j,k+1} f(jh, x(jh)), \quad \forall k \in \mathbb{N}_0, \\ x(0) = \xi_0 \in \mathbb{R}. \end{cases} \tag{1.13}$$

Replacing $x(kh+h)$ in the right hand of Eq. (1.12) by Eq. (1.13) gives

$$\begin{cases} x(kh+h) = \xi_0 + h^\alpha \sum_{j=0}^{k} b_{j,k+1} f(jh, x(jh)) \\ \qquad + \dfrac{h^\alpha}{\Gamma(2+\alpha)} f(kh+h, \tilde{x}(kh+h)), \quad \forall k \in \mathbb{N}_0, \\ x(0) = \xi_0 \in \mathbb{R}. \end{cases} \tag{1.14}$$

The method determined by Eqs. (1.13) and (1.14) is called the fractional Adams–Bashforth–Moulton difference, which mainly includes four steps as

(1) Predict: predict $\tilde{x}(\cdot)$ by Eq. (1.13);
(2) Evaluate: compute $f(\cdot, \tilde{x}(\cdot))$;
(3) Correct: correct $x(\cdot)$ by Eq. (1.14);
(4) Evaluate: compute $f(\cdot, x(\cdot))$ to prepare for the next loop iteration.

Hence, this method is commonly called the fractional predictor-corrector scheme, or fractional PECE (Predict, Evaluate, Correct, Evaluate) method. Fractional PECE method has been proposed and briefly studied in literature [38]. Some detailed mathematical analysis is provided in literature [39], and numerical experiments and comparisons with other methods are reported in literature [40, 41].

## 1.4 Mittag–Leffler Euler Difference

Motivated by exponential Euler differences, a question naturally arises: could the idea of exponential Euler differences be applied to studying semilinear problems of FODEs (1.9)? The excellent works in the literature [42–45] answer this question in the affirmative. For example, Garrappa and Popolizio [44, 45] studied a generalised exponential Euler integrator for FODEs in the sense of Caputo. Papers [42, 43] discussed the exponential Euler difference schemes for some stochastic FODEs in the Caputo sense. Remarkably, the literature [42–45] mainly considered the exponential Euler differences (also called Mittag–Leffler Euler differences) for FODEs from the point of view of the theory of numerical computations. Inspired by this point, the aim of this book is to establish the exponential Euler difference schemes for FODEs from the viewpoint of theory in difference equations. It should be noted that the work in this book extends and complements the existing results on this subject.

## 1.5 Fractional Calculus

This section introduces the concepts and properties of the *Gamma* function, Riemann–Liouville and Caputo fractional calculus, and the Mittag–Leffler function. These topics can be found in the literature [46–50].

The *Gamma* function $\Gamma(z)$ usually is defined by

$$\Gamma(z) = \int_0^\infty t^{z-1} e^{-t} \mathrm{d}t, \quad \Re(z) > 0.$$

It is analytic in the right half complex plane $\mathbb{C}_+ := \{z \in \mathbb{C} : \Re(z) > 0\}$ and

$$\Gamma(z+1) = z\Gamma(z).$$

Since $\Gamma(1) = 1$, therefore $\Gamma(n+1) = n!$ for $n \in \mathbb{N}$.

Also it holds

$$\Gamma(z)\Gamma(1-z) = \frac{\pi}{\sin(\pi z)}, \quad 0 < \Re(z) < 1 \text{ (Reflection formula)}$$

and

$$\Gamma(z)\Gamma(z+\frac{1}{2}) = \sqrt{\pi} 2^{1-2z} \Gamma(2z) \text{ (Legendre's duplication formula)}$$

are very useful in practice. Directly, $\Gamma\left(\frac{1}{2}\right) = \sqrt{\pi}$. For more knowledge of *Gamma* function, please refer to Chapter 1 of book [46].

### 1.5.1 Riemann–Liouville Fractional Calculus

This subsection mainly gives the definitions and properties of Riemann–Liouville fractional calculus, which can be found in books [46, 47].

The left-sided Riemann–Liouville fractional integral of $x \in L((a,b), \mathbb{R})$ with order $\alpha > 0$ based at $x = a$ is given by

$${}_aI_t^\alpha x(t) = \frac{1}{\Gamma(\alpha)} \int_a^t (t-s)^{\alpha-1} x(s) \mathrm{d}s$$

and the right-sided Riemann–Liouville fractional integral of order $\alpha > 0$ based at $x = b$ is defined by

$${}_tI_b^\alpha x(t) = \frac{1}{\Gamma(\alpha)} \int_t^b (s-t)^{\alpha-1} x(s) \mathrm{d}s.$$

If $\alpha = k \in \mathbb{N}$, they recover the $k$-fold integrals ${}_aI_t^k x$ and ${}_tI_b^k x$.

**Lemma 1.1** ([46]) *Let $x \in L((a,b),\mathbb{R})$. Then for almost all $t \in [a,b]$, it holds*

$$\lim_{\alpha\to 0^+} {}_aI_t^\alpha x(t) = \lim_{\alpha\to 0^+} {}_tI_b^\alpha x(t) = x(t).$$

**Lemma 1.2** (Semigroup Property [46]) *For $x \in L((a,b),\mathbb{R})$, $\alpha, \beta \geq 0$, there holds*

$${}_aI_t^\alpha\, {}_aI_t^\beta x = {}_aI_t^\beta\, {}_aI_t^\alpha x = {}_aI_t^{\alpha+\beta} x, \quad {}_tI_b^\alpha\, {}_tI_b^\beta x = {}_tI_b^\beta\, {}_tI_b^\alpha x = {}_tI_b^{\alpha+\beta} x.$$

**Lemma 1.3** (Integration by Parts [46]) *For any $f \in L^p((a,b),\mathbb{R})$, $g \in L^q((a,b),\mathbb{R})$, $p, q \geq 1$ with $p^{-1} + q^{-1} \leq 1 + \alpha (p, q \neq 1$, when $p^{-1} + q^{-1} = 1 + \alpha)$, the following identity holds*

$$\int_a^b g(t)\, {}_aI_t^\alpha f(t)\mathrm{d}t = \int_a^b f(t)\, {}_tI_b^\alpha g(t)\mathrm{d}t.$$

**Lemma 1.4** (Mean Value Theorem [47]) *Let $\alpha > 0$, $x \in C([a,b],\mathbb{R})$, $g$ be Lebesgue integrable on $(a,b)$ and do not change its sign in $(a,b)$. Then for almost every $t \in (a,b]$, there exists some $\xi \in (a,t) \subset (a,b)$ such that ${}_aI_t^\alpha(fg)(t) = f(\xi)\, {}_aI_t^\alpha g(t)$. If additionally, $\alpha \geq 1$ or $g \in C([a,b],\mathbb{R})$, then this result holds for every $t \in (a,b]$.*

For $x \in L((a,b),\mathbb{R})$ and $m - 1 < \alpha \leq m$, $m \in \mathbb{N}$, its left-sided Riemann–Liouville fractional derivative of order $\alpha$ based at $x = a$ is defined by

$${}_a^RD_t^\alpha x(t) = \frac{\mathrm{d}^m}{\mathrm{d}t^m}\, {}_aI_t^{m-\alpha} x(t) = \frac{1}{\Gamma(m-\alpha)} \frac{\mathrm{d}^m}{\mathrm{d}t^m} \int_a^t (t-s)^{m-\alpha-1} x(s)\mathrm{d}s,$$

if the integral on the right-hand side exists. Its right-sided Riemann–Liouville fractional derivative of order $\alpha$ based at $x = b$ is defined by

$${}_t^RD_b^\alpha x(t) = (-1)^m \frac{\mathrm{d}^m}{\mathrm{d}t^m}\, {}_tI_b^{m-\alpha} x(t) = \frac{(-1)^m}{\Gamma(m-\alpha)} \frac{\mathrm{d}^m}{\mathrm{d}t^m} \int_t^b (s-t)^{m-\alpha-1} x(s)\mathrm{d}s,$$

if the integral on the right-hand side exists.

**Lemma 1.5** ([46]) *If $m - 1 < \alpha \leq m$, $m \in \mathbb{N}$, $x \in L((a,b),\mathbb{R})$, ${}_aI_t^{m-\alpha} x \in AC^m([a,b],\mathbb{R})$ and ${}_a^RD_t^\alpha x(t) = 0$ in $(a,b)$, then*

$$x(t) = \sum_{j=1}^m \frac{{}_a^RD_t^{\alpha-j} x(a^+)}{\Gamma(\alpha - j + 1)} (x-a)^{\alpha-j}.$$

**Lemma 1.6** ([46]) *For any $\alpha, \beta \geq 0$, it holds*

$${}_a^RD_t^\alpha\, {}_a^RD_t^\beta x = {}_a^RD_t^{\alpha+\beta} x, \quad \forall x \in {}_aI_t^{\alpha+\beta}(L((a,b),\mathbb{R})).$$

**Lemma 1.7** ([46]) *Let $m-1<\alpha\le m$, $m\in\mathbb{N}$. Then*

(1) *For any $x\in L((a,b),\mathbb{R})$, ${}_a^RD_t^\alpha\,{}_aI_t^\alpha x=x$.*
(2) *If ${}_aI_t^\alpha x\in AC^m([a,b],\mathbb{R})$, then*

$$ {}_aI_t^\alpha\,{}_a^RD_t^\alpha x=x-\sum_{j=0}^{m-1}{}_a^RD_t^{\alpha-j-1}x(a^+)\frac{(t-a)^{\alpha-j-1}}{\Gamma(\alpha-j)}. $$

**Lemma 1.8** ([46]) *Let $x\in L((a,b),\mathbb{R})$.*

(1) *If ${}_a^RD_t^{\alpha-\beta}x\in L((a,b),\mathbb{R})$ for $\alpha,\beta\ge 0$, then*

$$ {}_a^RD_t^\alpha\,{}_aI_t^\beta x(t)={}_a^RD_t^{\alpha-\beta}x(t). $$

(2) *If ${}_a^RD_t^\beta x\in L((a,b),\mathbb{R})$ with $0\le m-1\le\beta\le m$, then for any $\alpha\ge 0$, it holds*

$$ {}_aI_t^\alpha\,{}_a^RD_t^\beta x(t)={}_a^RD_t^{\beta-\alpha}x(t)-\sum_{j=1}^{m}{}_a^RD_t^{\beta-j}x(a)\frac{(t-a)^{\alpha-j}}{\Gamma(\alpha+1-j)}. $$

(3) *If $\alpha,\beta\in(0,1]$, then*

$$ {}_a^RD_t^\alpha\,{}_a^RD_t^\beta x(t)=\frac{\mathrm{d}}{\mathrm{d}t}\left[{}_a^RD_t^{\alpha+\beta-1}x-{}_aI_t^{1-\beta}x(a)\frac{(t-a)^{-\alpha}}{\Gamma(1-\alpha)}\right], $$

*provided that all the terms make sense in $L((a,b),\mathbb{R})$.*

**Lemma 1.9** (Integration by Parts [47]) *Let $\alpha>0$, $p,q\ge 1$ with $p^{-1}+q^{-1}\le 1+\alpha(p,q\ne 1$, when $p^{-1}+q^{-1}=1+\alpha)$. If $f\in{}_aI_t^\alpha(L^p((a,b),\mathbb{R}))$, $g\in{}_tI_b^\alpha(L^q((a,b),\mathbb{R}))$, then*

$$ \int_a^b g(t)\,{}_a^RD_t^\alpha f(t)\mathrm{d}t=\int_a^b f(t)\,{}_t^RD_b^\alpha g(t)\mathrm{d}t. $$

**Lemma 1.10** (Laplace Transform [46, 47]) *Let $\alpha>0$, $x\in L(0,b)$ for any $b>0$, $|x(t)|\le ce^{p_0t}$ for all $t>b>0$ and some positive constants $c$, $p_0$.*

(1) *The relation $\mathcal{L}[{}_0I_t^\alpha x](z)=z^{-\alpha}\mathcal{L}[x](z)$ holds for any $\Re(z)>p_0$.*
(2) *If $m-1<\alpha\le m$, $m\in\mathbb{N}$, $x\in AC^m([0,b],\mathbb{R})$ and*

$$ \lim_{t\to\infty}{}_0^RD_t^{\alpha+j-m}x(t)=0,\quad j=0,1,\ldots,m-1. $$

*Then*

$$ \mathcal{L}[{}_0^RD_t^\alpha x](z)=z^\alpha\mathcal{L}[x](z)-\sum_{j=0}^{m-1}z^{m-j-1}({}_0^RD_t^{\alpha+j-m}x)(0^+),\quad\forall\Re(z)>p_0. $$

**Lemma 1.11** (Fourier Transform [46]) *If* $0 < \alpha < 1$ *and* $x$ *is sufficiently regular, then*

$$\mathcal{F}[{}_{-\infty}I_t^{\alpha}x](\xi) = (\mathrm{i}\xi)^{-\alpha}\mathcal{F}[x](\xi), \quad \mathcal{F}[{}_tI_{\infty}^{\alpha}x](\xi) = (-\mathrm{i}\xi)^{-\alpha}\mathcal{F}[x](\xi),$$

*where* $(\pm\mathrm{i}\xi)^{\alpha} := |\xi|^{\alpha}e^{\pm\alpha\pi\mathrm{i}\frac{\mathrm{sign}(\xi)}{2}}$ *for* $\xi \in \mathbb{R}$ *and* $\mathrm{i} = \sqrt{-1}$. *Similarly, for any* $\alpha > 0$,

$$\mathcal{F}[{}_{-\infty}^{\ R}D_t^{\alpha}x](\xi) = (\mathrm{i}\xi)^{\alpha}\mathcal{F}[x](\xi), \quad \mathcal{F}[{}_t^RD_{\infty}^{\alpha}x](\xi) = (-\mathrm{i}\xi)^{\alpha}\mathcal{F}[x](\xi).$$

### 1.5.2 Caputo Fractional Calculus

Let us turn to the Caputo fractional derivatives in books [46, 47], which is one of the most popular fractional derivatives owing to its ability to handle initial conditions in a physically transparent way.

For $x \in L((a,b),\mathbb{R})$ and $m-1 < \alpha \leq m, m \in \mathbb{N}$, its left-sided Caputo fractional derivative of order $\alpha$ based at $x = a$ is defined by

$${}_a^CD_t^{\alpha}x(t) = {}_aI_t^{m-\alpha}x^{(m)}(t) = \frac{1}{\Gamma(m-\alpha)}\int_a^t (t-s)^{m-\alpha-1}x^{(m)}(s)\mathrm{d}s,$$

if the integral on the right-hand side exists. Likewise, its right-sided Caputo fractional derivative of order $\alpha$ based at $x = b$ is defined by

$${}_t^CD_b^{\alpha}x(t) = (-1)^m\,{}_tI_b^{m-\alpha}x^{(m)}(t) = \frac{(-1)^m}{\Gamma(m-\alpha)}\int_t^b (s-t)^{m-\alpha-1}x^{(m)}(s)\mathrm{d}s,$$

if the integral on the right-hand side exists.

**Lemma 1.12** ([46]) *If* $m-1 < \alpha \leq m$, $m \in \mathbb{N}$, $x \in AC^m([a,b],\mathbb{R})$ *and* ${}_a^CD_t^{\alpha}x(t) = 0$ *in* $(a,b)$, *then*

$$x(t) = \sum_{k=0}^{m-1}\frac{x^{(k)}(a^+)}{k!}(t-a)^k.$$

**Lemma 1.13** ([46]) *Let* $m-1 < \alpha \leq m$, $m \in \mathbb{N}$, $x \in AC^m([a,b],\mathbb{R})$. *Then*

$${}_a^RD_t^{\alpha}x(t) = {}_a^CD_t^{\alpha}x(t) + \sum_{j=0}^{m-1}\frac{(t-a)^{j-\alpha}}{\Gamma(j-\alpha+1)}x^{(j)}(a^+),$$

*and*

$${}_t^RD_b^{\alpha}x(t) = {}_t^CD_b^{\alpha}x(t) + \sum_{j=0}^{m-1}\frac{(-1)^j(b-t)^{j-\alpha}}{\Gamma(j-\alpha+1)}x^{(j)}(b^-).$$

Very often, Caputo fractional derivative has another definition as follows.

**Definition 1.1** ([47]) For $m-1<\alpha\le m, m\in\mathbb{N}, x\in C^{m-1}([a,b],\mathbb{R}), {}_aI_t^{m-\alpha}(x-T_{m-1}x)(t)\in AC^m([a,b],\mathbb{R})$, the left-sided Caputo fractional derivative is defined by

$$ {}_a^CD_t^{\alpha*}x(t)={}_a^RD_t^{\alpha}(x-T_{m-1}x)(t), $$

where

$$ T_{m-1}x(t)=\sum_{j=0}^{m-1}\frac{(t-a)^{j-\alpha}}{\Gamma(j-\alpha+1)}x^{(j)}(a^+),\quad \forall t\in(a,b]. $$

Similarly one can define the right-sided version by

$$ {}_t^CD_b^{\alpha*}x(t)={}_t^RD_b^{\alpha}\left(x(t)-\sum_{k=0}^{m-1}\frac{(-1)^k(b-t)^k}{k!}x^{(k)}(b^-)\right). $$

If $m-1<\alpha<m, m\in\mathbb{N}$, then ${}_a^CD_t^{\alpha*}x={}_a^CD_t^{\alpha}x$ for $x\in AC^m([a,b],\mathbb{R})$. Generally, ${}_a^CD_t^{\alpha}x$ and ${}_a^CD_t^{\alpha*}x$ can be different and the latter requires less regularity assumption.

**Lemma 1.14** ([46, 47]) *The following statements hold.*

(1) *Let* $m-1<\alpha<m$, $m\in\mathbb{N}$, $x\in C([a,b],\mathbb{R})$. *Then*

$$ {}_a^CD_t^{\alpha*}\,{}_aI_t^{\alpha}x(t)=x(t) $$

*for all* $t\in(a,b)$.

(2) *Let* $x\in C^m([a,b],\mathbb{R})$ *for some* $m\in\mathbb{N}$, $\alpha,\beta>0$. *Moreover, it exists some* $\ell\in\mathbb{N}$ *with* $\ell\le m$ *and* $\beta,\alpha+\beta\in[\ell-1,\ell]$. *Then*

$$ {}_a^CD_t^{\alpha*}\,{}_a^CD_t^{\beta*}x={}_a^CD_t^{(\alpha+\beta)*}x. $$

**Lemma 1.15** ([46, 47]) *Let* $m-1<\alpha<m$, $m\in\mathbb{N}$. *Then the following statements hold.*

(1) *If* $x\in L((a,b),\mathbb{R})$, ${}_aI_t^{\alpha-m+1}x\in AC([a,b],\mathbb{R})$ *with* ${}_aI_t^{\alpha-m+1}x(a)=0$, *then*

$$ {}_a^CD_t^{\alpha}\,{}_aI_t^{\alpha}x(t)=x(t),\quad \forall t\in(a,b),\ a.e.. $$

(2) *If* $x\in AC^m([a,b],\mathbb{R})$, *then*

$$ {}_aI_t^{\alpha}\,{}_a^CD_t^{\alpha}x(t)=x(t)-T_{m-1}x(t),\quad \forall t\in(a,b),\ a.e.. $$

*If* $x\in C^m([a,b],\mathbb{R})$, *then the identity holds for all* $t\in[a,b]$.

**Lemma 1.16** (Laplace Transform [46, 47]) *Let* $m-1<\alpha\le m$, $m\in\mathbb{N}$, $x\in C^m([0,+\infty),\mathbb{R})$ *and* $|x(t)|\le ce^{p_0 t}$ *for large* $t$ *and some* $p_0\in\mathbb{R}$, $x\in AC^m([0,b],\mathbb{R})$ *for any* $b>0$, *the Laplace transforms of* $x$ *and* $x^{(m)}$ *exist,*

$$\lim_{t\to\infty} x^{(k)}(t)=0,\quad k=0,1,\ldots,m-1.$$

*Then*

$$\mathcal{L}[{}_0^C D_t^\alpha x](z)=z^\alpha\mathcal{L}[x](z)-\sum_{k=0}^{m-1} z^{\alpha-k-1}x^{(k)}(0),\quad \Re(z)>p_0.$$

**Lemma 1.17** (Mean Value Theorem [47]) *Let* $m-1<\alpha<m$, $m\in\mathbb{N}$, $x\in C^m([0,+\infty),\mathbb{R})$ *and* ${}_0^C D_t^\alpha x\in C^m([0,+\infty),\mathbb{R})$. *Then there exists some* $\xi\in(a,b)$ *such that*

$$\frac{x(b)-\sum\limits_{k=0}^{m-1}\dfrac{(b-a)^k}{k!}x^{(k)}(a)}{(b-a)^\alpha}=\frac{{}_a^C D_t^\alpha x(\xi)}{\Gamma(\alpha+1)}.$$

### *1.5.3 Mittag–Leffler Function*

This text demonstrates the concepts of Mittag–Leffler functions and their properties, as found in literature [46–50].

The one-parameter and two-parameter Mittag–Leffler functions are defined by

$$\mathbb{E}_\alpha(z)=\sum_{k=0}^\infty\frac{z^k}{\Gamma(\alpha k+1)},\quad \mathbb{E}_{\alpha,\beta}(z)=\sum_{k=0}^\infty\frac{z^k}{\Gamma(\alpha k+\beta)},\quad \Re(\alpha)>0,\ \beta,z\in\mathbb{C}.$$

In particular, $\mathbb{E}_1(z)=e^z$; $\mathbb{E}_{\alpha,1}(z)=\mathbb{E}_\alpha(z)$; $\mathbb{E}_{1,2}(z)=\dfrac{e^z-1}{z}$.

Also, it could define matrix Mittag–Leffler functions as

$$\mathbb{E}_\alpha(A)=\sum_{k=0}^\infty\frac{A^k}{\Gamma(\alpha k+1)},\quad \mathbb{E}_{\alpha,\beta}(A)=\sum_{k=0}^\infty\frac{A^k}{\Gamma(\alpha k+\beta)}$$

for all $\Re(\alpha)>0$, $\beta\in\mathbb{C}$, $A\in\mathbb{C}^{n\times n}$.

The asymptotic properties for $\mathbb{E}_{\alpha,\beta}(z)$ with $\alpha>0$ are given below.

**Lemma 1.18** ([46]) *Let* $\alpha\in(0,2)$, $\beta\in\mathbb{R}$, $\varphi\in\left(\frac{\alpha}{2}\pi,\min(\pi,\alpha\pi)\right)$ *and* $N\in\mathbb{N}$. *Then for* $|\arg(z)|\le\varphi$ *with* $|z|\to\infty$, *it obtains*

$$\mathbb{E}_{\alpha,\beta}(z)=\frac{1}{\alpha}z^{\frac{1-\beta}{\alpha}}e^{z^{\frac{1}{\alpha}}}-\sum_{k=1}^N\frac{1}{\Gamma(\beta-\alpha k)}\frac{1}{z^k}+O\left(\frac{1}{z^{N+1}}\right),$$

*and for $\varphi \le |\arg(z)| \le \pi$ with $|z| \to \infty$*

$$\mathbb{E}_{\alpha,\beta}(z) = -\sum_{k=1}^{N} \frac{1}{\Gamma(\beta - \alpha k)} \frac{1}{z^k} + O\left(\frac{1}{z^{N+1}}\right).$$

**Lemma 1.19** ([46]) *Let $\alpha \ge 2$, $\beta \in \mathbb{R}$ and $N \in \mathbb{N}$. Then*

$$\mathbb{E}_{\alpha,\beta}(z) = \frac{1}{\alpha} \sum \left(z^{\frac{1}{\alpha}} e^{\frac{2m\pi \mathrm{i}}{\alpha}}\right)^{1-\beta} e^{z^{\frac{1}{\alpha}} e^{\frac{2m\pi \mathrm{i}}{\alpha}}} - \sum_{k=1}^{N} \frac{z^{-k}}{\Gamma(\beta - \alpha k)} + O\left(\frac{1}{z^{N+1}}\right),$$

*as $z \to \infty$, where the first sum is taken for integer $m$ satisfying the condition $|\arg(z) + 2m\pi| \le \frac{3\alpha\pi}{4}$.*

**Lemma 1.20** ([46]) *Let $\alpha, \lambda \in \mathbb{C}$, $\Re(\alpha) > 0$. Then*

$$\frac{\mathrm{d}}{\mathrm{d}z}\left[z^{\alpha} \mathbb{E}_{\alpha,\alpha+1}(\lambda z^{\alpha})\right] = z^{\alpha-1} \mathbb{E}_{\alpha,\alpha}(\lambda z^{\alpha}), \quad \forall z \in \mathbb{C}\backslash\{0\}.$$

**Lemma 1.21** ([48–50]) *If $\alpha \in (0, 1]$ and $\lambda \in \mathbb{R}$, then*

(1) $\mathbb{E}_{\alpha}(0) = 1$, $\mathbb{E}_{\alpha,\alpha}(0) = \dfrac{1}{\Gamma(\alpha)}$.
(2) $\mathbb{E}_{\alpha}(\lambda t^{\alpha}) \in (0, 1)$ *for* $\lambda < 0$ *and* $\mathbb{E}_{\alpha}(\lambda t^{\alpha}) \in (1, +\infty)$ *for* $\lambda > 0$, $\forall t > 0$.
(3) $\mathbb{E}_{\alpha}(\lambda t_1^{\alpha}) > \mathbb{E}_{\alpha}(\lambda t_2^{\alpha})$ *and* $\mathbb{E}_{\alpha,\alpha}(\lambda t_1^{\alpha}) > \mathbb{E}_{\alpha,\alpha}(\lambda t_2^{\alpha})$ *for* $\lambda < 0$ *and* $t_1 < t_2$.
(4) $\mathbb{E}_{\alpha}(\lambda t_1^{\alpha}) < \mathbb{E}_{\alpha}(\lambda t_2^{\alpha})$ *and* $\mathbb{E}_{\alpha,\alpha}(\lambda t_1^{\alpha}) < \mathbb{E}_{\alpha,\alpha}(\lambda t_2^{\alpha})$ *for* $\lambda > 0$ *and* $t_1 < t_2$.

Considering the linear FODEs below

$$\begin{cases} {}_a^C D_t^{\alpha} x(t) = \lambda x(t) + f(t), & t \in (a, +\infty), \\ x(a) = \xi_0 \in \mathbb{R}. \end{cases} \tag{1.15}$$

**Lemma 1.22** ([46]) *If $\alpha \in (0, 1]$ and $\lambda \ne 0$, then Eq. (1.15) has a unique solution given by*

$$x(t) = \mathbb{E}_{\alpha}\left[\lambda(t-a)^{\alpha}\right]\xi_0 + \int_a^t (t-s)^{\alpha-1} \mathbb{E}_{\alpha,\alpha}[\lambda(t-s)^{\alpha}] f(s) \mathrm{d}s, \quad t \in \mathbb{R}_a.$$

**Lemma 1.23** *If $\alpha \in (0, 1]$ and $\lambda \in \mathbb{R}$, then*

$$t^{\alpha} \mathbb{E}_{\alpha,\alpha+1}(\lambda t^{\alpha}) = [\mathbb{E}_{\alpha}(\lambda t^{\alpha}) - 1]\lambda^{-1}, \quad \forall t \in \mathbb{R}.$$

***Proof*** Consider the initial value problem below

$${}_a^C D_t^{\alpha} x(t) = \lambda x(t) - \lambda, \quad x(a) = 1, \quad \forall t \in [a, +\infty). \tag{1.16}$$

By Lemma 1.22, Eq. (1.16) has a unique solution in the form of

$$x(t) = \mathbb{E}_\alpha(\lambda t^\alpha)x(a) - \int_a^t (t-s)^{\alpha-1}\mathbb{E}_{\alpha,\alpha}[\lambda(t-s)^\alpha]\lambda \mathrm{d}s$$
$$= \mathbb{E}_\alpha(\lambda t^\alpha) - t^\alpha \mathbb{E}_{\alpha,\alpha+1}(\lambda t^\alpha)\lambda, \tag{1.17}$$

where $t \in [a, +\infty)$. Lemma 1.20 is employed in the calculation of Eq. (1.17). Apparently, $x = 1$ is a solution of Eq. (1.16). Together with Eq. (1.17), one gains $1 = \mathbb{E}_\alpha(\lambda t^\alpha) - t^\alpha \mathbb{E}_{\alpha,\alpha+1}(\lambda t^\alpha)\lambda$, $t \in (a, +\infty)$. So Lemma 1.23 is obtained by the arbitrariness of $a \in \mathbb{R}$. The proof is finished. □

Consider the nonlinear FODEs

$$\begin{cases} {}_a^C D_t^\alpha x(t) = \lambda x(t) + f(t, x(t)), & \forall t \in [a, +\infty), \\ x(a) = \xi_0 \in \mathbb{R}, \end{cases} \tag{1.18}$$

where $x \in C([a, +\infty), \mathbb{R})$, $\alpha \in (0, 1]$, $\lambda \in \mathbb{R}\backslash\{0\}$, $f \in C([a, +\infty) \times \mathbb{R}, \mathbb{R})$ meets

(Lip) It exists a constant $L_f > 0$ such that $|f(t, x) - f(t, x)| \leq L_f|x - y|, \forall x, y \in \mathbb{R}, t \in [a, +\infty)$.

**Lemma 1.24** *If (Lip) holds, then Eq. (1.18) has a unique solution which can be described as*

$$x(t) = \mathbb{E}_\alpha\big[\lambda(t-a)^\alpha\big]\xi_0$$
$$+ \int_a^t (t-s)^{\alpha-1}\mathbb{E}_{\alpha,\alpha}[\lambda(t-s)^\alpha]f(s, x(s))\mathrm{d}s, \quad \forall t \in \mathbb{R}_a. \tag{1.19}$$

***Proof*** Owing to $\lim_{s\to 0} \mathbb{E}_\alpha(\lambda s^\alpha) = 1$, it exists a positive constant $h_\diamond$ small enough causing $0 < h_\diamond < b - a$ and $L_f \max_{s\in[0,h_\diamond]} \left\|[\mathbb{E}_\alpha(\lambda s^\alpha) - 1]\lambda^{-1}\right\| < 1$. Let $\mathbb{B}_{h_\diamond} = \{x \in C([a, a + h_\diamond], \mathbb{R}) : x(a) = \xi_0\}$ with norm $\|x\|_1 = \max_{s\in[a,a+h_\diamond]} |x(s)|$, $\forall x \in \mathbb{B}_{h_\diamond}$.

For each $\varphi \in \mathbb{B}_{h_\diamond}$, consider the auxiliary linear FODEs

$${}_a^C D_t^\alpha x(t) = \lambda x(t) + f(t, \varphi(t)), \quad t \in [a, a + h_\diamond], \quad x(a) = \xi_0 \in \mathbb{R},$$

From Lemma 1.22, define operator $\mathbf{T} : \varphi \to x^\varphi$, where

$$x^\varphi = \mathbf{T}\varphi = \mathbb{E}_\alpha(\lambda t^\alpha)\xi_0 + \int_a^t (t-s)^{\alpha-1}\mathbb{E}_{\alpha,\alpha}[\lambda(t-s)^\alpha]f(s, \varphi(s))\mathrm{d}s$$

for $\varphi \in \mathbb{B}_{h_\diamond}$, $t \in [a, a + h_\diamond]$. Apparently, $\mathbf{T} : \mathbb{B}_{h_\diamond} \to \mathbb{B}_{h_\diamond}$ and the unique fixed point $\varphi^* \in \mathbb{B}_{h_\diamond}$ of $\mathbf{T}$ is the unique solution of Eq. (1.18) on $[a, a + h_\diamond]$. For any $\varphi, \psi \in \mathbb{B}_{h_\diamond}$, from Lemmas 1.21, 1.20 and 1.23, it has

$$
\begin{aligned}
\|\mathbf{T}\varphi(t) - \mathbf{T}\psi(t)\|_1 &\le \left\| \int_a^t (t-s)^{\alpha-1}\mathbb{E}_{\alpha,\alpha}[\lambda(t-s)^\alpha]\mathrm{d}s \right\|_1 L_f \|\varphi - \psi\|_1 \\
&\le \left\| (t-s)^\alpha \mathbb{E}_{\alpha,\alpha+1}[\lambda(t-s)^\alpha] \Big|_{s=t}^{s=a} \right\| L_f \|\varphi - \psi\|_1 \\
&= \left\| (\mathbb{E}_\alpha[\lambda(t-a)^\alpha] - 1)\lambda^{-1} \right\|_1 L_f \|\varphi - \psi\|_1 \\
&< \|\varphi - \psi\|_1, \quad t \in [a, a+h_\diamond].
\end{aligned}
$$

Thus $\mathbf{T}$ has a unique fixed point $\varphi^* \in C([a, a+h_\diamond], \mathbb{R})$ and $\varphi^*$ is a unique solution of Eq. (1.18) for $t \in [a, a+h_\diamond]$. By the extension method utilized in Theorem 3.3 of [46], it easily extends the existence interval of $\varphi^*$ to $\mathbb{R}_a$. The proof is achieved. □

## 1.6 Caputo–Fabrizio Operator

Let $M$ be a given function with $M(0) = M(1) = 1$. The left-sided and right-sided Caputo–Fabrizio operators of $x \in C^1([a, b], \mathbb{R})$ with order $\alpha \in [0, 1)$ are respectively defined as

$$
{}^{CF}_{\ a}D^\alpha_t x(t) = \frac{M(\alpha)}{1-\alpha}\int_a^t e^{\bar{\alpha}(t-s)}\frac{\mathrm{d}x(s)}{\mathrm{d}s}\mathrm{d}s, \tag{1.20}
$$

$$
{}^{CF}_{\ t}D^\alpha_b x(t) = \frac{M(\alpha)}{1-\alpha}\int_t^b e^{\bar{\alpha}(t-s)}\frac{\mathrm{d}x(s)}{\mathrm{d}s}\mathrm{d}s, \tag{1.21}
$$

where $t \in [a, b]$ and $\bar{\alpha} = \frac{\alpha}{\alpha-1}$. Please see literature [51–55] for more details.

**Lemma 1.25** ([51]) *Let $\alpha \in [0, 1)$ and $x \in C^1([a, b], \mathbb{R})$. Then*

(1) ${}^{CF}_{\ a}D^0_t x(t) = x(t) - x(a)$ *and* ${}^{CF}_{\ t}D^0_b x(t) = x(b) - x(t)$.

(2) $\lim_{\alpha\to 1^-} {}^{CF}_{\ a}D^\alpha_t x(t) = \frac{\mathrm{d}x(t)}{\mathrm{d}t}$ *and* $\lim_{\alpha\to 1^-} {}^{CF}_{\ t}D^\alpha_b x(t) = \frac{\mathrm{d}x(t)}{\mathrm{d}t}$.

**Remark 1.1** Caputo–Fabrizio operators has some limitations. It does not fit the usual concepts neither for fractional nor for integer derivative/integral, see [52] for details.

Consider the linear CFOEs as below

$$
\begin{cases}
{}^{CF}_{\ 0}D^\alpha_t x(t) = \lambda x(t) + f(t) + H_\alpha(t), & \forall t \in \mathbb{R}_0, \\
x(0) = \xi_0 \in \mathbb{R},
\end{cases} \tag{1.22}
$$

where $H_\alpha(t) = -e^{\bar{\alpha}t}[\lambda\xi_0 + f(0)]$, $\forall t \in \mathbb{R}_0$.

**Lemma 1.26** *Let $\alpha \in [0, 1)$, $M(\alpha) = 1$ and $f \in C^1(\mathbb{R}_0, \mathbb{R})$. Then Eq. (1.22) has a unique solution, which can be described by*

$$x(t) = e^{p_\alpha t}\xi_0 + \frac{1-\alpha}{K_\alpha}[f(t) - e^{p_\alpha t} f(0)] + \frac{\alpha}{K_\alpha^2}\int_0^t e^{p_\alpha (t-s)} f(s)\,\mathrm{d}s, \quad (1.23)$$

*where $t \in \mathbb{R}_0$,*

$$K_\alpha = 1 - \lambda(1-\alpha) \neq 0, \quad p_\alpha = \frac{\lambda\alpha}{K_\alpha}.$$

***Proof*** Equation (1.22) can be rewritten as

$$\frac{1}{1-\alpha}\int_a^t e^{-\bar{\alpha}s}\frac{\mathrm{d}x(s)}{\mathrm{d}s}\mathrm{d}s = e^{-\bar{\alpha}t}[\lambda x(t) + \tilde{f}(t)], \quad \forall t \in \mathbb{R}_0,$$

where $\tilde{f}(t) = f(t) + H_\alpha(t)$ for all $t \in \mathbb{R}_0$. Next, it deduces

$$\frac{\mathrm{d}x(t)}{\mathrm{d}t} = (1-\alpha)\left[\lambda\frac{\mathrm{d}x(t)}{\mathrm{d}t} + \frac{\mathrm{d}\tilde{f}(t)}{\mathrm{d}t} - \lambda\bar{\alpha}x(t) - \bar{\alpha}\tilde{f}(t)\right], \quad \forall t \in \mathbb{R}_0,$$

or equivalently,

$$\frac{\mathrm{d}x(t)}{\mathrm{d}t} = p_\alpha x(t) + \frac{1-\alpha}{K_\alpha}\frac{\mathrm{d}\tilde{f}(t)}{\mathrm{d}t} + \frac{\alpha}{K_\alpha}\tilde{f}(t), \quad \forall t \in \mathbb{R}_0.$$

By the constant variation formula of ordinary differential equations, it yields

$$x(t) = e^{p_\alpha t}\xi_0 + \frac{1-\alpha}{K_\alpha}[\tilde{f}(t) - e^{p_\alpha t}\tilde{f}(0)] + \frac{\alpha}{K_\alpha^2}\int_0^t e^{p_\alpha(t-s)}\tilde{f}(s)\mathrm{d}s, \quad \forall t \in \mathbb{R}_0.$$

Substituting $\tilde{f}(t) = f(t) + H_\alpha(t)$ into the above equation, it induces Eq. (1.23).

On the other hand, from Eq. (1.23), it has

$$e^{-p_\alpha t}x(t) = \xi_0 + \frac{1-\alpha}{K_\alpha}[e^{-p_\alpha t}f(t) - f(0)] + \frac{\alpha}{K_\alpha^2}\int_0^t e^{-p_\alpha s}f(s)\mathrm{d}s, \quad \forall t \in \mathbb{R}_0.$$

Differentiating both sides of the above equation, it derives

$$\frac{\mathrm{d}x(t)}{\mathrm{d}t} = p_\alpha x(t) + \frac{\alpha}{K_\alpha}f(t) + \frac{1-\alpha}{K_\alpha} \times \frac{\mathrm{d}f(t)}{\mathrm{d}t}, \quad \forall t \in \mathbb{R}_0,$$

which is equivalent to

$$e^{-\bar{\alpha}t}\frac{\mathrm{d}x(t)}{\mathrm{d}t} = \alpha e^{-\bar{\alpha}t}[\lambda x(t) + f(t)] + (1-\alpha)e^{-\bar{\alpha}t}\left[\lambda\frac{\mathrm{d}x(t)}{\mathrm{d}t} + \frac{\mathrm{d}f(t)}{\mathrm{d}t}\right]$$

$$= (1-\alpha)\frac{\mathrm{d}}{\mathrm{d}t}\{e^{-\bar{\alpha}t}[\lambda x(t) + f(t)]\}, \quad \forall t \in \mathbb{R}_0.$$

Integrating both sides of the above equation from 0 to $t$, it infers

$$\int_0^t e^{-\bar{\alpha}s}\frac{\mathrm{d}x(s)}{\mathrm{d}s}\mathrm{d}s = (1-\alpha)\big\{e^{-\bar{\alpha}t}[\lambda x(t) + f(t)] - [\lambda x(0) + f(0)]\big\}, \quad \forall t \in \mathbb{R}_0,$$

that is,

$${}^{CF}_{\ 0}D^{\alpha}_{t}x(t) = \frac{1}{1-\alpha}\int_0^t e^{\bar{\alpha}(t-s)}\frac{\mathrm{d}x(s)}{\mathrm{d}s}\mathrm{d}s = \lambda x(t) + f(t) + H_\alpha(t), \quad \forall t \in \mathbb{R}_0.$$

This completes the proof. □

If $\lambda = 0$ in Eq. (1.22), then it achieves

**Corollary 1.1** *Let* $\alpha \in [0, 1)$, $M(\alpha) = 1$, $\lambda = 0$ *and* $f \in C^1(\mathbb{R}_0, \mathbb{R})$. *Then Eq. (1.22) has a unique solution, which can be described by*

$$x(t) = \xi_0 + (1-\alpha)[f(t) - f(0)] + \alpha\int_0^t f(s)\,\mathrm{d}s, \quad \forall t \in \mathbb{R}_0. \tag{1.24}$$

**Remark 1.2** In Eq. (1.22), $H_\alpha$ is called the remainder term caused by the Caputo–Fabrizio operator. According to the proof of equation Lemma 1.22, the remainder term is necessary for the existence of a unique solution to Eq. (1.22). If $\alpha \to 1^-$, the remainder term becomes zero and Eq. (1.22) is converted to

$$\frac{\mathrm{d}x(t)}{\mathrm{d}t} = \lambda x(t) + f(t), \quad \forall t \in \mathbb{R}_0, \quad x(0) = \xi_0 \in \mathbb{R}.$$

**Remark 1.3** Up to now, it has few papers focusing on the existence of a unique solution to Eq. (1.22), please see literature [53–55]. In 2015, Losada and Nieto in literature [54] firstly acquired the following linear CFOEs

$${}^{CF}_{\ 0}D^{\alpha}_{t}x(t) = f(t), \quad \forall t \in \mathbb{R}_0, \quad x(0) = \xi_0 \in \mathbb{R} \tag{1.25}$$

has a unique solution given by Eq. (1.24).

Remarkably, Corollary 1.1 shows that the solution given by Eq. (1.24) satisfies Eq. (1.22) *(*$\lambda = 0$*)*, i.e., it is not a solution of Eq. (1.25) unless $f(0) = 0$. In 2019, Yépez-Martínez and Gómez-Aguilar in paper [53] had pointed out this contradiction. Taking an example as illustration, if $f(t) \equiv 1$, then $x(t) = x_0 + \alpha t$ is a solution of

Eq. (1.24), but it is not a solution of Eq. (1.25), $\forall t \in \mathbb{R}_0$. Because of this, Yépez-Martínez and Gómez-Aguilar modified the definition of Caputo–Fabrizio operator and gave a new definition as follows

$$^{modified-CF}_{\quad 0}D_t^{\alpha} x(t) = \frac{1}{1-\alpha}\frac{\mathrm{d}}{\mathrm{d}t}\int_0^t e^{\bar{\alpha}(t-s)}[x(s)-x(0)]\mathrm{d}s, \quad \forall t \in \mathbb{R}_0.$$

Yépez-Martínez and Gómez-Aguilar [53] employed the method of Laplace transform to

$$^{modified-CF}_{\quad 0}D_t^{\alpha} x(t) = 1, \quad \forall t \in \mathbb{R}_0, \tag{1.26}$$

and obtained its solution

$$x(t) = x(0) + (1-\alpha) + \alpha t, \quad \forall t \in \mathbb{R}_0.$$

Although the above solution satisfies Eq. (1.26),

$$\lim_{t \to 0^+} x(t) = x(0) + (1-\alpha) \neq x(0)$$

when $\alpha \neq 1$. Consequently, the remainder term $H_\alpha$ is necessary for the existence of a unique solution to Eq. (1.22). Apparently, the result of Lemma 1.22 makes up for the lacks in literature [53–55].

## 1.7 Outline of the Book

The book is structured into **seven** chapters.

**This chapter** discusses the research background and significance of Euler, fractional Euler, exponential Euler and Mittag–Leffler Euler differences. Besides, it introduces the basic theory of fractional calculus, which involves Riemann–Liouville fractional integral, Riemann–Liouville fractional derivative, Caputo fractional derivative, and Caputo–Fabrizio operator.

**Chapter** 2 establishes the basic theory of Delta Mittag–Leffler Euler differences. This text presents the proposed Delta Mittag–Leffler Euler difference schemes for fractional-order differential equations in the sense of Caputo, along with their crucial features such as convergence. Additionally, Delta Mittag–Leffler Euler difference equations involving the Delta Mittag–Leffler Euler difference operator are introduced, and basic 0qualitative research is conducted, including global solution and global exponential stability analysis.

**Chapter** 3 presents an application of the theory discussed in Chap. 2. This chapter considers a fuzzy Cohen-Grossberg neural networks (CGNNs) of fractional order in the sense of Caputo. A Mittag–Leffler Euler difference for CGNNs is constructed by using the Delta Mittag–Leffler Euler difference. The chapter then considers the global solution, equilibrium point, global exponential stability, and synchronization of the proposed difference model.

**Chapter** 4 discusses the theory of Nabla Mittag–Leffler Euler differences. It establishes the corresponding Nabla Mittag–Leffler Euler difference schemes for fractional-order differential equations in the sense of Caputo. The obtained difference model falls within the scope of implicit Euler differences. Therefore, fractional PECE algorithm is proposed to effectively solve these implicit differences in numerical calculations. Further, this chapter introduces Nabla Mittag–Leffler Euler difference equations involving the Nabla Mittag–Leffler Euler difference operator. It then proceeds to study the existence of a unique globally exponentially stable solution for this model.

**Chapter** 5 establishes the fundamental structure of the exponential Euler difference form for differential equations with Caputo–Fabrizio operator and time lags. The research indicates that the obtained difference form also falls under the category of implicit Euler differences. Consequently, the fractional PECE algorithm is proposed to solve this implicit difference. Finally, the global exponential synchronization of the discrete-time model is discussed from the perspective of control theory.

**Chapter** 6 presents an application of the theory in Chap. 5 for fuzzy BAM neural networks (BAMNNs) with Caputo–Fabrizio operator. It establishes the exponential Euler difference scheme for a fuzzy BAMNNs with Caputo–Fabrizio operator. Additionally, it investigates the existence of a unique bounded asymptotically almost automorphic sequence, global exponential stability, and synchronization of the proposed exponential Euler difference model.

**Chapter** 7 summarises the main works of this book and provides future research perspectives.

# Chapter 2
# The Theory of Delta MLEDs

Exponential Euler differences have undergone rapid development in recent years, and a variety of exponential Euler difference methods have become significant research topics. This chapter introduces a novel difference method called the Mittag–Leffler Euler difference for semi-linear problems of fractional order, which includes the exponential Euler difference. A set of theories and methods for studying Mittag–Leffler Euler difference equations has been constructed. This provides a new avenue for discussing discrete-time models of fractional order in science and engineering.

## 2.1 Introduction

The theory of continuous-time fractional calculus [46–50, 60–62] has a long history within a wide field of mathematical analysis, which is almost as old as the classical theory of integer-order. A distinct feature of fractional derivative is nonlocal and the future status not only depends on the current status but also on the history of its old status. Although a sizeable mathematical theory in continuous-time fractional calculus and fractional-order differential equations (FODEs) have been existing, it has no extensive parallel theory for discrete-time fractional calculus and fractional-order difference equations (FODCEs) until recently. During the past decades, nevertheless, a surge of discrete configurations for fractional calculus were proposed and developed in the study of FODCEs. These works implied that the study of FODCEs has massive unexpected difficulties and technical complications compared with FODEs. For instance, Samko et al. [63] put forward Grünwald-Letnikov like operator and it had been applied widely in the research of engineering area [64–67]. The authors in literature [68–71] introduced on their own the basic Delta and Nabla FODCEs in the sense of Reimann–Liouville and Caputo, which were diffusely used in the discussion of FODCEs during these years, see [72–76]. Articles [77, 78] were devoted to the



T. Zhang et al., *Mittag-Leffler Euler Difference Techniques*,
https://doi.org/10.1007/978-981-96-6228-9_2

investigation of discrete-time fractional Laplacian equations, especially in homoclinic solutions, regularity and applications, etc. Ortigueira and Tenreiro Machado in literature [79] recently proposed a new discrete-time fractional derivatives based on the bilinear transformation and some basic properties were discussed. Meanwhile, the Delta and Nabla FODCEs with $h$-differences were described in literature [80]. The other types of FODCEs could be read in literature [81–85].

Over the years, various forms of numerical methods [15] were suggested to solve differential equations and various types of difference equations had been generated simultaneously from mathematics to computer science to engineering to biology. Such as, Euler differences [15], linear multi-step differences [15], Runge–Kutta differences [15], Taylor series methods [15], multi-value multi-stage difference methods [15], exponential Euler difference methods [19–29], finite difference methods [86] and finite element methods [86], etc.

In the light of variation-of-constants formulas, exponential Euler differences have got rapid development recently and a variety of exponential Euler difference methods have become very significant researching topics in literature [19–29], e.g., exponential Euler integrators [19–21, 26–29], exponential Runge–Kutta methods [22], multi-step exponential integrators [23], exponential general linear methods [24] and exponential Rosenbrock-type methods [25], etc. These numerical methods possess perfect linear stability and can be accomplished effectively. Consider semi-linear problems as described below

$$\begin{cases} \dfrac{\mathrm{d}X(t)}{\mathrm{d}t} = AX(t) + F(t, X(t)), \quad \forall t \in (a, +\infty), \\ X(a) = \xi \in \mathbb{R}^n. \end{cases} \tag{2.1}$$

Here, $X = (x_1, \ldots, x_n)^T \in \mathbb{R}^n$, $a \in \mathbb{R}$, $A$ is a nonsingular $n$-order real matrix and $F \in C([a, +\infty) \times \mathbb{R}^n, \mathbb{R}^n)$ satisfy

($C_1$) $|A|_{\det} \neq 0$ and $A$ is a diagonalization matrix, i.e., it exists a diagonal matrix $D = \mathrm{diag}\{d_1, \ldots, d_n\}$ and $n$-order invertible real matrix $P$ such that $P^{-1}AP = D$.

($C_2$) It exists a $n$-order constant matrix $L_F > \mathbb{O}_n$ such that

$$\|F(t, X) - F(t, Y)\| \leq \|L_F\|\|X - Y\|, \quad \forall X, Y \in \mathbb{R}^n, t \in [a, +\infty).$$

Let $h > 0$ be a given time step length. Adapting exponential Euler integrators in [19–21, 26–29] to Eq. (2.1) yields the exponential Euler difference equations as noted below

$$\begin{cases} X(t+h) = e^{Ah}X(t) + (e^{Ah} - \mathbb{I}_n)A^{-1}F(t, X(t)), \quad \forall t \in (h\mathbb{N}_0)_a, \\ X(a) = \xi \in \mathbb{R}^n, \end{cases} \tag{2.2}$$

where $(h\mathbb{N}_0)_a = \{a + kh : k = 0, 1, \ldots\}$. On the basis of Eq. (2.2), literature [19–21, 26–29] were devoted to discussing the various dynamical behaviours for a variety of exponential Euler difference models, e.g., periodicity [19, 28], almost periodicity [26, 27, 29], stability [21, 28, 29] and randomness [21, 27], etc.

Motivated by exponential Euler difference methods for semi-linear differential equations of first order, one question naturally comes to mind: could the thought of exponential Euler differences be applied in the research of semi-linear problems of FODEs as noted below?

$$\begin{cases} {}_a^C D_t^\alpha X(t) = AX(t) + F(t, X(t)), \quad \forall t \in (a, +\infty), \\ X(a) = \xi \in \mathbb{R}^n, \end{cases} \tag{2.3}$$

where ${}_a^C D_t^\alpha X = ({}_a^C D_t^{\alpha_1} x_1, \ldots, {}_a^C D_t^{\alpha_n} x_n)^T$ is Caputo fractional derivative with lower limit at $a \in \mathbb{R}, \alpha = (\alpha_1, \ldots, \alpha_n)^T \in \mathbb{R}^n, 0 < \alpha_i \leq 1 \ (i = 1, 2, \ldots, n)$ and the meanings of other symbols in Eq. (2.3) are the same as those in Eq. (2.1). The main aim of this chapter will show that the thought of exponential Euler differences is indeed able to be applied in the research of semi-linear problems of FODEs. Correspondingly, it is called Mittag–Leffler Euler differences in this book. On the basis of the variation-of-constants formulas in FODEs and some important properties of Mittag–Leffler functions, the numerical solution of FODEs (2.3) achieved by Mittag–Leffler Euler differences is subsequently acquired and its convergence should be discussed. On the other hand, some novel Mittag–Leffler Euler difference operators should be defined stimulated by Mittag–Leffler Euler differences. Further, the original difference equations involving Mittag–Leffler Euler difference operators are proposed and some basic issues (e.g., the existence of a unique global bounded solution and global exponential stability) will be investigated. This chapter is to construct a set of theories and methods in studying Mittag–Leffler Euler difference equations, which should provide a new idea and approach to discuss the discrete-time models of fractional order in areas of science and engineering in the future.

The structure of this chapter is as follows. In Sect. 2.2, Delta Mittag–Leffler Euler difference is set up for a FODEs in the sense of Caputo and the advantages of this difference are analyzed in Sect. 2.3. In Sect. 2.4, the convergence is studied. In Sect. 2.5, a novel Mittag–Leffler Euler difference equations with Mittag–Leffler or polynomial kernel is introduced. Further, the well-posed problems of the proposed difference equations are investigated in Sect. 2.6. An explicative example is used to visually elaborate the researching findings in Sect. 2.7.

## 2.2 Mittag–Leffler Euler Difference

Let

$$\mathcal{T}_a = \left\{ t_p \in [a, +\infty) : t_0 = a, \lim_{p \to +\infty} t_p = +\infty, t_p \le t_{p+1}, p \in \mathbb{N}_0 \right\}$$

and $[a, +\infty) = \bigcup_{p=0}^{+\infty} [t_p, t_{p+1})$ for some $a \in \mathbb{R}$. Choosing a set

$$\mathcal{N}_0 = \left\{ N_p \in \mathbb{Z} : N_0 = 0, N_p < N_{p+1}, p \in \mathbb{N}_0 \right\}$$

ensures that

$$C_a = \{ c_k \in [a, +\infty) : k \in \mathbb{N}_0 \}$$

satisfies the conditions below

(P1) $c_{N_p} = t_p, \forall p \in \mathbb{N}_0$.

(P2) $c_{N_p+j} = c_{N_p} + \dfrac{(c_{N_{p+1}} - c_{N_p})j}{N_{p+1} - N_p}, \forall j \in [0, N_{p+1} - N_p]_{\mathbb{Z}}, p \in \mathbb{N}_0$.

$C_a$ is called a $\mathcal{T}_a$-$\mathcal{N}_0$ sequence if $C_a$ satisfies conditions (P1)–(P2) with $\mathcal{T}_a$ and $\mathcal{N}_0$. Obviously, $C_a$ is a partition of $[a, +\infty)$ and $[a, +\infty) = \bigcup_{k=0}^{+\infty} [c_k, c_{k+1})$. Clearly, $N_{p+1} - N_p - 1$ stands for the number of all split points in $[t_p, t_{p+1}]$, $\forall p \in \mathbb{N}_0$. Besides, $\Delta c_k = c_{k+1} - c_k$ depicts the corresponding step length of $[t_p, t_{p+1}]$ for $k \in [N_p, N_{p+1})_{\mathbb{Z}}$, $\forall p \in \mathbb{N}_0$. This paper always assumes that $C_a$ is a $\mathcal{T}_a$-$\mathcal{N}_0$ sequence and $(h\mathbb{N}_0)_a = \{a + kh : k \in \mathbb{N}_0\}$ for some $h > 0$.

This chapter is devoted to studying the semi-linear FODEs with piecewise feature as noted below

$$\begin{cases} {}_{t_p}^{C}D_t^{\alpha} X(t) = AX(t) + F(t, X(t)), & \forall t \in (t_p, t_{p+1}], \\ X(a) = \xi \in \mathbb{R}^n, \end{cases} \tag{2.4}$$

where $p = 0, 1, \ldots, \alpha = (\alpha_1, \ldots, \alpha_n)^T \in \mathbb{R}^n$, $\mathbf{0}_n < \alpha \le \mathbf{1}_n$,

$$X = (x_1, \ldots, x_n)^T \in C([a, +\infty), \mathbb{R}^n), \quad {}_{t_p}^{C}D_t^{\alpha} X = ({}_{t_p}^{C}D_t^{\alpha_1} x_1, \ldots, {}_{t_p}^{C}D_t^{\alpha_n} x_n)^T,$$

$A \in \mathbb{R}^{n \times n}$ and $F \in C([a, +\infty) \times \mathbb{R}^n, \mathbb{R}^n)$ meet the assumptions ($C_1$)-($C_2$).

By ($C_1$), FODEs (2.4) can be turned into

$$\begin{cases} {}_{t_p}^{C}D_t^{\alpha} Y(t) = DY(t) + P^{-1}F(t, PY(t)), & \forall t \in (t_p, t_{p+1}], \\ Y(a) = P^{-1}\xi \in \mathbb{R}^n, \end{cases} \tag{2.5}$$

where $p = 0, 1, \ldots, Y = P^{-1}X$, $X$ is solution of FODEs (2.4).

Let $h_k = \Delta c_k = c_{k+1} - c_k$,

$$\mu_k = \max\{N_p \in \mathcal{N}_0 : N_p \le k,\, p \in \mathbb{N}_0\}, \quad \nu_k = k - \mu_k, \quad \forall k \in \mathbb{N}_0.$$

Supposing that ($C_1$)-($C_2$) hold and $C_a$ is a $\mathcal{T}_a$-$\mathcal{N}_0$ sequence. Let

$$\begin{cases} J_t^{\alpha} = \operatorname{diag}\left(t^{\alpha_1 - 1}, \ldots, t^{\alpha_n - 1}\right), \\ \mathcal{E}_{\alpha}(Dt^{\alpha}) = \operatorname{diag}\left(\mathbb{E}_{\alpha_1}(d_1 t^{\alpha_1}), \ldots, \mathbb{E}_{\alpha_n}(d_n t^{\alpha_n})\right), \\ \mathcal{E}_{\alpha,\alpha}(Dt^{\alpha}) = \operatorname{diag}\left(\mathbb{E}_{\alpha_1,\alpha_1}(d_1 t^{\alpha_1}), \ldots, \mathbb{E}_{\alpha_n,\alpha_n}(d_n t^{\alpha_n})\right) \end{cases} \tag{2.6}$$

for all $t \in [a, +\infty)$.

By Lemma 1.24, Eq. (2.5) possesses a unique solution given by

$$\begin{aligned} Y(t) = {} & \mathcal{E}_{\alpha}\big[D(t - t_p)^{\alpha}\big]Y(t_p) \\ & + \int_{t_p}^{t} J_{t-s}^{\alpha}\mathcal{E}_{\alpha,\alpha}[D(t-s)^{\alpha}]P^{-1}F(s, PY(s))\mathrm{d}s, \end{aligned} \tag{2.7}$$

where $t \in (t_p, t_{p+1}]$, $p \in \mathbb{N}_0$. Discussing Eq. (2.7) with piecewise constant arguments as noted below

$$\begin{aligned} Y(t) = {} & \mathcal{E}_{\alpha}\big[D(t - t_p)^{\alpha}\big]Y(t_p) \\ & + \int_{t_p}^{t} J_{t-s}^{\alpha}\mathcal{E}_{\alpha,\alpha}[D(t-s)^{\alpha}]P^{-1}F(\lfloor s \rfloor, PY(\lfloor s \rfloor))\mathrm{d}s, \end{aligned} \tag{2.8}$$

where $\lfloor s \rfloor = \max\{c_k : s \ge c_k, k \in \mathbb{N}_0\}$, $t \in (t_p, t_{p+1}]$, $p = 0, 1, \ldots$. For arbitrary $t \ge a$, there exists $k \in \mathbb{N}_0$ ensuring that $t \in (c_k, c_{k+1}] \subseteq (t_p, t_{p+1}]$, then Eq. (2.8) is transformed into

$$\begin{aligned} Y(t) = {} & \mathcal{E}_{\alpha}\big[D(t - t_p)^{\alpha}\big]Y(t_p) \\ & + \int_{c_k}^{t} J_{t-s}^{\alpha}\mathcal{E}_{\alpha,\alpha}[D(t-s)^{\alpha}]\mathrm{d}s\, P^{-1}F(c_k, PY(c_k)) \\ & + \sum_{\iota=N_p}^{k-1}\int_{c_\iota}^{c_{\iota+1}} J_{t-s}^{\alpha}\mathcal{E}_{\alpha,\alpha}[D(t-s)^{\alpha}]\mathrm{d}s\, P^{-1}F(c_\iota, PY(c_\iota)) \\ = {} & \mathcal{E}_{\alpha}\big[D(t - c_{\mu_k})^{\alpha}\big]Y(c_{\mu_k}) \\ & + \big[\mathcal{E}_{\alpha}(D(t-c_k)^{\alpha}) - \mathbb{I}_n\big]D^{-1}P^{-1}F(c_k, PY(c_k)) \\ & + \sum_{l=1}^{k-\mu_k}\Big\{\mathcal{E}_{\alpha}[D(t-c_{k-l})^{\alpha}] - \mathcal{E}_{\alpha}[D(t-c_{k-l+1})^{\alpha}]\Big\} \\ & \times D^{-1}P^{-1}F(c_{k-l}, PY(c_{k-l})), \end{aligned} \tag{2.9}$$

where $t \in (c_k, c_{k+1}] \subseteq (t_p, t_{p+1}]$, $p, k \in \mathbb{N}_0$. In the calculation of (2.9), Lemma 1.23 is employed.

Letting $t \to (c_k)^+$ and $t \to (c_{k+1})^-$ in (2.9) respectively, it derives

$$\begin{cases} Y(c_k) &= \mathcal{E}_\alpha\big[D(c_k - c_{\mu_k})^\alpha\big]Y(c_{\mu_k}) \\ &\quad + \sum_{l=0}^{\nu_k} W_\alpha^{[l-1,k]} D^{-1}P^{-1}F(c_{k-l}, PY(c_{k-l})), \\ Y(c_{k+1}) &= \mathcal{E}_\alpha\big[D(c_{k+1} - c_{\mu_k})^\alpha\big]Y(c_{\mu_k}) \\ &\quad + \sum_{l=0}^{\nu_k} W_\alpha^{[l,k]} D^{-1}P^{-1}F(c_{k-l}, PY(c_{k-l})), \quad \forall k \in \mathbb{N}_0. \end{cases} \tag{2.10}$$

In the above derivation, the facts are used: $c_k - c_s = (k-s)h_k$ and $c_{k+1} - c_s = (k+1-s)h_k$ for $s \in [\mu_k, k]_{\mathbb{Z}}$ and $(c_k, c_{k+1}] \subseteq (t_p, t_{p+1}]$, $k \in \mathbb{N}_0$. By using cancellation law, it yields from (2.10) that

$$Y(c_{k+1}) = \Lambda_\alpha^{[k]} Y(c_k) + \sum_{l=0}^{\nu_k} \Theta_\alpha^{[l,k]} P^{-1}F(c_{k-l}, PY(c_{k-l})), \quad \forall k \in \mathbb{N}_0. \tag{2.11}$$

So Mittag–Leffler Euler difference (MLED) of FODEs (2.5) is described as

$$\begin{cases} Y^{[k+1]} = \Lambda_\alpha^{[k]} Y^{[k]} + \sum_{l=0}^{\nu_k} \Theta_\alpha^{[l,k]} P^{-1} F^{[k-l]}(PY^{[k-l]}), \\ Y^{[0]} = P^{-1}\xi, \end{cases} \tag{PMLED$_\Delta$}$$

where

$$Y^{[k]} = Y(c_k) = P^{-1}X(c_k), \quad F^{[k]}(Z) = F(c_k, Z)$$

is an abbreviation for arbitrary $Z \in \mathbb{R}^n$,

$$\begin{cases} \Lambda_\alpha^{[k]} = \operatorname{diag}\left(\langle \Lambda_\alpha^{[k]}\rangle_1, \ldots, \langle \Lambda_\alpha^{[k]}\rangle_n\right), \\ \langle \Lambda_\alpha^{[k]}\rangle_i = \frac{\mathbb{E}_{\alpha_i}[d_i(\nu_k h_k + h_k)^{\alpha_i}]}{\mathbb{E}_{\alpha_i}[d_i(\nu_k h_k)^{\alpha_i}]}, \\ \Theta_\alpha^{[l,k]} = \left\{W_\alpha^{[l,k]} - \Lambda_\alpha^{[k]} W_\alpha^{[l-1,k]}\right\} D^{-1}, \\ W_\alpha^{[l,k]} = \operatorname{diag}\left(\langle W_\alpha^{[l,k]}\rangle_1, \ldots, \langle W_\alpha^{[l,k]}\rangle_n\right), \quad \langle W_\alpha^{[-1,k]}\rangle_i \equiv 0, \\ \langle W_\alpha^{[l,k]}\rangle_i = \mathbb{E}_{\alpha_i}[d_i(lh_k + h_k)^{\alpha_i}] - \mathbb{E}_{\alpha_i}[d_i(lh_k)^{\alpha_i}] \end{cases} \tag{2.12}$$

for $l \in [0, \nu_k]_{\mathbb{Z}}$, $k \in \mathbb{N}_0$.

Let

$$\underline{h} = \inf_{k \in \mathbb{N}_0} h_k, \quad \bar{h} = \sup_{k \in \mathbb{N}_0} h_k, \quad N^* = \sup_{p \in \mathbb{N}_0} (N_{p+1} - N_p), \quad t_\diamond = \sup_{p \in \mathbb{N}_0} (t_{p+1} - t_p).$$

Throughout this paper, supposing that $t_\diamond < +\infty$.

**Remark 2.1** In $(\text{PMLED})_\Delta$, it exists some indetectable facts as noted below.

(1) $\nu_k \in [0, N_{p+1} - N_p)_\mathbb{Z} \subseteq [0, N^*)_\mathbb{Z}$ for $k \in [N_p, N_{p+1})_\mathbb{Z}$, $p \in \mathbb{N}_0$.
(2) $(\nu_k + 1)h_k \in (0, t_\diamond]$ for all $k \in \mathbb{N}_0$.
(3) $\left|\Lambda_\alpha^{[k]}\right|_{\det} > 0$ for $\alpha \in (0, 1]$ and $k \in \mathbb{N}_0$.
(4) $\langle \Lambda_\alpha^{[k]} \rangle_i \in (0, 1)$ for $d_i < 0$ and $\langle \Lambda_\alpha^{[k]} \rangle_i \in (1, +\infty)$ for $d_i > 0$, $k \in \mathbb{N}_0$, $i = 1, 2, \ldots, n$.
(5) $\langle W_\alpha^{[l,k]} D^{-1} \rangle_i = \dfrac{\langle W_\alpha^{[l,k]} \rangle_i}{d_i} > 0$ for $d_i \neq 0$, $l \in \mathbb{N}_0$, $i = 1, 2, \ldots, n$.
(6) If $\alpha = \mathbf{1}_n$, then $\Theta_\alpha^{[l,k]} \equiv \mathbb{O}_n$ for all $l \in [1, \nu_k]_\mathbb{Z}$, $k \in \mathbb{N}_0$.
(7) $\Lambda_\alpha$ and $\Theta_\alpha$ are abbreviate forms, which are dependent on both $\mathcal{N}_0$ and $\mathcal{H} = \{h_k : k \in \mathbb{N}_0\}$.

**Remark 2.2** Apparently, $X(c_k) = PY(c_k)$ $(k \in \mathbb{N}_0)$ stands for numerical solution of FODEs (2.4). Further, $(\text{PMLED})_\Delta$ is equivalent to piecewise difference equations

$$\begin{cases} Y^{[k+1]} = \Lambda_\alpha^{[k]} Y^{[k]} + \sum_{l=0}^{k-N_p} \Theta_\alpha^{[l,k]} P^{-1} F^{[k-l]}(PY^{[k-l]}), \\ Y^{[0]} = P^{-1}\xi, \end{cases}$$

where $k \in [N_p, N_{p+1})_\mathbb{Z}$, $p \in \mathbb{N}_0$.

**Remark 2.3** Some literature recently were devoted to investigating FODEs with short-term memory (i.e., piecewise feature), which just discuss the behaviors of state variables in the recent past, see [87–90]. To authors' knowledge, almost no research focuses on the study of FODCEs with short-term memory to date. By Fig. 2.1 and the relative definitions as described above, $t_p = c_{N_p}$ and $t_{p+1} - t_p = c_{N_{p+1}} - c_{N_p}$ portray the starting point and the length of each piece (i.e., $(t_p, t_{p+1}]$) of short-term memory in FODEs (2.4) or (2.5), respectively, $p \in \mathbb{N}_0$. As a consequence, $(\text{PMLED})_\Delta$ is a fractional-order difference equations with short-term memory and

$$\sum_{l=1}^{\nu_k} \Theta_\alpha^{[l,k]} P^{-1} F^{[k-l]}(PY^{[k-l]})$$

stands for the trait of short-term memory of $(\text{PMLED})_\Delta$, $k \in \mathbb{N}_0$.

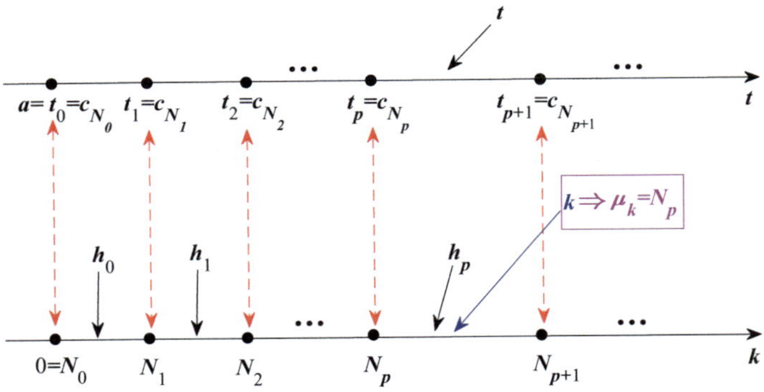


**Fig. 2.1** Graphic illustrations of sequences $\mathcal{T}_a$, $\mathcal{N}_0$, $C_a$ and $\mathcal{H}$

**MLED with Constant Step Length** $h$

In $(\text{PMLED})_\Delta$, taking

$$\mathcal{T}_a = \left\{t_p : t_0 = a, t_{p+1} = t_p + h(N_{p+1} - N_p), \forall p \in \mathbb{N}_0\right\}$$

for some $h > 0$. Then $(h\mathbb{N}_0)_a = \{a + kh : k \in \mathbb{N}_0\}$ is a $\mathcal{T}_a$-$\mathcal{N}_0$ sequence satisfying conditions (P1)–(P2). Consequently, $(\text{PMLED})_\Delta$ is reformed to

$$\begin{cases} Y^{[k+1]} = \Lambda_\alpha^{[k]} Y^{[k]} + \sum_{l=0}^{\nu_k} \Theta_\alpha^{[l,k]} P^{-1} F^{[k-l]}(PY^{[k-l]}), \\ Y^{[0]} = P^{-1}\xi, \end{cases} \tag{2.13}$$

where $Y^{[k]} = Y(a + kh)$, $\Lambda_\alpha^{[k]}$ and $\Theta_\alpha^{[l,k]}$ are defined as those in (2.12) with $h_k \equiv h$, $l \in [0, \nu_k]_{\mathbb{Z}}$, $k \in \mathbb{N}_0$.

Equation (2.13) is equivalent to piecewise $h$-difference equations as follows

$$\begin{cases} Y^{[k+1]} = \Lambda_\alpha^{[k]} Y^{[k]} + \sum_{l=0}^{k-N_p} \Theta_\alpha^{[l,k]} P^{-1} F^{[k-l]}(PY^{[k-l]}), \\ Y^{[0]} = P^{-1}\xi, \end{cases}$$

where $k \in [N_p, N_{p+1})_{\mathbb{Z}}$, $p \in \mathbb{N}_0$.

**Remark 2.4** It should be noted that $(\text{PMLED})_\Delta$ is fractional-order difference equations with different step length $h_p$ in each segment $[N_p, N_{p+1}]_{\mathbb{Z}}$ (i.e., aperiodic time scale), but Eq. (2.13) has a constant step length $h$ in every segment $[N_p, N_{p+1}]_{\mathbb{Z}}$ (i.e., periodic time scale), $p \in \mathbb{N}_0$. In addition, if $A$ is a diagonal matrix with $A = \text{diag}\{a_1, \ldots, a_n\}$, then the above discussions are apparently valid by taking $D = A$ and $P = \mathbb{I}_n$.

**Long-term Memory Model**

This example considers Eq. (2.3). For the purpose of gaining MLED of Eq. (2.3) and by the derivation of $(\mathrm{PMLED})_\Delta$, $\mu_k$ should be redefined as $\mu_k \equiv 0$ and $h_k \equiv h$ for all $k \in \mathbb{N}_0$. Then $(\mathrm{PMLED})_\Delta$ is reformed to

$$\begin{cases} Y^{[k+1]} = \Lambda_\alpha^{[k]} Y^{[k]} + \sum_{l=0}^{k} \Theta_\alpha^{[l,k]} P^{-1} F^{[k-l]}(PY^{[k-l]}), \\ Y^{[0]} = P^{-1}\xi, \end{cases} \qquad (\mathrm{MLED})_\Delta$$

where $Y^{[k]} = Y(a+kh)$, $X(a+kh) = PY(a+kh)$ stands for the numerical solution of FODEs (2.3), $P$ is defined as that in $(\mathrm{C}_1)$, $\Lambda_\alpha^{[k]}$ and $\Theta_\alpha^{[l,k]}$ are defined as those in (2.12) with $\nu_k = k$ and $h_k \equiv h$, $l \in [0,k] \cap \mathbb{Z}$, $k \in \mathbb{N}_0$.

If $\alpha = \mathbf{1}_n = (1, \ldots, 1)^T \in \mathbb{R}^n$, then $((\mathrm{MLED})_\Delta)$ is simplified to Eq. (2.2). So Mittag–Leffler Euler difference method expands the approach of exponential Euler difference. Besides, the convergence of Mittag–Leffler Euler difference has been studied in Sect. 2.4.

**MLED with Arbitrary Matrix $A$**

In FODEs (2.4), $A$ is supposed to be a diagonalizable matrix. Hereby, assume that $A$ is an arbitrary matrix with $|A|_{\det} \neq 0$ and $\alpha \in (0,1]$ in FODEs (2.4). Let $C_a$ be a $\mathcal{T}_a$-$\mathcal{N}_0$ sequence, $h_k$, $\mu_k$ and $\nu_k$ be defined as the same as those in $(\mathrm{PMLED})_\Delta$, $\forall k \in \mathbb{N}_0$. Supposing that $(\mathrm{C}_2)$ holds. By Mittag–Leffler Euler difference method as that in $(\mathrm{PMLED})_\Delta$, it easily gets MLED of FODEs (2.4) as described below

$$\begin{cases} X^{[k+1]} = \hat{\Lambda}_\alpha^{[k]} X^{[k]} + \sum_{l=0}^{\nu_k} \hat{\Theta}_\alpha^{[l,k]} F^{[k-l]}(X^{[k-l]}), \\ X^{[0]} = \xi, \end{cases} \qquad (2.14)$$

where $X^{[k]} = X(c_k)$,

$$\hat{\Lambda}_\alpha^{[k]} = \mathbb{E}_\alpha[A(\nu_k h_k + h_k)^\alpha]\{\mathbb{E}_\alpha[A(\nu_k h_k)^\alpha]\}^{-1},$$

$$\hat{\Theta}_\alpha^{[l,k]} = \hat{W}_\alpha^{[l,k]} - \hat{\Lambda}_\alpha^{[k]} \hat{W}_\alpha^{[l-1,k]}, \quad \hat{W}_\alpha^{[-1,k]} \equiv 0,$$

$$\hat{W}_\alpha^{[l,k]} = (lh_k + h_k)^\alpha \mathbb{E}_{\alpha,\alpha+1}[A(lh_k + h_k)^\alpha] - (lh_k)^\alpha \mathbb{E}_{\alpha,\alpha+1}[A(lh_k)^\alpha]$$

for $l \in [0, \nu_k]_{\mathbb{Z}}$, $k \in \mathbb{N}_0$.

**Exponential Euler Difference Equations**
If $\alpha = \mathbf{1}_n$ in Eq. (2.13), then

$$\Lambda_\alpha^{[k]} \equiv e^{Dh}, \ \ \Theta_\alpha^{[0,k]} \equiv (e^{Dh} - \mathbb{I}_n)D^{-1},$$

$\Theta_\alpha^{[l,k]} \equiv \mathbb{O}_n$ (see Remark 2.1) for all $l \in [1, \nu_k]_{\mathbb{Z}}$, $k \in \mathbb{N}_0$.
So $(\text{PMLED})_\Delta$ is turned into

$$\begin{cases} Y^{[k+1]} = e^{Dh} Y^{[k]} + D^{-1}(e^{Dh} - \mathbb{I}_n)P^{-1}F^{[k]}(PY^{[k]}), \\ Y^{[0]} = P^{-1}\xi, \end{cases}$$

which is equal to Eq. (2.2), $Y^{[k]} = P^{-1}X(a + kh)$, $k \in \mathbb{N}_0$. Equation (2.2) is called exponential Euler analogue (EEA) corresponding to Eq. (2.1).

**Remark 2.5** In Eq. (2.2), the memory terms $F^{[s]}(X^{[s]})$ $(s = k-1, k-2, \ldots, k-\nu_k)$ in $(\text{PMLED})_\Delta$ have become zero. On the basis of Eq. (2.2), the authors [19–29] studied the dynamic behaviours of different forms of discrete applied models. Obviously, $(\text{PMLED})_\Delta$ of this chapter is an expansion of discrete analogues in literature [19–29].

Let

$$\bar{\Lambda}_\alpha = \text{diag}\left(\langle\bar{\Lambda}_\alpha\rangle_1, \ldots, \langle\bar{\Lambda}_\alpha\rangle_n\right), \quad \bar{\Theta}_\alpha = \sup_{k\in\mathbb{N}_0} \sum_{l=0}^{\nu_k} \left[\Theta_\alpha^{[l,k]}\right]_{\text{abs}},$$

where

$$\langle\bar{\Lambda}_\alpha\rangle_i = \sup_{k\in\mathbb{N}_0} \langle\Lambda_\alpha^{[k]}\rangle_i = \sup_{k\in\mathbb{N}_0} \frac{\mathbb{E}_{\alpha_i}[d_i(\nu_k h_k + h_k)^{\alpha_i}]}{\mathbb{E}_{\alpha_i}[d_i(\nu_k h_k)^{\alpha_i}]}, \quad i = 1, 2, \ldots, n.$$

**Lemma 2.1** *In* $(\text{PMLED})_\Delta$, *supposing that* $(\text{C}_1)$ *is fulfilled. Then*

(1) $0 < \langle\bar{\Lambda}_\alpha\rangle_i < 1$ *provided that* $[D]_{\text{dia}} < \mathbf{0}_n$, $\forall k \in \mathbb{N}_0$, $i = 1, 2, \ldots, n$.

(2) $\sum_{l=0}^{\nu_k} \Theta_\alpha^{[l,k]} = (\Lambda_\alpha^{[k]} - \mathbb{I}_n)D^{-1}$, $\forall k \in \mathbb{N}_0$.

(3) $\bar{\Theta}_\alpha \le D^{-1}(\bar{\Lambda}_\alpha + \mathbb{I}_n)\left[\mathcal{E}_\alpha(Dt_\diamond^\alpha) - \mathbb{I}_n\right]$, *where* $\mathcal{E}_\alpha$ *is defined as that in* (2.6).

***Proof*** Owing to $\nu_k h_k \in [0, t_\diamond]$ for all $k \in \mathbb{N}_0$, it yields from Lemma 1.21 that

$$d_i < 0 \Rightarrow \langle\bar{\Lambda}_\alpha\rangle_i = \sup_{k\in\mathbb{N}_0} \frac{\mathbb{E}_{\alpha_i}[d_i(\nu_k h_k + h_k)^{\alpha_i}]}{\mathbb{E}_{\alpha_i}[d_i(\nu_k h_k)^{\alpha_i}]} \le \max_{s\in[0,t_\diamond]} \frac{\mathbb{E}_{\alpha_i}[d_i(s + \underline{h})^{\alpha_i}]}{\mathbb{E}_{\alpha_i}[d_i s^{\alpha_i}]} \in (0, 1),$$

where $i = 1, 2, \ldots, n$. So (1) of Lemma 2.1 is proved. Further,

$$\begin{aligned}\sum_{l=0}^{\nu_k} \langle \Theta_\alpha^{[l,k]} \rangle_i &= \sum_{l=0}^{\nu_k} \left[ \langle W_\alpha^{[l,k]} \rangle_i - \langle \Lambda_\alpha^{[k]} \rangle_i \langle W_\alpha^{[l-1,k]} \rangle_i \right] d_i^{-1} \\ &= \left[ \left( \mathbb{E}_\alpha[d_i(\nu_k h_k + h_k)^\alpha] - 1 \right) - \langle \Lambda_\alpha^{[k]} \rangle_i \left( \mathbb{E}_\alpha[d_i(\nu_k h_k)^\alpha] - 1 \right) \right] d_i^{-1} \\ &= (\langle \Lambda_\alpha^{[k]} \rangle_i - 1) d_i^{-1}, \quad i = 1, 2, \ldots, n,\end{aligned}$$

which yields $\sum_{l=0}^{\nu_k} \Theta_\alpha^{[l,k]} = (\Lambda_\alpha^{[k]} - \mathbb{I}_n) D^{-1}$ for all $k \in \mathbb{N}_0$. Besides, it indicates from Remark 2.1 that

$$\begin{aligned}\sum_{l=0}^{\nu_k} \langle [\Theta_\alpha^{[l,k]}]_{\text{abs}} \rangle_i &\leq \sum_{l=0}^{\nu_k} \max \left\{ \frac{\langle W_\alpha^{[l,k]} \rangle_i}{d_i}, \langle \Lambda_\alpha^{[k]} \rangle_i \frac{\langle W_\alpha^{[l-1,k]} \rangle_i}{d_i} \right\} \\ &\leq \sum_{l=0}^{\nu_k} \left\{ \frac{\langle W_\alpha^{[l,k]} \rangle_i}{d_i} + \langle \Lambda_\alpha^{[k]} \rangle_i \frac{\langle W_\alpha^{[l-1,k]} \rangle_i}{d_i} \right\} \\ &\leq \frac{\{\mathbb{E}_{\alpha_i}[d_i t_\diamond^{\alpha_i}] - 1\}}{d_i} + \frac{\langle \bar{\Lambda}_\alpha \rangle_i \{\mathbb{E}_{\alpha_i}[d_i t_\diamond^{\alpha_i}] - 1\}}{d_i} \\ &\leq \frac{\left[1 + \langle \bar{\Lambda}_\alpha \rangle_i\right] \{\mathbb{E}_{\alpha_i}[d_i t_\diamond^{\alpha_i}] - 1\}}{d_i}, \quad i = 1, 2, \ldots, n.\end{aligned}$$

Consequently, the rest of Lemma 2.1 is proved. The proof is achieved. □

Let

$$\mathcal{O}_\alpha = \sup_{k \in \mathbb{N}_0} \Theta_\alpha^{[0,k]}, \quad \mathcal{R}_\alpha = \sup_{k \in \mathbb{N}_0} \sum_{l=1}^{\nu_k} \left[\Theta_\alpha^{[l,k]}\right]_{\text{abs}}.$$

If $\alpha = \mathbf{1}_n$ in $(\text{PMLED})_\Delta$, $(h\mathbb{N}_0)_a$ is a $\mathcal{T}_a$-$\mathcal{N}_0$ sequence, then it yields

$$\bar{\Lambda}_\alpha = e^{Dh}, \quad \bar{\Theta}_\alpha = (e^{Dh} - \mathbb{I}_n) D^{-1} = \mathcal{O}_\alpha, \quad \mathcal{R}_\alpha = \mathbb{O}_n.$$

Taking $\alpha$ and $D$ as two constants in $(\text{PMLED})_\Delta$. Furthermore, $\mathcal{N}_0 = \{N_p = 10p : p \in \mathbb{N}_0\}$ and $h_k \equiv 0.01$, $\forall k \in \mathbb{N}_0$. Some scalar information of $\bar{\Theta}_\alpha$, $\mathcal{O}_\alpha$ and $\mathcal{R}_\alpha$ are calculated in Tables 2.1 and 2.2.

**Remark 2.6** By observing the last columns of Tables 2.1 and 2.2, $\mathcal{R}_\alpha$ gradually becomes 0 as $\alpha$ tends to 1. In other words, the value of $\bar{\Theta}_\alpha$ is mainly dependent on the value of $\mathcal{O}_\alpha$ as $\alpha \to 1$. The reason mainly lies in the gradual disappearance of the memory terms in $(\text{PMLED})_\Delta$ as $\alpha \to 1$. If $\alpha = \mathbf{1}_n$, $\mathcal{R}_\alpha \equiv \mathbb{O}_n$ owing to Remark 2.1.

**Table 2.1** $D = -5$

| | $\bar{\Theta}_\alpha$ | $O_\alpha$ | $\mathcal{R}_\alpha$ | $\mathcal{R}_\alpha/\bar{\Theta}_\alpha$ (%) |
|---|---|---|---|---|
| $\alpha = 0.3$ | 0.2357 | 0.1203 | 0.1154 | 48.97 |
| $\alpha = 0.5$ | 0.1458 | 0.0769 | 0.0689 | 47.27 |
| $\alpha = 0.7$ | 0.0657 | 0.0381 | 0.0276 | 41.97 |
| $\alpha = 0.9$ | 0.0208 | 0.0158 | 0.0050 | 24.26 |
| $\alpha = 0.98$ | 0.0115 | 0.0108 | 0.0007 | 6.67 |
| $\alpha = 1$ | 0.0098 | 0.0098 | 0 | 0 |

**Table 2.2** $D = 5$

| | $\bar{\Theta}_\alpha$ | $O_\alpha$ | $\mathcal{R}_\alpha$ | $\mathcal{R}_\alpha/\bar{\Theta}_\alpha$ (%) |
|---|---|---|---|---|
| $\alpha = 0.3$ | 9.0804 | 5.2880 | 3.7924 | 41.76 |
| $\alpha = 0.5$ | 0.3230 | 0.1905 | 0.1326 | 41.03 |
| $\alpha = 0.7$ | 0.0794 | 0.0510 | 0.0284 | 35.79 |
| $\alpha = 0.9$ | 0.0216 | 0.0173 | 0.0043 | 19.99 |
| $\alpha = 0.98$ | 0.0120 | 0.0114 | 0.0006 | 5.36 |
| $\alpha = 1$ | 0.0103 | 0.0103 | 0 | 0 |

**Remark 2.7** From the proof of Lemma 2.1 and the definition of $\Theta_\alpha^{[l,k]}$, $\Lambda_\alpha^{[k]}$ and $\Theta_\alpha^{[l,k]}$ in $(\text{PMLED})_\Delta$ are mainly dependent on $lh_k$ ($l \in [0, \nu_k]_\mathbb{Z}$), which is included in a bounded interval $[0, t_\diamond]$, i.e., $lh_k \in [0, \nu_k h_k] \subseteq [0, t_\diamond]$, $l \in [0, \nu_k]_\mathbb{Z}$, $k \in \mathbb{N}_0$, please see Remark 2.1. So the bounded estimations $\bar{\Lambda}_\alpha$ and $\bar{\Theta}_\alpha$ in Lemma 2.1 are rational. Nevertheless, in $(\text{MLED})_\Delta$, $\mu_k \equiv 0$ and $\{kh : k \in \mathbb{N}_0\}$ is not bounded. Owing to the complexity of Mittag–Leffler function $\mathbb{E}$, it is hard to estimate the bounds of $\Lambda_\alpha^{[k]}$ and $\Theta^{[l,k]}$ in $(\text{MLED})_\Delta$, $l \in [0, k]_\mathbb{Z}$, $k \in \mathbb{N}_0$. Noting that the bounds $\bar{\Lambda}_\alpha$ and $\bar{\Theta}_\alpha$ are crucial to the study of $(\text{PMLED})_\Delta$ in this chapter.

By Lemma 2.1, it easily obtains

**Lemma 2.2** *If* $(\text{C}_1)$ *holds and* $[D]_{\text{dia}} < \mathbf{0}_n$ *in* $(\text{PMLED})_\Delta$, *then* $\|\mathcal{E}_\alpha[Dt_\diamond^\alpha - \mathbb{I}_n\| \in (0, 1)$, $\|\bar{\Lambda}_\alpha\| \in (0, 1)$ *and* $\|\bar{\Theta}_\alpha\| \leq 2\|D^{-1}\|$.

***Proof*** This lemma can be directly gained by Lemma 2.1, so we omit its proof. This completes the proof. □

**Remark 2.8** Supposing that $[D]_{\text{dia}} < \mathbf{0}_n$, and in view of Lemmas 2.1 and 2.2, $\|\bar{\Theta}_\alpha\|$ is finite as $N^* \to +\infty$.

## 2.3 Advantages of Mittag–Leffler Euler Difference

Some advantages of Mittag–Leffler Euler difference are summarized as follows.

(1) $(\text{PMLED})_\Delta$ possesses the property of short-term memory.

**($1_A$)** As is well-known, memory property is a representative feature of FODEs. The memory property of $(\text{PMLED})_\Delta$ is displayed in the form of

$$\sum_{l=1}^{\nu_k} \Theta_\alpha^{[l,k]} P^{-1} F^{[k-l]}(PY^{[k-l]}), \quad \nu_k \in [0, N^* - 1], k \in \mathbb{N}_0. \tag{2.15}$$

By Remark 2.1, $\alpha = \mathbf{1}_n \Rightarrow \Theta_\alpha^{[l,k]} = \mathbb{O}_n$ for all $l \in [1, \nu_k]_\mathbb{Z}$, $k \in \mathbb{N}_0$. Consequently, if $\alpha = \mathbf{1}_n$ in $(\text{PMLED})_\Delta$, the memory term of $(\text{PMLED})_\Delta$ (i.e., (2.15)) becomes zero and $(\text{PMLED})_\Delta$ is turned into EEDEs (2.2). Besides, since $\nu_k \in [0, N^* - 1]_\mathbb{Z}$ $(k \in \mathbb{N}_0)$, (2.15) is at most the sum of $N^* - 1$ terms and it displays the property of short-term memory.

**($1_B$)** Observing $(\text{PMLED})_\Delta$, the fractional features are partly exhibited in the parameters $\Lambda_\alpha^{[k]}$ and $\Theta_\alpha^{[l,k]}$ for all $l \in [0, \nu_k]_\mathbb{Z}$, $k \in \mathbb{N}_0$. The trait of short-term memory in $(\text{PMLED})_\Delta$ enables us to easily derive these parameters' estimates, please see Lemmas 2.1 and 2.2. From the proof of Lemma 2.1, the property of short-term memory is mainly displayed in $\nu_k h_k \in [0, t_\diamond]$ $(k \in \mathbb{N}_0)$ and $t_\diamond$ is the maximal memory length. Inversely, if FODEs (2.3) is considered, the estimate of upper bound of $\Lambda_\alpha^{[k]}$ for the corresponding MLED $(\text{MLED})_\Delta$ is transformed into $\bar{\Lambda}_\alpha = \text{diag}\left(\langle\bar{\Lambda}_\alpha\rangle_1, \ldots, \langle\bar{\Lambda}_\alpha\rangle_n\right)$, where

$$\langle\bar{\Lambda}_\alpha\rangle_i = \sup_{k \in \mathbb{N}_0} \langle\Lambda_\alpha^{[k]}\rangle_i = \sup_{k \in \mathbb{N}_0} \frac{\mathbb{E}_{\alpha_i}[d_i(kh_k + h_k)^{\alpha_i}]}{\mathbb{E}_{\alpha_i}[d_i(kh_k)^{\alpha_i}]}, \quad i = 1, 2, \ldots, n.$$

Since $\langle\Lambda_\alpha^{[k]}\rangle_i$ $(i = 1, 2, \ldots, n)$ have no determinate monotonicity and $k \in \mathbb{N}_0$ is infinite, then the above estimate is difficultly calculated in practical examinations. It also appears in the estimate of $\Theta_\alpha^{[l,k]}$ for $l \in [0, k]_\mathbb{Z}$, $k \in \mathbb{N}_0$.

(2) $(\text{PMLED})_\Delta$ provides a numerical approximation algorithm for FODEs (2.4). Let $c_k$ be defined as that in $(\text{PMLED})_\Delta$, $k \in \mathbb{N}_0$. Supposing that $X(t)$ is a solution of FODEs 2.4 for $t \in [a, +\infty)$ and $Y^{[k]} = Y(c_k)$ in Eq. (2.8) is a solution of $(\text{PMLED})_\Delta$ for $k \in \mathbb{N}_0$. If $N^* \to +\infty$, then

$$Y^{[k]} = Y(c_k) \to P^{-1}X(t), \quad \forall t \in (c_k, c_{k+1}], k \in \mathbb{N}_0.$$

For instance, taking into account Eq. (2.13) with $a = 0$. $c_k = kh$, $k \in \mathbb{N}_0$. If $h \to 0^+$ (i.e., $N^* \to +\infty$), then

$$Y^{[k]} = Y(c_k) = Y(kh) \to P^{-1}X(t), \quad \forall t \in (kh, kh + h], k \in \mathbb{N}_0.$$

As an illustration, consider the linear Caputo FODEs below

$$\mathbf{CF}: \quad {}_{0}^{C}D_{t}^{\alpha}x(t) = -ax(t) + b, \quad \forall t \in \mathbb{R}_0, \tag{2.16}$$

where $\alpha \in (0, 1]$, $x(0) = \xi_0 \in \mathbb{R}$, $a, b \in \mathbb{R}$. Corresponding to FODEs (2.4), $A = -a$. Let $\mathcal{T}_a$ be defined as that in Eq. (2.13). By Eq. (2.2), MLED of CF (2.16) is represented by

$$\mathbf{ML}: \quad x(kh+h) = \Lambda_{\alpha}^{[k]} x(kh) + \frac{b(1-\Lambda_{\alpha}^{[k]})}{a}, \quad x(0) = \xi_0, \tag{2.17}$$

where $\Lambda_{\alpha}^{[k]} = \frac{\mathbb{E}_{\alpha}[-a(kh+h)^{\alpha}]}{\mathbb{E}_{\alpha}[-a(kh)^{\alpha}]}$, $k \in \mathbb{N}_0$. Taking $a = 0.3$, $b = 1$, $\alpha \in \{0.3, 0.5, 0.7, 0.9\}$ and $h \in \{0.05, 0.1, 0.5, 1\}$, Figs. 2.2 and 2.3 image the graphs of solutions to (2.16)–(2.17) and implies that the solutions of ML (2.17) tend to the solutions of CF (2.16), as $h$ tends to 0.

(3) In allusion to FODEs, the most frequently used and effective difference methods are the fractional Euler differences (FEDs) based on the fractional integral, see literature [37, 91–93]. Compared with the classical fractional Euler differences

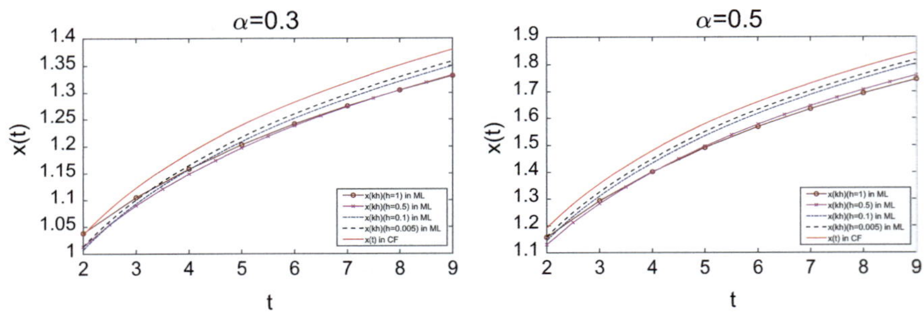


**Fig. 2.2** Trajectories of solutions to CF (2.16) and ML (2.17) with $\alpha = 0.3, 0.5$ and $h = 1, 0.5, 0.1, 0.005$

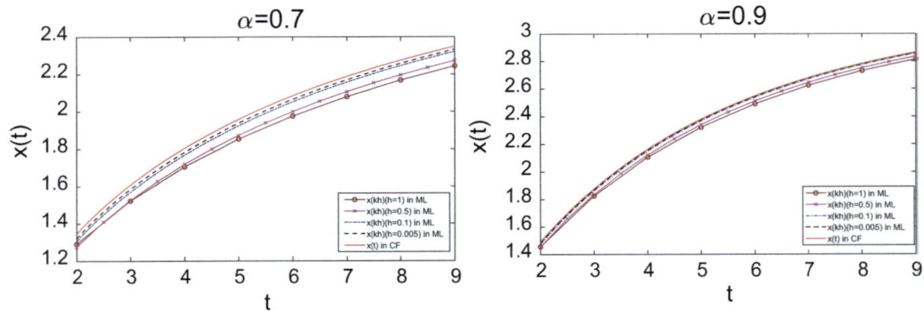


**Fig. 2.3** Trajectories of solutions to CF (2.16) and ML (2.17) with $\alpha = 0.7, 0.9$ and $h = 1, 0.5, 0.1, 0.005$

in literature [37, 91–93], Mittag–Leffler Euler differences can better depict and retain the dynamical characteristics of FODEs and they have a high level of implementation efficiency for some FODEs. Please see the numerical illustration below.

**Numerical example: Mittag–Leffler Euler and fractional Euler differences**
Considering FODEs as described below

$$
\begin{aligned}
{}^{c}D_{5p}^{\alpha}X(t) = -6X(t) + \sin^2(\sqrt{2}t)\Big[|X(t)+1| \\
-|X(t)-1|\Big] + 5\cos(\sqrt{3}t),
\end{aligned} \tag{2.18}
$$

where $X(0) = \xi \in \mathbb{R}$, $t \in (5p, 5p+5]$, $p \in \mathbb{N}_0$.

Taking

$$N \in \mathbb{N}, \quad h = 1/N, \quad \mathcal{T}_0 = \{t_p = 5p : p \in \mathbb{N}_0\},$$

$$\mathcal{N}_0 = \left\{N_p = 5pN : p \in \mathbb{N}_0\right\}, \quad C_0 = \{c_k = kh : k \in \mathbb{N}_0\}.$$

Then $C_0$ is a $\mathcal{T}_0$-$\mathcal{N}_0$ sequence. By using Mittag–Leffler Euler differences, MLED of FODEs (2.18) is depicted by

$$
\begin{aligned}
X^{[k+1]} = \lambda_{\alpha}^{[k]} X^{[k]} + \sum_{l=0}^{\nu_k} \theta_{\alpha}^{[l,k]} \Bigg\{ \sin^2(\sqrt{2}(k-l)h) \\
\times\Big[|X^{[k-l]}+1| - |X^{[k-l]}-1|\Big] + 5\cos(\sqrt{3}(k-l)h)\Bigg\},
\end{aligned} \tag{2.19}
$$

where $X^{[0]} = \xi \in \mathbb{R}$,

$$\nu_k = k - \mu_k, \quad \mu_k = \max_{p \in \mathbb{N}_0} \left\{5pN : 5pN \le k\right\},$$

$$\lambda_{\alpha}^{[k]} = \frac{\mathbb{E}_{\alpha}[-6(\nu_k h + h)^{\alpha}]}{\mathbb{E}_{\alpha}[-6(\nu_k h)^{\alpha}]}, \quad \theta_{\alpha}^{[l,k]} = -\frac{1}{6}\Big[w_{\alpha}^{[l]} - \lambda^{[k]} w_{\alpha}^{[l-1]}\Big],$$

$$w_{\alpha}^{[-1]} = 0, \quad w_{\alpha}^{[l]} = \mathbb{E}_{\alpha}[-6(lh+h)^{\alpha}] - \mathbb{E}_{\alpha}[-6(lh)^{\alpha}]$$

for $l \in [0, \nu_k]_{\mathbb{Z}}$, $k \in \mathbb{N}_0$.

Through the use of fractional Euler differences (also known as Adams–Bashforth differences) in literature [37, 91–93], it obtains another type of numerical solution of FODEs (2.18) as below

$$X_*^{[k+1]} = X_*^{[k]} + \sum_{l=0}^{\nu_k} \hat{\theta}_\alpha^{[l]} \bigg\{ -6X_*^{[k-l]} + \sin^2(\sqrt{2}(k-l)h) \times \Big[ |X_*^{[k-l]} + 1| - |X_*^{[k-l]} - 1| \Big] + 5\cos(\sqrt{3}(k-l)h) \bigg\}, \quad (2.20)$$

where $X_*^{[0]} = \xi \in \mathbb{R}$,

$$\hat{\theta}_\alpha^{[l]} = \frac{1}{\Gamma(\alpha+1)} \left[ \hat{w}_\alpha^{[l]} - \hat{w}_\alpha^{[l-1]} \right], \quad \hat{w}_\alpha^{[-1]} = 0, \quad \hat{w}_\alpha^{[l]} = \left[ (l+1)^\alpha - l^\alpha \right] h^\alpha$$

for $l \in [0, \nu_k]_{\mathbb{Z}}$, $k \in \mathbb{N}_0$.

The solutions of Eqs. (2.19) and (2.20) with different step lengths $h \in \{0.01, 0.1, 1\}$ and orders $\alpha \in \{0.3, 0.5, 0.8, 1\}$ are drawn in Figs. 2.4, 2.5, 2.6, 2.7, 2.8 and 2.9. By observing Figs. 2.4, 2.5, 2.6, 2.7, 2.8, 2.9 and 2.10, they imply that

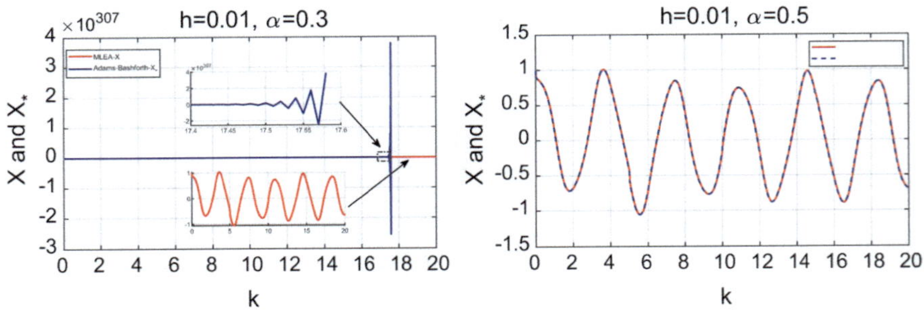


**Fig. 2.4** $X$ and $X_*$ of Eqs. (2.19) and (2.20) with $h = 0.01$ and $\alpha = 0.3, 0.5$

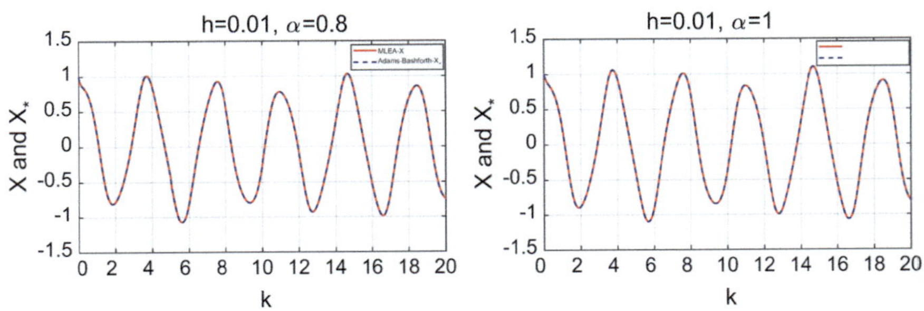


**Fig. 2.5** $X$ and $X_*$ Eqs. (2.19) and (2.20) with $h = 0.01$ and $\alpha = 0.8, 1$

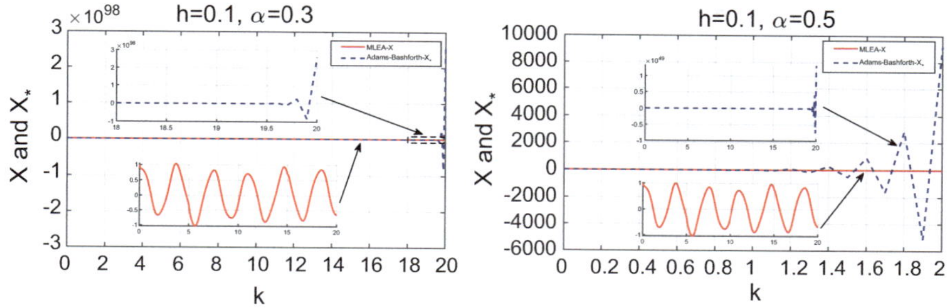


**Fig. 2.6** $X$ and $X_*$ of Eqs. (2.19) and (2.20) with $h = 0.1$ and $\alpha = 0.3, 0.5$

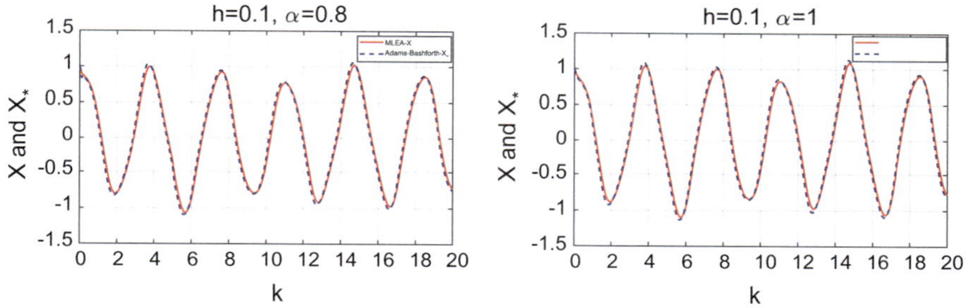


**Fig. 2.7** $X$ and $X_*$ of Eqs. (2.19) and (2.20) with $h = 0.1$ and $\alpha = 0.8, 1$

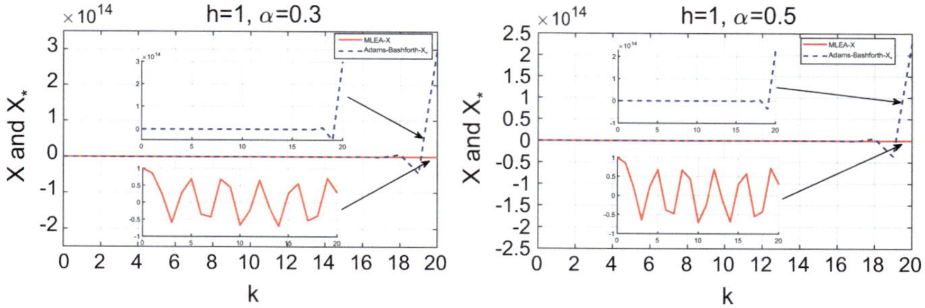


**Fig. 2.8** $X$ and $X_*$ of Eqs. (2.19) and (2.20) with $h = 1$ and $\alpha = 0.3, 0.5$

**(a)** Equation (2.20) produces non-periodic oscillations (divergence) for smaller orders $\alpha$ or bigger step lengths $h$, e.g., Figs. 2.4, 2.6, 2.8 and 2.9. Inversely, like continuous system (2.18), the solutions of Eq. (2.19) always keep the almost periodic oscillations (convergence) for arbitrary step lengths $h \in \{0.01, 0.1, 1\}$ and orders $\alpha \in \{0.3, 0.5, 0.8, 1\}$. Consequently, Mittag–Leffler Euler differences can depict and retain the dynamical characteristics of FODEs better than fractional Euler differences.

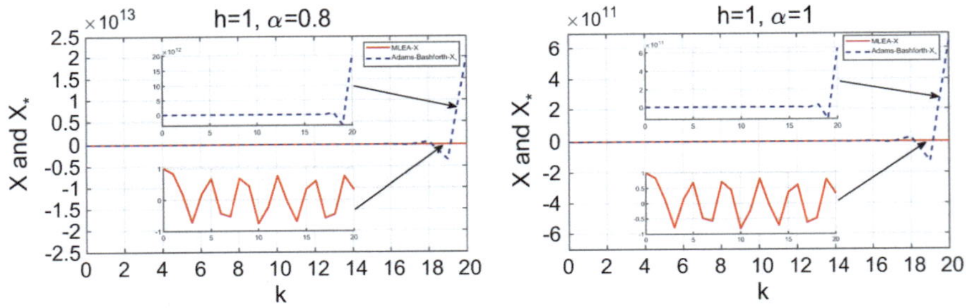


**Fig. 2.9** $X$ and $X_*$ of Eqs. (2.19) and (2.20) with $h = 1$ and $\alpha = 0.8, 1$

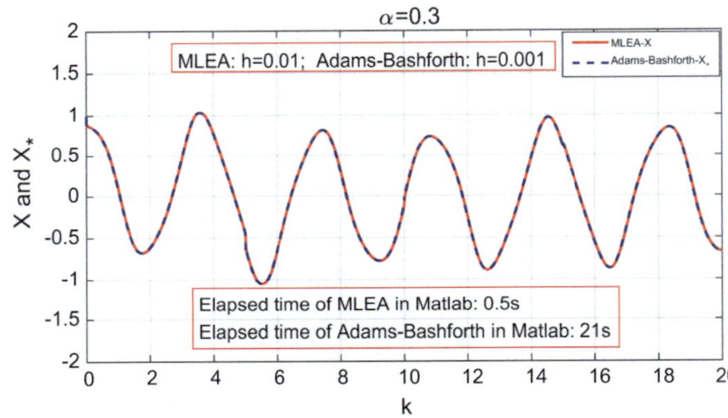


**Fig. 2.10** $X$ and $X_*$ of Eqs. (2.19) and (2.20) with $\alpha = 0.3$

**(b)** If fractional Euler differences are utilized to approximate FODEs, it is better to choose bigger order $\alpha$ and smaller step length $h$. However, it may result in a low of implementation efficiency. In Fig. 2.10, the degree of approximation of Mittag–Leffler Euler difference is about the same as fractional Euler difference, but the elapsed time of fractional Euler difference is about 42 times Mittag–Leffler Euler difference.

**(c)** For the same step length $h$, the bigger the order $\alpha$, the higher the degrees of approximations of Eqs. (2.19) and (2.20).

## 2.4 Convergence

Supposing that $Y$, $\mathbf{Y}$ and $\tilde{Y}$ are the solutions of Eq. (2.5), $(\text{PMLED})_\Delta$ and Eq. (2.8) with initial values $Y(a) = \mathbf{Y}(a) = \tilde{Y}(a) = \xi \in \mathbb{R}^n$, respectively. Let $P_0 = \|P\|_\text{r}\|P^{-1}\|_\text{r}$, $T > a$ be a constant, $C_a$ be a $\mathcal{T}_a$-$\mathcal{N}_0$ sequence and $\lfloor t \rfloor = \max\{c_k : t \geq c_k, k \in \mathbb{N}_0\}$ for all $t \in [a, T]$.

**Lemma 2.3** *Let* ($C_1$)-($C_2$) *hold. Then it exists a constant* $M_T > 0$ *such that*

$$\max_{s\in[a,T]}\{\|Y(s)\|, \|\tilde{Y}(s)\|\} \le M_T.$$

*Hereby,* $M_T$ *is independent of* $\bar{h}$.

***Proof*** Set $T_a = \max\{T, T-a\}$,

$$\bar{\mathcal{E}}_\alpha = \max_{t\in[a,T_a]}\|\mathcal{E}_\alpha(Dt^\alpha)\|, \quad F_T = \max_{t\in[a,T]}\|F(t, \mathbf{0}_n)\|,$$

where $\mathcal{E}_\alpha$ is defined as that in (2.6). Owing to $\lim_{s\to 0}\mathcal{E}_\alpha(Ds^\alpha) = \mathbb{I}_n$, it exists a constant $\breve{h} > 0$ small enough causing $a + 2\breve{h} \le \min\{t_1, T\}$ and

$$h_\diamond = P_0 L_F \max_{s\in[0,\breve{h}]}\left\|[\mathcal{E}_\alpha(Ds^\alpha) - \mathbb{I}_n]D^{-1}\right\| < 1,$$

where $t_1 \in \mathcal{T}_a$. Noting that $\breve{h}$ is a constant, which is independent of $\bar{h}$.

If $t \in [a, a+\breve{h}]$, then it gets from Eq. (2.8) that

$$\begin{aligned}
\|\tilde{Y}(t)\| &= \left\|\mathcal{E}_\alpha(Dt^\alpha)\tilde{Y}(a) + \int_a^t J_{t-s}^\alpha \mathcal{E}_{\alpha,\alpha}[D(t-s)^\alpha]P^{-1}F(\lfloor s\rfloor, P\tilde{Y}(\lfloor s\rfloor))\mathrm{d}s\right\| \\
&\le \bar{\mathcal{E}}_\alpha\|\tilde{Y}_a\| + \left\|\int_a^t J_{t-s}^\alpha \mathcal{E}_{\alpha,\alpha}[D(t-s)^\alpha]\mathrm{d}s\right\| \\
&\quad \times \max_{s\in[a,a+\breve{h}]}\left\|P^{-1}F(\lfloor s\rfloor, P\tilde{Y}(\lfloor s\rfloor)) - P^{-1}F(\lfloor s\rfloor, \mathbf{0}_n)\right\| \\
&\quad + \left\|\int_a^t J_{t-s}^\alpha \mathcal{E}_{\alpha,\alpha}[D(t-s)^\alpha]\mathrm{d}s\right\| \max_{s\in[a,a+\breve{h}]}\|P^{-1}F(\lfloor s\rfloor, \mathbf{0}_n)\| \\
&\le \bar{\mathcal{E}}_\alpha\|\tilde{Y}_a\| + \left\|[\mathcal{E}_\alpha(D(t-a)^\alpha) - \mathbb{I}_n]D^{-1}\right\| \\
&\quad \times \left[P_0 L_F \max_{s\in[a,a+\breve{h}]}\|\tilde{Y}(s)\| + \|P^{-1}\|_{\mathrm{r}} F_T\right] \\
&\le \bar{\mathcal{E}}_\alpha\|\tilde{Y}_a\| + \max_{s\in[a,a+\breve{h}]}\left\|[\mathcal{E}_\alpha(D(s-a)^\alpha) - \mathbb{I}_n]D^{-1}\right\| \\
&\quad \times \left[P_0 L_F \max_{s\in[a,a+\breve{h}]}\|\tilde{Y}(s)\| + \|P^{-1}\|_{\mathrm{r}} F_T\right],
\end{aligned}$$

which deduces

$$\max_{s\in[a,a+\breve{h}]}\|\tilde{Y}(s)\| \le \frac{\bar{\mathcal{E}}_\alpha\|\tilde{Y}_a\| + \left\|[\mathcal{E}_\alpha(D\breve{h}^\alpha) - \mathbb{I}_n]D^{-1}\right\| \|P^{-1}\|_{\mathrm{r}} F_T}{1 - h_\diamond} = \tilde{M}_0. \quad (2.21)$$

Let $F_{\tilde{M}_0} = \max_{t_1 \in [a,T], t_2 \in [0,\tilde{M}_0]} \|F(t_1, Pt_2)\|$. If $t \in [a+\breve{h}, a+2\breve{h}]$, taking a similar discussion as the above derivation yields

$$
\begin{aligned}
\|\tilde{Y}(t)\| \leq & \left\| \mathcal{E}_\alpha(Dt^\alpha)\tilde{Y}(a) + \int_a^{a+\breve{h}} J_{t-s}^\alpha \mathcal{E}_{\alpha,\alpha}[D(t-s)^\alpha] P^{-1} F(\lfloor s \rfloor, P\tilde{Y}(\lfloor s \rfloor)) \mathrm{d}s \right\| \\
& + \left\| \int_{a+\breve{h}}^t J_{t-s}^\alpha \mathcal{E}_{\alpha,\alpha}[D(t-s)^\alpha] P^{-1} F(\lfloor s \rfloor, P\tilde{Y}(\lfloor s \rfloor)) \mathrm{d}s \right\| \\
\leq & \ \bar{\mathcal{E}}_\alpha \|\tilde{Y}_a\| + \left\| [\mathcal{E}_\alpha(D(t-a)^\alpha) - \mathcal{E}_\alpha(D(t-a-\breve{h})^\alpha)] D^{-1} \right\| \|P^{-1}\|_{\mathrm{r}} F_{\tilde{M}_0} \\
& + \left\| [\mathcal{E}_\alpha(D(t-a-\breve{h})^\alpha) - \mathbb{I}_n] D^{-1} \right\| \left[ P_0 L_F \max_{s \in [a,a+2\breve{h}]} \|\tilde{Y}(s)\| + \|P^{-1}\|_{\mathrm{r}} F_T \right] \\
\leq & \ \bar{\mathcal{E}}_\alpha \left[ \|\tilde{Y}_a\| + 2\|D^{-1}\| \|P^{-1}\|_{\mathrm{r}} F_{\tilde{M}_0} \right] + \left\| [\mathcal{E}_\alpha(D\breve{h}^\alpha) - \mathbb{I}_n] D^{-1} \right\| \\
& \times \left[ P_0 L_F \max_{s \in [a,a+2\breve{h}]} \|\tilde{Y}(s)\| + \|P^{-1}\|_{\mathrm{r}} F_T \right],
\end{aligned}
$$

implies

$$
\begin{aligned}
& \max_{t \in [a,a+2\breve{h}]} \|\tilde{Y}(t)\| \\
\leq & \ \frac{\tilde{M}_0 + \bar{\mathcal{E}}_\alpha \left[ \|\tilde{Y}_a\| + 2\|D^{-1}\| \|P^{-1}\|_{\mathrm{r}} F_{\tilde{M}_0} \right] + \left\| [\mathcal{E}_\alpha(D\breve{h}^\alpha) - \mathbb{I}_n] D^{-1} \right\| \|P^{-1}\|_{\mathrm{r}} F_T}{1 - h_\diamond} \\
= & \ \tilde{M}_1. 
\end{aligned}
\tag{2.22}
$$

Noting that $\tilde{M}_0$ and $\tilde{M}_1$ are both independent of $\bar{h}$. By an extension technique as those in (2.21)–(2.22), there exists a constant $\tilde{M}_T > 0$ such that $\max_{s \in [a,T]} \|\tilde{Y}(s)\| \leq \tilde{M}_T$. By the same derivation as $\tilde{Y}$, it also exists a constant $\widehat{M}_T > 0$ such that $\max_{s \in [a,T]} \|Y(s)\| \leq \widehat{M}_T$. Taking $M_T = \max\{\widehat{M}_T, \tilde{M}_T\}$, so

$$
\max_{s \in [a,T]} \{\|Y(s)\|, \|\tilde{Y}(s)\|\} \leq M_T
$$

and $M_T$ is independent of $\bar{h}$. The proof is achieved. □

**Lemma 2.4** *Let* (C$_1$) *hold. Then*

$$
\lim_{\bar{h} \to 0} \|\tilde{Y}(t) - \tilde{Y}(\lfloor t \rfloor)\| = 0
$$

*uniformly for* $t \in [a, T]$.

***Proof*** From Lemma 2.3, it exists a constant $M_T > 0$ such that $\max_{s\in[a,T]} \|\tilde{Y}(s)\| \le M_T$. Set

$$\bar{\alpha} = \max_{i=1,2,\ldots,n} \alpha_i, \quad \underline{\alpha} = \min_{i=1,2,\ldots,n} \alpha_i,$$

$$F_{M_T} = \max_{t_1\in[a,T], t_2\in[0,M_T]} \|F(t_1, Pt_2)\|, \quad \bar{\mathcal{E}}_{\alpha,\alpha} = \max_{t\in[a,T]} \|\mathcal{E}_{\alpha,\alpha}(Dt^{\alpha})\|,$$

where $\mathcal{E}_{\alpha,\alpha}$ is defined as that in (2.6). According to the uniform continuity of $\mathcal{E}_{\alpha}(Dt^{\alpha})$ and $\mathcal{E}_{\alpha,\alpha}(Dt^{\alpha})$ for $t \in [a, T]$, for arbitrary $\epsilon > 0$, there exists a constant $\delta_1 > 0$ small enough such that

$$\|\mathcal{E}_{\alpha}(Dt_1^{\alpha}) - \mathcal{E}_{\alpha}(Dt_2^{\alpha})\| < \epsilon, \quad \|\mathcal{E}_{\alpha,\alpha}(Dt_1^{\alpha}) - \mathcal{E}_{\alpha,\alpha}(Dt_2^{\alpha})\| < \epsilon, \tag{2.23}$$

if $|t_1 - t_2| < \delta_1$, $t_1, t_2 \in [a, T]$. Meanwhile, it exists a very small constant $\delta_2 > 0$ such that if $\bar{h} < \delta_2$, then

$$\max_{i=1,2,\ldots,n} \{(t - c_q)^{\alpha_i}, (t - c_{N_p})^{\alpha_i} - (c_q - c_{N_p})^{\alpha_i}\} < \epsilon, \tag{2.24}$$

where $t \in (c_q, c_{q+1}] \subseteq (c_{N_p}, c_{N_{p+1}}]$, $q \in [N_p, N_{p+1}]_{\mathbb{Z}}$ and $p \in \mathbb{N}_0$.

Let $\bar{h} \le \min\{\delta_1, \delta_2\}$. Taking arbitrary $t' \in [a, T]$. In view of (P1)–(P2), it exists $p' \in \mathbb{N}_0$ and $q' \in [N_{p'}, N_{p'+1}]_{\mathbb{Z}}$ such that $t' \in (c_{q'}, c_{q'+1}] \subseteq (c_{N_{p'}}, c_{N_{p'+1}}]$. In the light of Eq. (2.8), Lemma 2.3, (2.23) and (2.24) in turn, it follows

$$\begin{aligned}
\|\tilde{Y}(t') - \tilde{Y}(\lfloor t' \rfloor)\| = {} & \|\tilde{Y}(t') - \tilde{Y}(c_{q'})\| \\
\le {} & \Big\| \mathcal{E}_{\alpha}\big[D(t' - c_{N_{p'}})^{\alpha}\big]\tilde{Y}(c_{N_{p'}}) \\
& - \mathcal{E}_{\alpha}\big[D(c_{q'} - c_{N_{p'}})^{\alpha}\big]\tilde{Y}(c_{N_{p'}}) \Big\| \\
& + \Big\| \int_{c_{N_{p'}}}^{t'} J_{t'-s}^{\alpha} \mathcal{E}_{\alpha,\alpha}[D(t' - s)^{\alpha}] P^{-1} F(\lfloor s \rfloor, P\tilde{Y}(\lfloor s \rfloor)) \mathrm{d}s \\
& - \int_{c_{N_{p'}}}^{c_{q'}} J_{c_{q'}-s}^{\alpha} \mathcal{E}_{\alpha,\alpha}[D(c_{q'} - s)^{\alpha}] P^{-1} F(\lfloor s \rfloor, P\tilde{Y}(\lfloor s \rfloor)) \mathrm{d}s \Big\| \\
\le {} & M_T \epsilon + \Bigg\{ \Big\| \int_{c_{N_{p'}}}^{c_{q'}} \Big[ J_{t'-s}^{\alpha} \mathcal{E}_{\alpha,\alpha}[D(t' - s)^{\alpha}] \\
& - J_{c_{q'}-s}^{\alpha} \mathcal{E}_{\alpha,\alpha}[D(c_{q'} - s)^{\alpha}] \Big]_{\mathrm{abs}} \mathrm{d}s \Big\| \\
& + \Big\| \int_{c_{q'}}^{t'} J_{t'-s}^{\alpha} \mathcal{E}_{\alpha,\alpha}[D(t' - s)^{\alpha}] \mathrm{d}s \Big\| \Bigg\} \|P^{-1}\|_{\mathrm{r}} F_{M_T}
\end{aligned}$$

$$
\begin{aligned}
&\le M_T\epsilon + \Bigg\{ \left\| \int_{c_{N_{p'}}}^{c_{q'}} \left[ J_{t'-s}^{\alpha} - J_{c_{q'}-s}^{\alpha} \right]_{\text{abs}} \mathcal{E}_{\alpha,\alpha}[D(t'-s)^{\alpha}]\mathrm{d}s \right\| \\
&\quad + \left\| \int_{c_{N_{p'}}}^{c_{q'}} J_{c_{q'}-s}^{\alpha} \Big[ \mathcal{E}_{\alpha,\alpha}[D(t'-s)^{\alpha}] \right. \\
&\quad \left. - \mathcal{E}_{\alpha,\alpha}[D(c_{q'}-s)^{\alpha}] \Big]_{\text{abs}} \mathrm{d}s \right\| + \Big\| \big[ \mathcal{E}_{\alpha}(D(t'-c_{q'})^{\alpha}) \\
&\quad - \mathcal{E}_{\alpha}(D(t'-t')^{\alpha}) \big]_{\text{abs}} D^{-1} \Big\| \Bigg\} \|P^{-1}\|_{\text{r}} F_{M_T} \\
&\le M_T\epsilon + \Bigg\{ \left\| \int_{c_{N_{p'}}}^{c_{q'}} \left[ J_{t'-s}^{\alpha} - J_{c_{q'}-s}^{\alpha} \right]_{\text{abs}} \mathrm{d}s \right\| \bar{\mathcal{E}}_{\alpha,\alpha} \\
&\quad + \left\| \int_{c_{N_{p'}}}^{c_{q'}} J_{c_{q'}-s}^{\alpha} \mathrm{d}s \right\| \epsilon + \|D^{-1}\|\epsilon \Bigg\} \|P^{-1}\|_{\text{r}} F_{M_T} \\
&\le \Bigg\{ \max_{i=1,2,\ldots,n} \Bigg[ \frac{(t'-c_{q'})^{\alpha_i}}{\alpha_i} \\
&\quad + \frac{(t'-c_{N_{p'}})^{\alpha_i} - (c_{q'}-c_{N_{p'}})^{\alpha_i}}{\alpha_i} \Bigg] \bar{\mathcal{E}}_{\alpha,\alpha} \\
&\quad + \frac{(T-a)^{\bar{\alpha}}}{\underline{\alpha}} \epsilon + \|D^{-1}\|\epsilon \Bigg\} \|P^{-1}\|_{\text{r}} F_{M_T} \\
&\le \left( \frac{2\bar{\mathcal{E}}_{\alpha,\alpha}}{\underline{\alpha}} + \frac{(T-a)^{\bar{\alpha}}}{\underline{\alpha}} + \|D^{-1}\| \right) \|P^{-1}\|_{\text{r}} F_{M_T} \epsilon.
\end{aligned}
$$

By the arbitrariness of $\epsilon \in (0, 1)$ and $t' \in [a, T]$, Lemma 2.4 is valid. The proof is achieved. □

It needs a simplified version of Gronwall inequality in fractional calculus studied in literature [94].

**Lemma 2.5** (Gronwall Inequality in Fractional Calculus [94]) *Let $\beta > 0$, $T > a$ be a constant, $c$ and $d$ be nonnegative constants. If $v$ is nonnegative and locally integrable for $t \in [a, T]$ satisfying*

$$
v(t) \le c + d \int_a^t (t-s)^{\beta-1} v(s)\mathrm{d}s, \quad \forall t \in [a, T],
$$

*then $v(t) \le c\mathbb{E}_{\beta}(d\Gamma(\beta)(t-a)^{\beta})$ for $t \in [a, T]$.*

The following theorem illustrates the solution $\mathbf{Y}$ of $(\text{PMLED})_\Delta$ approximates the solution $Y$ of Eq. (2.5) as the maximum step length $\bar{h}$ tends to 0. In other words, numerical solution $\mathbf{X} = P\mathbf{Y}$ of Eq. (2.4) approximates the exact solution $X = PY$ of Eq. (2.4) as the maximum step length $\bar{h}$ tends to 0.

**Theorem 2.1** *Let* $(C_1)$*-*$(C_2)$ *hold. Then*

$$\lim_{\bar{h}\to 0} \|Y(t) - \mathbf{Y}(\lfloor t \rfloor)\| = 0$$

*uniformly for* $t \in [a, T]$*. In other words,* $\lim_{\bar{h}\to 0} \|X(t) - \mathbf{X}(\lfloor t \rfloor)\| = 0$ *uniformly for* $t \in [a, T]$*, where* $X$ *and* $\mathbf{X}$ *are the exact solution and numerical solution of Eq.* (2.4)*, respectively.*

***Proof*** Define $\Bbbk_0(t) = 1$ if $t \geq 0$, otherwise it is 0. On account of Lemma 2.4, for arbitrary $\epsilon > 0$, there exists $\delta_3 > 0$ small enough such that

$$\max_{s\in[a,T]} \|\tilde{Y}(s) - \tilde{Y}(\lfloor s \rfloor)\| < \epsilon$$

if $\bar{h} < \delta_3$.

From Eqs. (2.5) and (2.8), if $t \in [a, t_1] \cap [a, T]$ and $\bar{h} < \delta_3$, it induces

$$\begin{aligned}
\|Y(t) - \tilde{Y}(t)\| = & \left\| \int_a^t J_{t-s}^\alpha \mathcal{E}_{\alpha,\alpha}[D(t-s)^\alpha] P^{-1} \right. \\
& \left. \times \Big[ F(s, PY(s)) - F(\lfloor s \rfloor, P\tilde{Y}(\lfloor s \rfloor)) \Big] \mathrm{d}s \right\| \\
\leq & \left\| \int_a^t J_{t-s}^\alpha \mathcal{E}_{\alpha,\alpha}[D(t-s)^\alpha] P^{-1} \right. \\
& \left. \times \Big[ F(s, P\tilde{Y}(s)) - F(\lfloor s \rfloor, P\tilde{Y}(\lfloor s \rfloor)) \Big] \mathrm{d}s \right\| \\
& + \left\| \int_a^t J_{t-s}^\alpha \mathcal{E}_{\alpha,\alpha}[D(t-s)^\alpha] P^{-1} \right. \\
& \left. \times \Big[ F(s, PY(s)) - F(s, P\tilde{Y}(s)) \Big] \mathrm{d}s \right\| \\
\leq & \left\| \int_a^t J_{t-s}^\alpha \mathcal{E}_{\alpha,\alpha}[D(t-s)^\alpha] \|\tilde{Y}(s) - \tilde{Y}(\lfloor s \rfloor)\| \mathrm{d}s \right\| P_0 L_F \\
& + \left\| \int_a^t J_{t-s}^\alpha \mathcal{E}_{\alpha,\alpha}[D(t-s)^\alpha] \|Y(s) - \tilde{Y}(s)\| \mathrm{d}s \right\| P_0 L_F \\
\leq & \left\| [\mathcal{E}_\alpha(D(T-a)^\alpha) - \mathbb{I}_n] D^{-1} \right\| P_0 L_F \epsilon \\
& + \left\| \int_a^t J_{t-s}^\alpha \|Y(s) - \tilde{Y}(s)\| \mathrm{d}s \right\| P_0 L_F \bar{\mathcal{E}}_{\alpha,\alpha}
\end{aligned}$$

$$
\begin{aligned}
&\leq Q_1\epsilon + \int_a^{t-1} \Bbbk_0(s)(t-s)^{\bar{\alpha}-1}\|Y(s)-\tilde{Y}(s)\|\mathrm{d}s\, P_0 L_F \bar{\mathcal{E}}_{\alpha,\alpha} \\
&\quad + \int_{t-1}^{t} \Bbbk_0(s)(t-s)^{\underline{\alpha}-1}\|Y(s)-\tilde{Y}(s)\|\mathrm{d}s\, P_0 L_F \bar{\mathcal{E}}_{\alpha,\alpha} \\
&\leq Q_1\epsilon + \int_a^t \left[(t-s)^{\bar{\alpha}-1} + (t-s)^{\underline{\alpha}-1}\right] \\
&\quad \times\|Y(s)-\tilde{Y}(s)\|\mathrm{d}s\, P_0 L_F \bar{\mathcal{E}}_{\alpha,\alpha} \\
&\leq Q_1\epsilon + Q_2\int_a^t (t-s)^{\underline{\alpha}-1}\|Y(s)-\tilde{Y}(s)\|\mathrm{d}s,
\end{aligned}
\tag{2.25}
$$

where $t \in [a, t_1] \subseteq [a, T]$,

$$
Q_1 = \left\|[\mathcal{E}_\alpha(D(T-a)^\alpha) - \mathbb{I}_n]D^{-1}\right\| P_0 L_F, \quad Q_2 = P_0 L_F \bar{\mathcal{E}}_{\alpha,\alpha}[(T-a)^{\bar{\alpha}-\underline{\alpha}} + 1].
$$

Applying Lemma 2.5 into (2.25) deduces

$$
\max_{s\in[a,t_1]} \|Y(s)-\tilde{Y}(s)\| \leq Q_1\mathbb{E}_{\underline{\alpha}}[Q_2\Gamma(\underline{\alpha})(T-a)^{\underline{\alpha}}]\epsilon. \tag{2.26}
$$

If $t \in (t_1, t_2] \cap [a, T]$ and $\bar{h} < \delta_3$, utilizing a similar derivation as that in (2.25) leads to

$$
\begin{aligned}
\|Y(t)-\tilde{Y}(t)\| \leq\ & \left\|\mathcal{E}_\alpha\left[D(t-t_1)^\alpha\right]\right\| \|Y(t_1)-\tilde{Y}(t_1)\| \\
& + \left\| \int_{t_1}^t J_{t-s}^{\alpha}\mathcal{E}_{\alpha,\alpha}[D(t-s)^\alpha]P^{-1} \right. \\
& \left. \times\left[F(s, PY(s)) - F(\lfloor s\rfloor, P\tilde{Y}(\lfloor s\rfloor))\right]\mathrm{d}s \right\| \\
\leq\ & \bar{\mathcal{E}}_\alpha Q_1\mathbb{E}_{\underline{\alpha}}[Q_2\Gamma(\underline{\alpha})(T-a)^{\underline{\alpha}}]\epsilon \\
& + Q_1\epsilon + Q_2\int_a^t (t-s)^{\underline{\alpha}-1}\|Y(s)) - \tilde{Y}(s)\|\mathrm{d}s,
\end{aligned}
\tag{2.27}
$$

where $t \in (t_1, t_2] \cap [a, T]$. In view of (2.25), (2.27) is also valid for $t \in [a, t_1]$. By employing Gronwall inequality in Lemma 2.5, it obtains

$$
\begin{aligned}
\max_{s\in[a,t_2]} \|Y(s)-\tilde{Y}(s)\| \leq\ & Q_1\left\{\bar{\mathcal{E}}_\alpha\mathbb{E}_{\underline{\alpha}}[Q_2\Gamma(\underline{\alpha})(T-a)^{\underline{\alpha}}] + 1\right\} \\
& \times\mathbb{E}_{\underline{\alpha}}[Q_2\Gamma(\underline{\alpha})(T-a)^{\underline{\alpha}}]\epsilon.
\end{aligned}
\tag{2.28}
$$

By an extension technique as those in (2.26) and (2.28), it must exist a constant $\mathcal{M}_T > 0$ such that

$$
\max_{s\in[a,T]} \|Y(s)-\tilde{Y}(s)\| \leq \mathcal{M}_T\epsilon
$$

if $\bar{h} < \delta_3$. So

$$\lim_{\bar{h}\to 0} \|Y(t) - \tilde{Y}(t)\| = 0$$

uniformly for $t \in [a, T]$. By Lemma 2.4,

$$\lim_{\bar{h}\to 0} \|Y(t) - \tilde{Y}(\lfloor t \rfloor)\| = 0$$

uniformly for $t \in [a, T]$. By the derivation of $(\text{PMLED})_\Delta$, $\tilde{Y}(\lfloor t \rfloor) = \mathbf{Y}(\lfloor t \rfloor)$, $\forall t \in [a, T]$. The proof is achieved. □

## 2.5 Delta Mittag–Leffler Euler Difference Equations

Let $\mathrm{S}(C_a, \mathbb{R})$ and $\mathrm{S}((h\mathbb{N}_0)_a, \mathbb{R})$ denote the totality of all mappings $x : C_a \to \mathbb{R}$ and $x : (h\mathbb{N}_0)_a \to \mathbb{R}$, respectively. Discussing the FODEs with piecewise feature as below

$$\begin{cases} {}_{t_p}^{C}D_t^{\alpha} x(t) = f(t, x(t)) = \lambda x(t) + [f(t, x(t)) - \lambda x(t)], \\ x(a) = \xi_0 \in \mathbb{R}, \end{cases} \tag{2.29}$$

where $t \in (t_p, t_{p+1}]$, $p = 0, 1, \ldots$, $\alpha \in (0, 1]$, $\lambda \in \mathbb{R}\backslash\{0\}$, $f \in C([a, +\infty) \times \mathbb{R}, \mathbb{R})$ satisfies Lipschitz condition.

According to $(\text{PMLED})_\Delta$, numerical solution of Eq. (2.29) is given by

$$x(c_{k+1}) = \lambda_\alpha^{[k]} x(c_k) + \sum_{l=0}^{\nu_k} \theta_\alpha^{[l,k]} \left[ f(c_{k-l}, x(c_{k-l})) - \lambda x(c_{k-l}) \right], \tag{2.30}$$

where $x(a) = \xi_0 \in \mathbb{R}$, $c_k \in C_a$,

$$\begin{cases} \lambda_\alpha^{[k]} = \dfrac{\mathbb{E}_\alpha[\lambda(\nu_k h_k + h_k)^\alpha]}{\mathbb{E}_\alpha[\lambda(\nu_k h_k)^\alpha]}, \\ \theta_\alpha^{[l,k]} = \left\{ w_\alpha^{[l,k]} - \lambda_\alpha^{[k]} w_\alpha^{[l-1,k]} \right\} \lambda^{-1}, \quad w_\alpha^{[-1,k]} = 0, \\ w_\alpha^{[l,k]} = \mathbb{E}_\alpha[\lambda(l h_k + h_k)^\alpha] - \mathbb{E}_\alpha[\lambda(l h_k)^\alpha] \end{cases} \tag{2.31}$$

for $l \in [0, \nu_k]_{\mathbb{Z}}$, $k \in \mathbb{N}_0$.

Define

$$\underline{t} = \max\{N_p \in N_0 : t \geq c_{N_p}, p \in \mathbb{N}_0\}, \quad t^\circ = \underline{t} + \frac{t - c_{\underline{t}}}{h_{\underline{t}}}, \quad \forall t \in C_a. \tag{2.32}$$

**Remark 2.9** In (2.32), it has some indetectable relations as below.

(1) $t^\circ = k$ if $t = c_k$, $k \in \mathbb{N}_0$.
(2) $t^\circ = N_p + \dfrac{t - c_{N_p}}{h_{N_p}}$ if $t \in [c_{N_p}, c_{N_{p+1}})_{C_a}$, $p \in \mathbb{N}_0$.
(3) $\underline{t} \equiv N_p$ and $c_{\underline{t}} \equiv c_{N_p}$ if $t \in [c_{N_p}, c_{N_{p+1}})_{C_a}$, $p \in \mathbb{N}_0$.
(4) $c_{\underline{t}+1} - c_{\underline{t}} = h_{\underline{t}} = h_{t^\circ} = h_k \equiv h_{N_p}$ if $t = c_k$ and $k \in [N_p, N_{p+1})_{\mathbb{Z}}$, $p \in \mathbb{N}_0$.
(5) $t + h_{\underline{t}} = c_k + h_k = c_{k+1}$ if $t = c_k$, $k \in \mathbb{N}_0$.

By Remark 2.9, Eq. (2.30) is transformed into

$$x(t + h_{\underline{t}}) = \lambda_\alpha^{[t^\circ]} x(t) + \sum_{l=0}^{\nu_{t^\circ}} \theta_\alpha^{[l,t^\circ]} \Big[ f(t - lh_{\underline{t}}, x(t - lh_{\underline{t}})) - \lambda x(t - lh_{\underline{t}}) \Big], \quad (2.33)$$

where $t \in C_a$, $x(a) = \xi_0$.

### 2.5.1 Mittag–Leffler Euler Difference Operator

Define an operator $\mathbf{M}_\lambda^{-\alpha} : \mathrm{S}(C_a, \mathbb{R}) \to \mathrm{S}(C_a, \mathbb{R})$ by

$$\mathbf{M}_\lambda^{-\alpha} y(t) = \sum_{l=0}^{\nu_{t^\circ}} \theta_\alpha^{[l,t^\circ]} y(t - lh_{\underline{t}}), \quad \forall y \in \mathrm{S}(C_a, \mathbb{R}), t \in C_a. \quad (2.34)$$

Obviously, $\mathbf{M}_\lambda^{-\alpha} : \mathrm{S}(C_a, \mathbb{R}) \to \mathrm{S}(C_a, \mathbb{R})$ is a linear operator. By Cramer's rule in linear algebra, $\mathbf{M}_\lambda^{-\alpha} : \mathrm{S}(C_a, \mathbb{R}) \to \mathrm{S}(C_a, \mathbb{R})$ defined in Eq. (2.34) has a unique inverse operator denoted by $\mathbf{M}_\lambda^{\alpha} : \mathrm{S}(C_a, \mathbb{R}) \to \mathrm{S}(C_a, \mathbb{R})$, which can be calculated by a sequence of iterations as noted below

$$\mathbf{M}_\lambda^{\alpha} z(t) = \frac{\lambda}{\mathbb{E}_\alpha(\lambda h_{\underline{t}}^\alpha) - 1} \Big[ z(t) - \sum_{j=1}^{\nu_{t^\circ}} \theta_\alpha^{[j,t^\circ]} \mathbf{M}_\lambda^{\alpha} z(t - jh_{\underline{t}}) \Big], \quad (2.35)$$

where $z \in \mathrm{S}(C_a, \mathbb{R})$ and $t$ is successively equal to $c_0 (= a), c_1, c_2, \ldots$.

By employing operators $\mathbf{M}_\lambda^{-\alpha}$ and $\mathbf{M}_\lambda^{\alpha}$ defined in Eqs. (2.34)–(2.35), (2.33) can be reformulated by

$$x(t + h_{\underline{t}}) = \lambda_\alpha^{[t^\circ]} x(t) + \mathbf{M}_\lambda^{-\alpha} \big[ f(t, x(t)) - \lambda x(t) \big]$$

is equivalent to

$$\mathbf{M}_\lambda^{\alpha} \big[ x(t + h_{\underline{t}}) - \lambda_\alpha^{[t^\circ]} x(t) \big] + \lambda x(t) = f(t, x(t)), \quad \forall t \in C_a. \quad (2.36)$$

It should be noted that the solution $x$ of Eq. (2.36) is a numerical solution of Eq. (2.29) by an approach of Mittag–Leffler Euler difference. Motivated by Eq. (2.36), piecewise Mittag–Leffler Euler difference operator (PMLEDO) with parameter $\lambda \in \mathbb{R}\backslash\{0\}$ can be defined as below.

**Definition 2.1** Let $\alpha \in (0, 1]$, $\lambda \in \mathbb{R}\backslash\{0\}$, $\underline{t}$, $t^\circ$ and $\mathbf{M}_\lambda^\alpha$ be defined as those in (2.32) and (2.35), $\forall t \in C_a$. The $\alpha$-order Caputo piecewise Mittag–Leffler Euler Delta $h_{\underline{t}}$-difference of $x : C_a \to \mathbb{R}$ with lower limit $c_{\underline{t}} \in C_a$ and parameter $\lambda$ is defined by

$$ {}_{\lambda}^{C}\Delta_{c_{\underline{t}},h_{\underline{t}}}^{\alpha}x(t) = \mathbf{M}_\lambda^\alpha\left[x(t+h_{\underline{t}}) - \lambda_\alpha^{[t^\circ]}x(t)\right] + \lambda x(t), \quad \forall t \in C_a. \qquad (\text{PMLEDO})_\Delta^\lambda $$

Noting that $(\text{PMLEDO})_\Delta^\lambda$ in Definition 2.1 possesses piecewise feature and variable step length $h_{N_p}$ in each segment $[c_{N_p}, c_{N_{p+1}})_{C_a}$, $\forall p \in \mathbb{N}_0$. That is, $(\text{PMLEDO})_\Delta^\lambda$ has an equivalent piecewise form given by

$$ {}_{\lambda}^{C}\Delta_{c_{N_p},h_{N_p}}^{\alpha}x(t) = \mathbf{M}_\lambda^\alpha\left[x(t+h_{N_p}) - \lambda_\alpha^{[t^\circ]}x(t)\right] + \lambda x(t) \qquad (2.37) $$

for $t \in [c_{N_p}, c_{N_{p+1}})_{C_a}$ and $p \in \mathbb{N}_0$. In all calculations of (2.37),

$$ h_{\underline{t}} \equiv h_{N_p}, \quad t^\circ = N_p + \frac{t - c_{N_p}}{h_{N_p}}, \quad \nu_{t^\circ} = \frac{t - c_{N_p}}{h_{N_p}}, $$

for all $t \in [c_{N_p}, c_{N_{p+1}})_{C_a}$, $p \in \mathbb{N}_0$.

In $(\text{PMLEDO})_\Delta^\lambda$ or (2.37), let

$$ z_k = \mathbf{M}_\lambda^\alpha z(c_k), \quad z(c_k) = x(c_{k+1}) - \lambda_\alpha^{[k]}x(c_k) = z(t) $$

if $t = c_k \in C_a$, and $y_k = z_k + \lambda x(c_k)$, $k \in \mathbb{N}_0$. Obviously,

$$ {}_{\lambda}^{C}\Delta_{c_{\underline{t}},h_{\underline{t}}}^{\alpha}x(t) = y_k \text{ for } t = c_k, k \in \mathbb{N}_0. $$

More clearly, $y_k$ and $z_k$ can be piecewise calculated by

$$ \begin{cases} y_k = z_k + \lambda x(c_k), \quad \forall k \in \mathbb{N}_0, \\ z_k = \dfrac{\lambda}{\mathbb{E}_\alpha(\lambda h_{N_p}^\alpha) - 1}\left[z\left(c_{N_p} + (k - N_p)h_{N_p}\right) - \displaystyle\sum_{j=1}^{k-N_p} \theta_\alpha^{[j,k]} z_{k-j}\right], \end{cases} $$

where $k \in [N_p, N_{p+1})_{\mathbb{Z}}$, $p = 0, 1 \ldots$.

By employing $(\text{PMLEDO})_\Delta^\lambda$, Eqs. (2.30) and (2.33) are reformulated as

$$ {}_{\lambda}^{C}\Delta_{c_{\underline{t}},h_{\underline{t}}}^{\alpha}x(t) = f(t, x(t)), \quad \forall t \in C_a, \qquad (2.38) $$

where $x(a) = \xi_0$.

Also, $(\text{PMLED})_\Delta$ can be rewritten to the difference equations involving $(\text{PMLEDO})_\Delta^\lambda$ as noted below

$$ {}_D^C\Delta_{c_{\underline{t}},h_{\underline{t}}}^\alpha Y(t) = DY(t) + P^{-1}F(t, PY(t)), \quad \forall t \in C_a, \tag{2.39} $$

where $Y(a) = P^{-1}\xi \in \mathbb{R}^n$ and

$$ {}_D^C\Delta_{c_{\underline{t}},h_{\underline{t}}}^\alpha Y = \left( {}_{d_1}^C\Delta_{c_{\underline{t}},h_{\underline{t}}}^{\alpha_1} y_1, \ldots, {}_{d_n}^C\Delta_{c_{\underline{t}},h_{\underline{t}}}^{\alpha_n} y_n \right)^T, \quad \forall t \in C_a. $$

Equations (2.38) and (2.39) include $(\text{PMLEDO})_\Delta^\lambda$, which are called piecewise Mittag–Leffler Euler difference equations (PMLEDEs) in this book.

The following property is easily obtained by Eqs. (2.38) and (2.39).

**Property 2.1** *The solution of* PMLEDEs (2.38) *denotes the numerical solution of* FODEs (2.29). *Meanwhile, if $Y$ solves* PMLEDEs (2.39)*, then $PY$ is the numerical solution of* FODEs (2.4).

By Definition 2.1, non-piecewise Mittag–Leffler Euler difference operator (MLEDO) with lower limit $a \in \mathbb{R}$ and parameter $\lambda \in \mathbb{R}\backslash\{0\}$ can be defined as

**Definition 2.2** Let $\alpha \in (0, 1]$, $\lambda \in \mathbb{R}\backslash\{0\}$, $\lambda_\alpha^{[k]}$ and $\theta_\alpha^{[l,k]}$ be defined as those in (2.31) with $\nu_k = k$ and $h_k \equiv h$, $\forall l, k \in \mathbb{N}_0$. The $\alpha$-order Caputo Mittag–Leffler Euler Delta $h$-difference of $x : (h\mathbb{N}_0)_a \to \mathbb{R}$ with lower limit $a \in \mathbb{R}$ and parameter $\lambda$ is defined by

$$ {}_\lambda^C\Delta_{a,h}^\alpha x(t) = \mathbf{M}_\lambda^\alpha \left[ x(t+h) - \lambda_\alpha^{[t^\circ]} x(t) \right] + \lambda x(t), \quad \forall t \in (h\mathbb{N}_0)_a, \tag{$(\text{MLEDO})_\Delta^\lambda$} $$

where operator $\mathbf{M}_\lambda^\alpha$ is defined by an iterative sequence

$$ \mathbf{M}_\lambda^\alpha z(t) = \frac{\lambda}{\mathbb{E}_\alpha(\lambda h^\alpha) - 1} \left[ z(t) - \sum_{j=1}^{t^\circ} \theta_\alpha^{[j,t^\circ]} \mathbf{M}_\lambda^\alpha z(t - jh) \right], \tag{2.40} $$

where $z \in \mathrm{S}((h\mathbb{N}_0)_a, \mathbb{R})$, $t$ is successively equal to $a, a+h, a+2h, \ldots$. In all calculations of Definition 2.2 and Eq. (2.40),

$$ t^\circ = \frac{t-a}{h} \in \mathbb{N}_0, \quad h_{t^\circ} \equiv h, \quad \nu_{t^\circ} = t^\circ, \quad \forall t \in (h\mathbb{N}_0)_a. $$

In $(\text{MLEDO})_\Delta^\lambda$, let

$$ z(a+kh) = x(a+(k+1)h) - \lambda_\alpha^{[k]} x(a+kh) = z(t) $$

if $t = a + kh \in (h\mathbb{N}_0)_a$,

$$z_k = \mathbf{M}_\lambda^\alpha z(a + kh), \quad y_k = z_k + \lambda x(a + kh), \quad \forall k \in \mathbb{N}_0.$$

Obviously, ${}_\lambda^C\Delta_{a,h}^\alpha x(t) = y_k$ for $t = a + kh$, $k \in \mathbb{N}_0$. More clearly, $y_k$ and $z_k$ can be calculated by

$$\begin{cases} y_k = z_k + \lambda x(a + kh), \\ z_k = \dfrac{\lambda}{\mathbb{E}_\alpha(\lambda h^\alpha) - 1}\left[z(a + kh) - \displaystyle\sum_{j=1}^{k} \theta_\alpha^{[j,k]} z_{k-j}\right], \end{cases} \tag{2.41}$$

where $k = 0, 1, \ldots$.

By Definition 2.2, it has the following properties.

**Property 2.2** *The solution of MLEDEs given by*

$${}_\lambda^C\Delta_{a,h}^\alpha x(t) = f(t, x(t)), \quad \forall t \in (h\mathbb{N}_0)_a, \quad x(a) = \xi_0 \in \mathbb{R}, \tag{2.42}$$

*is the numerical solution of* FODEs *as below*

$${}_a^C D_t^\alpha x(t) = f(t, x(t)), \quad \forall t \in (a, +\infty), \quad x(a) = \xi_0 \in \mathbb{R}.$$

**Property 2.3** *Let $\alpha \in (0, 1]$, $\lambda \in \mathbb{R}\backslash\{0\}$ and $K \in \mathbb{R}$ be a constant. Then*

$${}_\lambda^C\Delta_{c_{\underline{t}},h_{\underline{t}}}^\alpha K = 0, \quad \forall t \in C_a.$$

*Also, ${}_\lambda^C\Delta_{a,h}^\alpha K = 0$, $\forall t \in (h\mathbb{N}_0)_a$.*

***Proof*** Let $x(t) \equiv K$, $\forall t \in (h\mathbb{N}_0)_a$. If $t = a$, then $k = 0$ in (2.41). By Eq. (2.31), it calculates $\lambda_\alpha^{[0]} = \mathbb{E}_\alpha(\lambda h^\alpha)$. By (2.41), $z(a) = x(a + h) - \lambda_\alpha^{[0]} x(a) = K - \lambda_\alpha^{[0]} K$ and

$$z_0 = \frac{\lambda z(a)}{\mathbb{E}_\alpha(\lambda h^\alpha) - 1} = -\lambda K,$$

which implies that ${}_\lambda^C\Delta_{a,h}^\alpha x(t) = y_0 = z_0 + \lambda x(a) = 0$ for $t = a$.

If $t = a + h$, then $k = 1$ in (2.41). Equation (2.31) results in

$$\lambda_\alpha^{[1]} = \frac{\mathbb{E}_\alpha[\lambda(2h)^\alpha]}{\mathbb{E}_\alpha(\lambda h^\alpha)}, \quad \theta_\alpha^{[1,1]} = \left\{\frac{\mathbb{E}_\alpha[\lambda(2h^\alpha]}{\mathbb{E}_\alpha(\lambda h^\alpha)} - \mathbb{E}_\alpha(\lambda h^\alpha)\right\}\lambda^{-1}.$$

By (2.41), $z(a + h) = x(a + 2h) - \lambda_\alpha^{[1]} x(a + h) = K - \lambda_\alpha^{[1]} K$ and

$$z_1 = \frac{\lambda}{\mathbb{E}_\alpha(\lambda h^\alpha) - 1}\left[z(a + h) - \theta_\alpha^{[1,1]} z_0\right] = -\lambda K,$$

which implies that

$$ {}^{C}_{\lambda}\Delta^{\alpha}_{a,h}x(t) = y_1 = z_1 + \lambda x(a) = 0 \text{ for } t = a + h. $$

By the mathematical induction, ${}^{C}_{\lambda}\Delta^{\alpha}_{a,h}x(t) = 0$, $\forall t \in (h\mathbb{N}_0)_a$. Similarly, it easily obtains ${}^{C}_{\lambda}\Delta^{\alpha}_{c_{\underline{t}},h_{\underline{t}}}K = 0$, $\forall t \in C_a$. The proof is ended. □

According to $(\text{PMLEDO})^{\lambda}_{\Delta}$ and $(\text{MLEDO})^{\lambda}_{\Delta}$, it easily gets the following properties.

**Property 2.4** *Let* $\alpha \in (0, 1]$ *and* $\lambda \in \mathbb{R}\backslash\{0\}$. *Then*

$$ {}^{C}_{\lambda}\Delta^{\alpha}_{c_{\underline{t}},h_{\underline{t}}}[m_1x(t) + m_2y(t)] = m_1\, {}^{C}_{\lambda}\Delta^{\alpha}_{c_{\underline{t}},h_{\underline{t}}}x(t) + m_2\, {}^{C}_{\lambda}\Delta^{\alpha}_{c_{\underline{t}},h_{\underline{t}}}y(t), \quad \forall t \in C_a, $$

*where* $m_1, m_2 \in \mathbb{R}$ *and* $x, y : C_a \to \mathbb{R}$. *Also,*

$$ {}^{C}_{\lambda}\Delta^{\alpha}_{a,h}[m_1x(t) + m_2y(t)] = m_1\, {}^{C}_{\lambda}\Delta^{\alpha}_{a,h}x(t) + m_2\, {}^{C}_{\lambda}\Delta^{\alpha}_{a,h}y(t), \quad \forall t \in (h\mathbb{N}_0)_a, $$

*where* $x, y : (h\mathbb{N}_0)_a \to \mathbb{R}$.

**Property 2.5** *Let* $\alpha \in (0, 1]$, $\lambda \in \mathbb{R}\backslash\{0\}$ *and* $\mathbf{M}^{-\alpha}_{\lambda}$ *be defined in Eq.* (2.34). *Then*

$$ \mathbf{M}^{-\alpha}_{\lambda}\, {}^{C}_{\lambda}\Delta^{\alpha}_{c_{\underline{t}},h_{\underline{t}}}x(t) = x(t + h_{\underline{t}}) - \lambda^{[t^\circ]}_{\alpha}x(t) + \lambda\sum_{j=0}^{\nu_{t^\circ}}\theta^{[j,t^\circ]}_{\alpha}x(t - jh_{\underline{t}}), \quad \forall t \in C_a, $$

*where* $x : C_a \to \mathbb{R}$. *Additionally,*

$$ \mathbf{M}^{-\alpha}_{\lambda}\, {}^{C}_{\lambda}\Delta^{\alpha}_{a,h}x(t) = x(t + h) - \lambda^{[t^\circ]}_{\alpha}x(t) + \lambda\sum_{j=0}^{t^\circ}\theta^{[j,t^\circ]}_{\alpha}x(t - jh), \quad \forall t \in (h\mathbb{N}_0)_a, $$

*where* $x : (h\mathbb{N}_0)_a \to \mathbb{R}$. *In all calculations of the last equation, taking* $\underline{t} = 0$, $h_{t^\circ} = h_{\underline{t}} \equiv h$, $t^\circ = \frac{t-a}{h}$ *and* $\nu_{t^\circ} = t^\circ$, $\forall t \in (h\mathbb{N}_0)_a$.

By employing difference operator $(\text{MLEDO})^{\lambda}_{\Delta}$, $(\text{MLED})_{\Delta}$ can be rewritten by the difference equations named Mittag–Leffler Euler difference equations (MLEDEs) as described below

$$ {}^{C}_{D}\Delta^{\alpha}_{a,h}Y(t) = DY(t) + P^{-1}F(t, PY(t)), \quad \forall t \in (h\mathbb{N}_0)_a, \tag{2.43} $$

where $D = P^{-1}AP = \text{diag}\{d_1, \ldots, d_n\}$ is defined as that in $(\text{C}_1)$, $Y = (y_1, \ldots, y_n)^T \in \mathbb{R}^n$ and ${}^{C}_{D}\Delta^{\alpha}_{a,h}Y = \left({}^{C}_{d_1}\Delta^{\alpha_1}_{a,h}y_1, \ldots, {}^{C}_{d_n}\Delta^{\alpha_n}_{a,h}y_n\right)^T$.

**Remark 2.10** There is a one-to-one correlation between MLEDEs and FODEs. For example, the solution of FODEs (2.3) can be approximated by the solution of the corresponding MLEDEs (2.43), see Properties 2.1 and 2.2. In literature [63–85], various concepts of fractional-order differences are defined and relative basic theories were established. However, the FODCEs discussed in [63–85] almost have no such one-to-one relation with FODEs. So MLEDEs starts a new avenue for the research of FODCEs.

**Exponential Euler and Euler Differences**
Let $\alpha = 1$ in $(\text{MLEDO})_{\Delta}^{\lambda}$. In view of Remark 2.1, then

$$ {}_{\lambda}\Delta_{a,h}x(t) = {}_{\lambda}^{C}\Delta_{a,h}^{1}x(t) = \frac{\lambda[x(t+h)-x(t)]}{e^{\lambda h}-1}, \quad \forall t \in (h\mathbb{N}_0)_a, \qquad (\text{EEDO})_{\Delta}^{\lambda} $$

where $x : (h\mathbb{N}_0)_a \to \mathbb{R}$. ${}_{\lambda}\Delta_{a,h}$ in $(\text{EEDO})_{\Delta}^{\lambda}$ is called exponential Euler difference operator (EEDO), which had been extensively discussed in literature [19–32].

Letting $\lambda \to 0$ in $(\text{EEDO})_{\Delta}^{\lambda}$, $(\text{EEDO})_{\Delta}^{\lambda}$ approaches the famous Euler difference operator (EDO) as below

$$ \Delta_h x(t) = \frac{x(t+h)-x(t)}{h}, \quad \forall t \in (h\mathbb{N}_0)_a, \qquad (\text{EDO})_{\Delta} $$

where $x : (h\mathbb{N}_0)_a \to \mathbb{R}$.

### 2.5.2 Fractional Euler Difference Operator

Letting $\lambda \to 0$ in (2.31) and using Lemma 1.23, it results in

$$ \lambda_{\alpha}^{[t^\circ]} \to 1, \quad \theta_{\alpha}^{[0,t^\circ]} \to \zeta_{\alpha}^{[0,t^\circ]} = \frac{h_{t^\circ}^{\alpha}}{\Gamma(\alpha+1)}, $$

$$ \theta_{\alpha}^{[j,t^\circ]} \to \zeta_{\alpha}^{[j,t^\circ]} = \frac{h_{t^\circ}^{\alpha}}{\Gamma(\alpha+1)}\left[(j+1)^{\alpha} - 2j^{\alpha} + (j-1)^{\alpha}\right] $$

for all $j \in [1, \nu_{t^\circ}]_{\mathbb{Z}}$ and $t \in C_a$.

By Eq. (2.33), another type of numerical solution for FODEs (2.29) by the approach of fractional Euler differences (also called fractional Adams–Bashforth differences) in [37] is described as

$$x(t+h_{\underline{t}}) = x(t) + \sum_{l=0}^{\nu_{t^\circ}} \zeta_\alpha^{[l,t^\circ]} f(t - lh_{\underline{t}}, x(t - lh_{\underline{t}})), \quad \forall t \in C_a. \tag{2.44}$$

Besides, if $\lambda \to 0$, $\mathbf{M}_\lambda^{-\alpha}$ and $\mathbf{M}_\lambda^\alpha$ defined in Eqs. (2.34)–(2.35) are turned into

$$\mathbf{M}_0^{-\alpha} y(t) = \sum_{l=0}^{\nu_{t^\circ}} \zeta_\alpha^{[l,t^\circ]} y(t - lh_{\underline{t}}) \tag{2.45}$$

and its inverse operator $\mathbf{M}_0^\alpha$ calculated by a sequence of iterations as below

$$\mathbf{M}_0^\alpha z(t) = \frac{\Gamma(\alpha+1)}{h_{\underline{t}}^\alpha}\left[z(t) - \sum_{j=1}^{\nu_{t^\circ}} \zeta_\alpha^{[j,t^\circ]} \mathbf{M}_0^\alpha z(t - jh_{\underline{t}})\right] \tag{2.46}$$

for $t$ is successively equal to $c_0(=a), c_1, c_2, \ldots, y, z \in \mathrm{S}(C_a, \mathbb{R})$.

Similar to Definition 2.1, piecewise fractional Euler difference operator (PFEDO) is defined as follows.

**Definition 2.3** Let $\alpha \in (0,1]$, $\underline{t}$, $t^\circ$ and $\mathbf{M}_0^\alpha$ be defined as those in (2.32) and (2.46), $\forall t \in C_a$. The $\alpha$-order Caputo piecewise fractional Euler Delta $h_{\underline{t}}$-difference of $x : C_a \to \mathbb{R}$ with lower limit $c_{\underline{t}} \in C_a$ is defined by

$$ {}_0^C\Delta_{c_{\underline{t}},h_{\underline{t}}}^\alpha x(t) = \mathbf{M}_0^\alpha \left[h_{\underline{t}} \Delta_{h_{\underline{t}}} x(t)\right], \quad \forall t \in C_a. \tag{PFEDO$_\Delta$}$$

Hereby, $\Delta_{h_{\underline{t}}} x(t) = \frac{x(t+h_{\underline{t}})-x(t)}{h_{\underline{t}}}$, $\forall t \in C_a$.

(PFEDO)$_\Delta$ has an equivalent piecewise form given by

$$ {}_0^C\Delta_{c_{N_p},h_{N_p}}^\alpha x(t) = \mathbf{M}_0^\alpha \left[h_{N_p} \Delta_{h_{N_p}} x(t)\right] \tag{2.47}$$

for all $t \in [c_{N_p}, c_{N_{p+1}})_{C_a}$, $p \in \mathbb{N}_0$. In all calculations of (2.47),

$$h_{\underline{t}} \equiv h_{N_p}, \quad t^\circ = N_p + \frac{t - c_{N_p}}{h_{N_p}}, \quad \nu_{t^\circ} = \frac{t - c_{N_p}}{h_{N_p}}$$

for all $t \in [c_{N_p}, c_{N_{p+1}})_{C_a}$, $p \in \mathbb{N}_0$.

In (PFEDO)$_\Delta$ or (2.47), let

$$z_k = \mathbf{M}_0^\alpha z(c_k), \quad z(c_k) = x(c_{k+1}) - x(c_k) = z(t)$$

if $t = c_k \in C_a$, $k \in \mathbb{N}_0$. Obviously,

$$ {}_0^C\Delta_{c_{\underline{t}},h_{\underline{t}}}^\alpha x(t) = z_k \text{ for } t = c_k, k \in \mathbb{N}_0.$$

More clearly, $z_k$ can be piecewise calculated by

$$z_k = \frac{\Gamma(\alpha+1)}{h_{N_p}^{\alpha}}\left[z\left(c_{N_p} + (k - N_p)h_{N_p}\right) - \sum_{j=1}^{k-N_p} \zeta_{\alpha}^{[j,k]} z_{k-j}\right],$$

where $k \in [N_p, N_{p+1})_{\mathbb{Z}}$, $p = 0, 1, \ldots$.

By Definition 2.3, Eq. (2.44) can be rewritten by the equation with (PFEDO)$_\Delta$

$$ {}_0^C\Delta_{c_{\underline{t}},h_{\underline{t}}}^{\alpha} x(t) = f(t, x(t)), \quad \forall t \in C_a, \quad x(a) = \xi_0 \in \mathbb{R}. \tag{2.48}$$

Equation (2.48) includes (PFEDO)$_\Delta$, which is called piecewise fractional Euler difference equations (PFEDEs).

By Definition 2.3, non-piecewise fractional Euler difference operator (FEDO) can be defined as below.

**Definition 2.4** Let $\alpha \in (0, 1]$. The $\alpha$-order Caputo fractional Euler Delta $h$-difference of $x : (h\mathbb{N}_0)_a \to \mathbb{R}$ with lower limit $a \in \mathbb{R}$ is defined by

$$ {}_0^C\Delta_{a,h}^{\alpha} x(t) = \mathbf{M}_0^{\alpha}\left[h\Delta_h x(t)\right], \quad \forall t \in (h\mathbb{N}_0)_a. \tag{FEDO$_\Delta$}$$

In (FEDO)$_\Delta$, operator $\mathbf{M}_0^{\alpha} : \mathrm{S}((h\mathbb{N}_0)_a, \mathbb{R}) \to \mathrm{S}((h\mathbb{N}_0)_a, \mathbb{R})$ can be calculated by

$$\mathbf{M}_0^{\alpha} z(t) = \frac{\Gamma(\alpha+1)}{h^{\alpha}}\left[z(t) - \sum_{j=1}^{t^\circ} \zeta_{\alpha}^{[j,t^\circ]} \mathbf{M}_0^{\alpha} z(t - jh)\right]$$

for $t$ is successively equal to $a, a+h, a+2h, \ldots$, $z \in \mathrm{S}((h\mathbb{N}_0)_a, \mathbb{R})$. In all calculations of this definition,

$$\Delta_h x(t) = \frac{x(t+h) - x(t)}{h}, \quad t^\circ = \frac{t-a}{h}, \quad h_{t^\circ} \equiv h, \quad \nu_{t^\circ} = t^\circ,$$

where $t \in (h\mathbb{N}_0)_a$.

In (FEDO)$_\Delta$, let

$$z_k = \mathbf{M}_0^{\alpha} z(a + kh), \quad z(a + kh) = x(a + (k+1)h) - x(a + kh) = z(t)$$

if $t = a + kh \in (h\mathbb{N}_0)_a$, $k \in \mathbb{N}_0$. Obviously,

$$ {}_0^C\Delta_{a,h}^{\alpha} x(t) = z_k \text{ for } t = a + kh, k \in \mathbb{N}_0.$$

More clearly, $z_k$ can be calculated by

$$z_k = \frac{\lambda}{\mathbb{E}_\alpha(\lambda h^\alpha) - 1}\left[z(a+kh) - \sum_{j=1}^{k} \zeta_\alpha^{[j,k]} z_{k-j}\right], \quad k = 0, 1, \ldots.$$

If $(\text{PFEDO})_\Delta$ is turned into $(\text{FEDO})_\Delta$ in Eq. (2.48), it is called fractional Euler difference equations (FEDEs), correspondingly.

Similar to MLEDO, some properties for FEDO are listed without any need of proof as follows.

**Property 2.6** *The solution of* PFEDEs (2.48) *denotes the numerical solution of* FODEs (2.29).

**Property 2.7** *The solution of FEDEs given by*

$$ {}_0^C\Delta_{a,h}^{\alpha} x(t) = f(t, x(t)), \quad \forall t \in (h\mathbb{N}_0)_a, \quad x(a) = \xi_0 \in \mathbb{R}$$

*is the numerical solution of* FODEs *as below*

$$ {}_a^C D_t^{\alpha} x(t) = f(t, x(t)), \quad \forall t \in [a, +\infty), \quad x(a) = \xi_0 \in \mathbb{R}.$$

**Property 2.8** *Let* $\alpha \in (0, 1]$ *and* $K \in \mathbb{R}$ *be a constant. Then*

$$ {}_0^C\Delta_{c_{\underline{t}},h_{\underline{t}}}^{\alpha} K = 0, \quad \forall t \in C_a.$$

*Also,* ${}_0^C\Delta_{a,h}^{\alpha} K = 0$, $\forall t \in (h\mathbb{N}_0)_a$.

**Property 2.9** *Let* $\alpha \in (0, 1]$. *Then*

$$ {}_0^C\Delta_{c_{\underline{t}},h_{\underline{t}}}^{\alpha} [m_1 x(t) + m_2 y(t)] = m_1 \, {}_0^C\Delta_{c_{\underline{t}},h_{\underline{t}}}^{\alpha} x(t) + m_2 \, {}_0^C\Delta_{c_{\underline{t}},h_{\underline{t}}}^{\alpha} y(t), \quad \forall t \in C_a,$$

*where* $m_1, m_2 \in \mathbb{R}$ *and* $x, y : C_a \to \mathbb{R}$. *Also,*

$$ {}_0^C\Delta_{a,h}^{\alpha} [m_1 x(t) + m_2 y(t)] = m_1 \, {}_0^C\Delta_{a,h}^{\alpha} x(t) + m_2 \, {}_0^C\Delta_{a,h}^{\alpha} y(t), \quad \forall t \in (h\mathbb{N}_0)_a,$$

*where* $x, y : (h\mathbb{N}_0)_a \to \mathbb{R}$.

**Property 2.10** *Let* $\alpha \in (0, 1]$ *and* $\mathbf{M}_0^{-\alpha}$ *be defined as that in Eq.* (2.45). *Then*

$$\mathbf{M}_0^{-\alpha} \, {}_0^C\Delta_{c_{\underline{t}},h_{\underline{t}}}^{\alpha} x(t) = x(t + h_{\underline{t}}) - x(t), \quad \forall t \in C_a,$$

*where* $x : C_a \to \mathbb{R}$. *Additionally,*

$$\mathbf{M}_0^{-\alpha} \, {}_0^C\Delta_{a,h}^{\alpha} x(t) = x(t + h) - x(t), \quad \forall t \in (h\mathbb{N}_0)_a,$$

*where* $x : (h\mathbb{N}_0)_a \to \mathbb{R}$. *In all calculations of the last equation, taking* $\underline{t} = 0$, $h_{t^\circ} = h_{\underline{t}} \equiv h$, $t^\circ = \frac{t-a}{h}$ *and* $\nu_{t^\circ} = t^\circ$, $\forall t \in (h\mathbb{N}_0)_a$.

**Remark 2.11** Letting $\alpha = 1$ in $(\text{FEDO})_\Delta$, then $\zeta_\alpha^{[j,k]} = 0\,(j, k \in \mathbb{N})$ and $(\text{FEDO})_\Delta$ is changed into $(\text{EDO})_\Delta$.

**MLEDO includes FEDO, EEDO and EDO**

See Fig. 2.11.

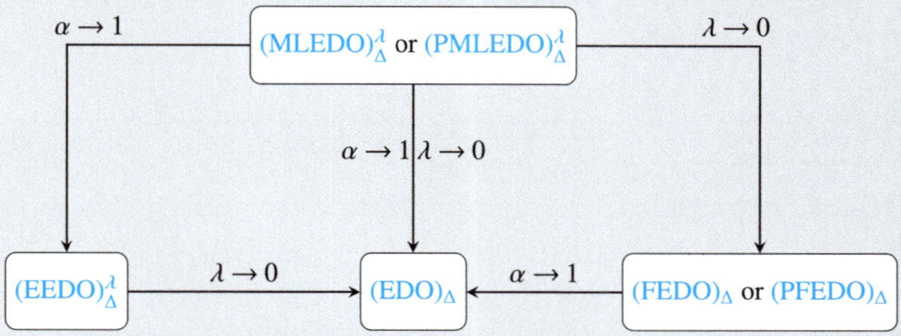


**Fig. 2.11** Relations of MLEDO, FEDO, EEDO and EDO

Figures 2.12, 2.13, 2.14 and 2.15 image MLEDO, FEDO, EEDO and EDO of some elementary functions $t$, $e^t$, $\ln t$ and $\sin t$ for $t \in 0.01\mathbb{N}_0 = \{0.01k : k \in \mathbb{N}_0\}$. By observing Figs. 2.12, 2.13, 2.14, 2.15, 2.16, 2.17, 2.18 and 2.19, they imply that

- **(a)** Figures 2.14 and 2.15 imply the fact that EEDO tends to EDO if $\lambda \to 0$.
- **(b)** Figures 2.16 and 2.17 imply the fact that MLEDO tends to FEDO if $\lambda \to 0$.
- **(c)** Figures 2.18 and 2.19 imply the fact that FEDO tends to EDO if $\alpha \to 1$.

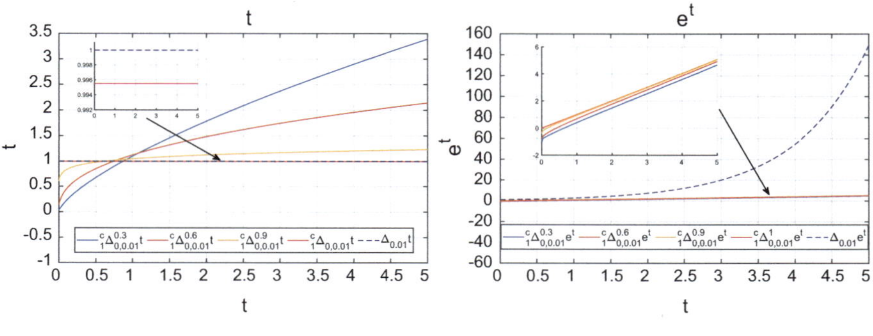


**Fig. 2.12** $(\text{MLEDO})_\Delta^\lambda$, $(\text{EEDO})_\Delta^\lambda$ and $(\text{EDO})_\Delta$ of elementary functions $t$ and $e^t$ with $\lambda = 1$, $h = 0.01$ and $\alpha = 0.3, 0.6, 0.9, 1$

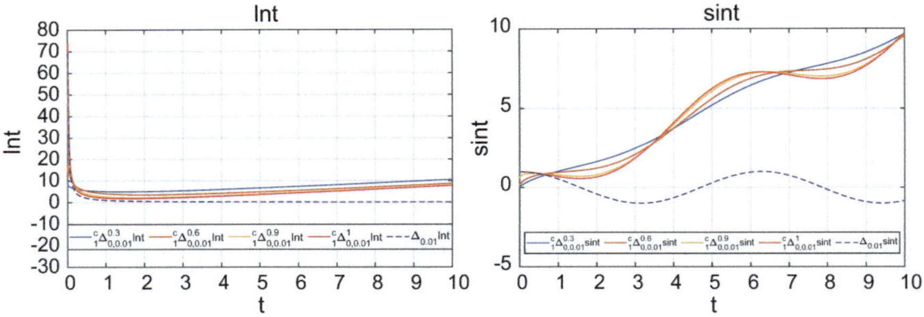


**Fig. 2.13** $(\text{MLEDO})^{\lambda}_{\Delta}$, $(\text{EEDO})^{\lambda}_{\Delta}$ and $(\text{EDO})_{\Delta}$ of elementary functions $\ln t$ and $\sin t$ with $\lambda = 1$, $h = 0.01$ and $\alpha = 0.3, 0.6, 0.9, 1$

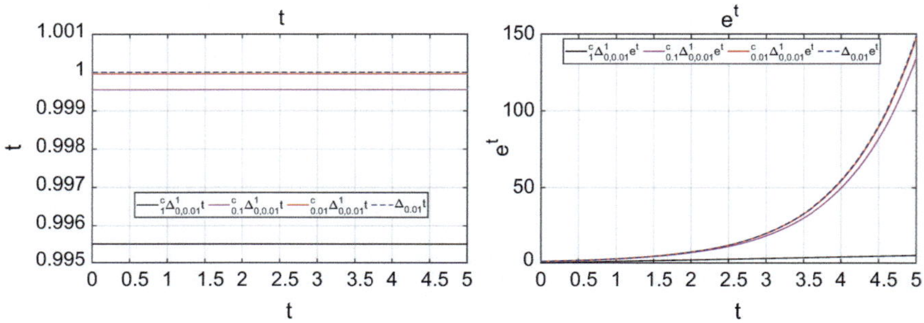


**Fig. 2.14** $(\text{EEDO})^{\lambda}_{\Delta}$ and $(\text{EDO})_{\Delta}$ of elementary functions $t$ and $e^t$ with $\lambda = 1, 0.1, 0.01$, $h = 0.01$ and $\alpha = 1$

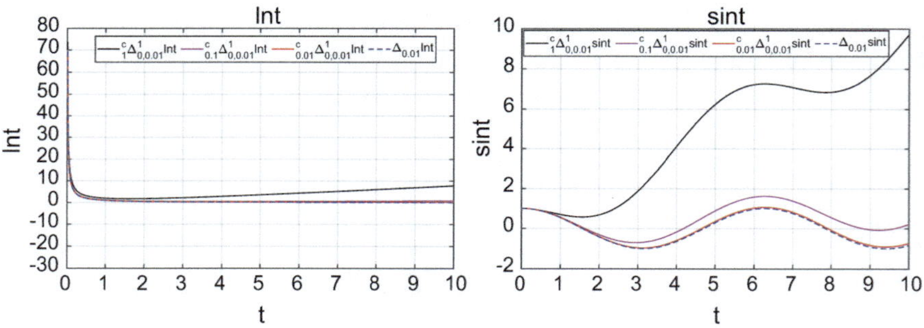


**Fig. 2.15** $(\text{EEDO})^{\lambda}_{\Delta}$ and $(\text{EDO})_{\Delta}$ of elementary functions $\ln t$ and $\sin t$ with $\lambda = 1, 0.1, 0.01$, $h = 0.01$ and $\alpha = 1$

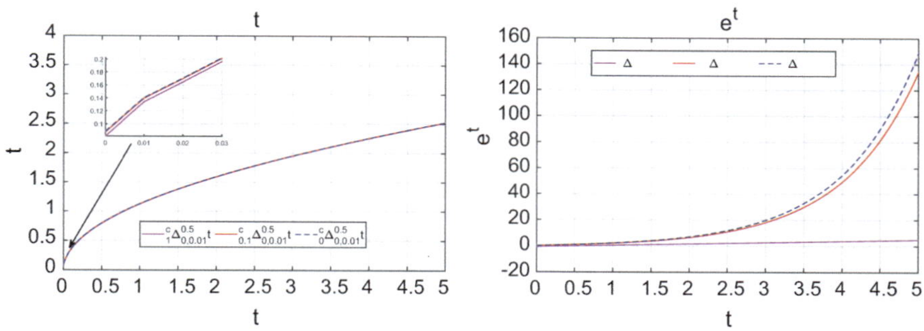


**Fig. 2.16** $(\mathrm{MLEDO})_\Delta^\lambda$ and $(\mathrm{FEDO})_\Delta$ of elementary functions $t$ and $e^t$ with $\lambda = 1, 0.1$, $h = 0.01$ and $\alpha = 0.5$

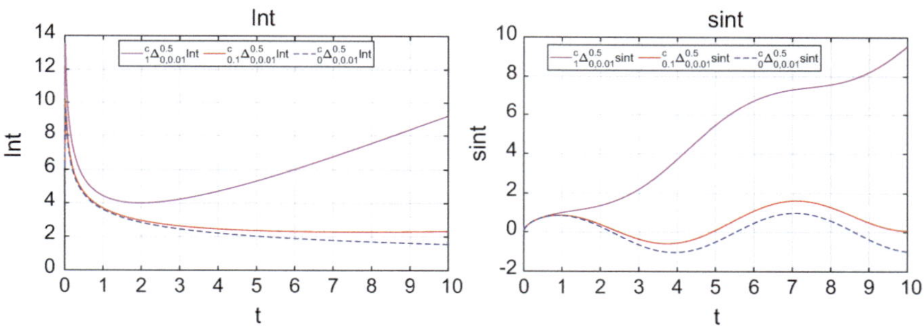


**Fig. 2.17** $(\mathrm{MLEDO})_\Delta^\lambda$ and $(\mathrm{FEDO})_\Delta$ of elementary functions $\ln t$ and $\sin t$ with $\lambda = 1, 0.1$, $h = 0.01$ and $\alpha = 0.5$

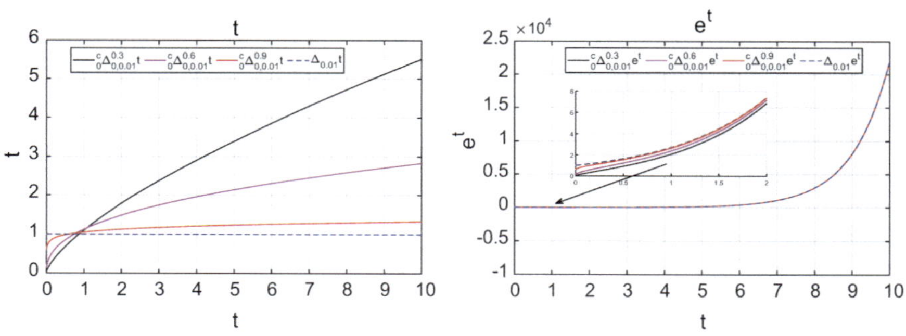


**Fig. 2.18** $(\mathrm{FEDO})_\Delta$ and $(\mathrm{EDO})_\Delta$ of elementary functions $t$ and $e^t$ with $h = 0.01$ and $\alpha = 0.3, 0.6, 0.9, 1$

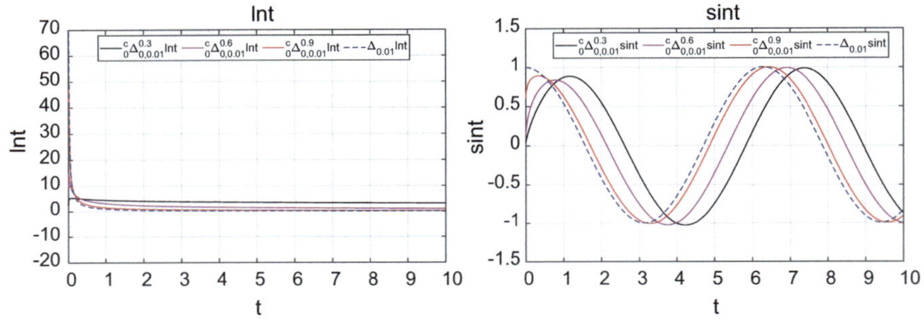


**Fig. 2.19** $(\text{FEDO})_\Delta$ and $(\text{EDO})_\Delta$ of elementary functions $\ln t$ and $\sin t$ with $h = 0.01$ and $\alpha = 0.3, 0.6, 0.9, 1$

### 2.5.3 Variation-of-Constants Formula

Let $f_\lambda(\cdot, x(\cdot)) = f(\cdot, x(\cdot)) - \lambda x(\cdot)$, Eq. (2.33) is changed into

$$x(t + h_{t^\circ}) = \lambda_\alpha^{[t^\circ]} x(t) + \sum_{l=0}^{\nu_{t^\circ}} \theta_\alpha^{[l,t^\circ]} f_\lambda(t - l h_{t^\circ}, x(t - l h_{t^\circ})), \tag{2.49}$$

where $t \in C_a$.

By employing the variation-of-constants formula on time scales in Theorem 2.27 of literature [95], the solution of Eq. (2.49) has the form of

$$x(t) = \prod_{\iota=t_0^\circ}^{t^\circ-1} \lambda_\alpha^{[\iota]} x(t_0) + \sum_{q=t_0^\circ}^{t^\circ-1} \prod_{\iota=q+1}^{t^\circ-1} \lambda_\alpha^{[\iota]} \sum_{l=0}^{\nu_q} \theta_\alpha^{[l,q]} f_\lambda(c_{q-l}, x(c_{q-l})), \tag{2.50}$$

where $t \in [t_0, +\infty)_{C_a}$, $t_0 \in C_a$ is a fixed point.

Since PMLEDEs (2.38) is equal to Eq. (2.33), then it obtains

**Lemma 2.6** *The solutions of* PMLEDEs (2.38) *satisfy Eq.* (2.50). *If* $\alpha = 1$ *and* $\lambda \neq 0$, *then Eq.* (2.50) *is turned into*

$$x(t) = \prod_{\iota=t_0^\circ}^{t^\circ-1} e^{\lambda h_\iota} x(t_0) + \sum_{q=t_0^\circ}^{t^\circ-1} \prod_{\iota=q+1}^{t^\circ-1} e^{\lambda h_\iota} \frac{(e^{\lambda h_q} - 1)}{\lambda} f_\lambda(c_q, x(c_q)),$$

*where* $t \in [t_0, +\infty)_{C_a}$, $t_0 \in C_a$ *is a fixed point.*

Similarly,

**Lemma 2.7** *If* $\alpha = 1$ *and* $\lambda \neq 0$*, then the solutions of* MLEDEs (2.42) *satisfy*

$$x(t) = \prod_{\iota=t_0^\circ}^{t^\circ-1} \lambda_\alpha^{[\iota]} x(t_0) + \sum_{q=t_0^\circ}^{t^\circ-1} \prod_{\iota=q+1}^{t^\circ-1} \lambda_\alpha^{[\iota]} \times \sum_{l=0}^{q} \theta_\alpha^{[l,q]} f_\lambda\big(a + (q-l)h, x(a + (q-l)h)\big), \tag{2.51}$$

*where* $f_\lambda(\cdot, x(\cdot)) = f(\cdot, x(\cdot)) - \lambda x(\cdot)$, $t_0 \in (h\mathbb{N}_0)_a$ *is a fixed point,* $t^\circ = \frac{t-a}{h}$ *and* $t \in [t_0, +\infty) \cap (h\mathbb{N}_0)_a$.

*If* $\alpha = 1$*, Eq.* (2.51) *is converted into*

$$x(t) = \prod_{\iota=t_0^\circ}^{t^\circ-1} e^{\lambda h} x(t_0) + \sum_{q=t_0^\circ}^{t^\circ-1} \frac{e^{\lambda(t^\circ-q-1)h}(e^{\lambda h}-1)}{\lambda} f_\lambda(a+qh, x(a+qh)), \tag{2.52}$$

*where* $t \in [t_0, +\infty)_{(h\mathbb{N}_0)_a}$, $t_0 \in (h\mathbb{N}_0)_a$ *is a fixed point.*

Compared with the previous FODCEs in literature [63–85], MLEDEs (e.g., Eq. (2.42)) possess the variation-of-constants formulas based on continuous-time Mittag–Leffler functions, which are more easier to be studied and calculated in practical applications. As everyone knows, variation-of-constants formula plays an important role in the discussion of semi-linear differential or difference equations, particularly in the existence and uniqueness of various solutions, e.g., periodic solutions [49, 50, 96], almost periodic solutions [27, 96, 97] and almost automorphic solutions [98–100], etc. Until now, several variation-of-constants formulas for FODCEs had been proposed involving various discrete Mittag–Leffler functions. For instance, the authors [68, 71] achieved the variation-of-constants formulas of Nabla Caputo FODCEs. The authors [69, 70, 76] investigated the variation-of-constants formulas for Delta Caputo FODCEs. Giving an example, literature [69, 70, 76] discussed a Delta FODCEs in the form of

$$\Delta_C^\alpha x(t) = \lambda x(t+\alpha-1) + F(t), \quad \forall t \in \mathbb{N}_0, \quad x(0) = \xi_0 \in \mathbb{R}, \tag{2.53}$$

where $\Delta_C^\alpha x$ denotes Delta Caputo fractional difference defined in Definition 13 of [69], $\alpha \in (0, 1]$, $a = \alpha - 1$, $\lambda \in \mathbb{R}$, $F : \mathbb{N}_0 \to \mathbb{R}$, $t \in \mathbb{N}_0$. A variation-of-constants formula corresponding to Eq. (2.53) was introduced in [69, 70, 76] as below

$$x(t) = E_{(\alpha,1)}(\lambda, t)\xi_0 + \sum_{q=0}^{t-\alpha} E_{(\alpha,\alpha)}(\lambda, t-q-1)F(q),$$

where $E$ is discrete Mittag–Leffler function:

$$E_{(\alpha,\beta)}(\lambda, t) = \sum_{j=0}^{\infty} \frac{(t + (j-1)(\alpha-1))^{(j\alpha)}(t + j(\alpha-1))^{(\beta-1)}}{\Gamma(\alpha j + \beta)} \lambda^j, \quad (2.54)$$

where $t \in \mathbb{N}_0$, $\lambda \in \mathbb{R}$, $\alpha, \beta \in (0, +\infty)$ and $t^{(r)} = \frac{\Gamma(t+1)}{\Gamma(t+1-r)}$ for $t \in \mathbb{N}_0$ and $r \in [0, +\infty)$.

In Eq. (2.51), $\lambda_\alpha^{[k]}$ and $\theta_\alpha^{[l,k]}$ are defined as those in (2.31) with $\nu_k = k$ and $h_k \equiv h$, $\forall l, k \in \mathbb{N}_0$. Obviously, the values of the kernel functions

$$\prod_{\iota=t_0^\circ}^{t^\circ-1} \lambda_\alpha^{[\iota]} \text{ and } \prod_{\iota=s^\circ+1}^{t^\circ-1} \lambda_\alpha^{[\iota]} \sum_{l=0}^{s^\circ} \theta_\alpha^{[l,s^\circ]}, \quad t_0, s, t \in (h\mathbb{N}_0)_a$$

of Eq. (2.51) are mainly dependent on continuous Mittag–Leffler functions. Some fine properties (e.g., Lemmas 1.21 and 1.23) of continuous Mittag–Leffler functions can be used to discuss Eq. (2.51). Meanwhile, many properties of discrete Mittag–Leffler function $E_{(\alpha,\beta)}(\lambda, \cdot)$ in Eq. (2.54) still need to further research and some properties of continuous Mittag–Leffler functions may not be valid for discrete Mittag–Leffler functions. For instance, continuous Mittag–Leffler function $\mathbb{E}_{\alpha,1}(z)$ ($\alpha > 0$) is convergent for $z \in \mathbb{C}$, but $E_{(\alpha,1)}(\lambda, \cdot)$ in Eq. (2.54) needs a strict condition $|\lambda| < \alpha$ to guarantee its convergence in the discussions of literatures [70, 76]. These reasons cause that the variation-of-constants formulas in [69, 70, 76] have more difficulty than Eq. (2.51) in the study of FODCEs.

**Remark 2.12** The form of Eq. (2.52) is consistent to variation-of-constants formulas studied in literature [26–29]. So Eq. (2.52) is an expansion of the corresponding results in literature [26–29].

## 2.6 Qualitative Analysis

Let us discuss a Cauchy problem of PMLEDEs as described below

$$\begin{cases} {}_{D}^{C}\Delta_{c_\iota, h_\iota}^{\alpha} X(t) = G(t, X(t)), \quad \forall t \in C_a, \\ X(a) = \varphi(a) \in \mathbb{R}^n, \end{cases} \quad \text{(PMLEDEs)}$$

where $X = (x_1, \ldots, x_n)^T \in \mathbb{R}^n$, $\alpha = (\alpha_1, \ldots, \alpha_n)^T \in \mathbb{R}^n$, $\mathbf{0}_n < \alpha \leq \mathbf{1}_n$,

$${}_{D}^{C}\Delta_{c_\iota, h_\iota}^{\alpha} X = \left( {}_{d_1}^{C}\Delta_{c_\iota, h_\iota}^{\alpha_1} x_1, \ldots, {}_{d_n}^{C}\Delta_{c_\iota, h_\iota}^{\alpha_n} x_n \right)^T,$$

${}_{d_i}^{C}\Delta_{c_t,h_t}^{\alpha_i}$ stands for PMLEDO defined in Definition 2.1 for $t \in C_a$ and $i = 1, 2, \ldots, n$, $G : C_a \times \mathbb{R}^n \to \mathbb{R}^n$ and $D = \text{diag}\{d_1, \ldots, d_n\} \in \mathbb{R}^{n\times n}$ meet

(H$_1$) There exists a constant $G_D^0 > 0$ such that $\|G_D(t, \mathbf{0}_n)\| \le G_D^0$, $\forall t \in C_a$, where $G_D(t, Y) := G(t, Y) - DY$ for arbitrary $(t, Y) \in C_a \times \mathbb{R}^n$.

(H$_2$) It exists a $n$-order constant matrix $L_{G_D} > \mathbb{O}_n$ such that

$$\|G_D(t, X) - G_D(t, Y)\| \le \|L_{G_D}\|\|X - Y\|, \quad \forall t \in C_a, X, Y \in \mathbb{R}^n.$$

(H$_3$) $[D]_{\text{dia}} < \mathbf{0}_n$ and $\xi_1 = \|\bar{\Theta}_\alpha(\mathbb{I}_n - \bar{\Lambda}_\alpha)^{-1}\|\|L_{G_D}\| < 1$.

By Lemma 2.6, the solution of (PMLEDEs) can be depicted by

$$X(t) = \prod_{\iota=0}^{t^\circ-1} \Lambda_\alpha^{[\iota]}\varphi(a) + \sum_{q=0}^{t^\circ-1}\prod_{\iota=q+1}^{t^\circ-1} \Lambda_\alpha^{[\iota]}\sum_{l=0}^{\nu_q} \Theta_\alpha^{[l,q]}G_D(c_{q-l}, X(c_{q-l})), \tag{2.55}$$

where $\Lambda_\alpha^{[\iota]}$ and $\Theta_\alpha^{[l,t^\circ]}$ are defined as those in (2.12), $l \in [0, t^\circ]_{\mathbb{Z}}$, $t \in C_a$.

### 2.6.1 Global Solution

Let

$$\mathbb{X}_L = \left\{X : [c_0, c_L]_{C_a} \to \mathbb{R}^n | X(a) = \varphi(a)\right\}$$

for arbitrary positive integer $L$. Then $\mathbb{X}_L$ is a Banach space with the norm $\|X\|_L = \sup_{t\in[c_0,c_L]_{C_a}} \|X(t)\|$, $\forall X \in \mathbb{X}_L$. Likewise,

$$\mathbb{X}_\infty = \left\{X : C_a \to \mathbb{R}^n | X(a) = \varphi(a)\right\}$$

is a Banach space with the norm $\|X\|_\infty = \sup_{t\in C_a} \|X(t)\|$, $\forall X \in \mathbb{X}_\infty$.

Let $\bar{\Lambda}_\alpha$ and $\bar{\Theta}_\alpha$ be defined as those in Lemma 2.1.

**Theorem 2.2** *Let the assumptions* (H$_1$)*,* (H$_2$) *and* (H$_3$) *be fulfilled. Then*

**(1)** (PMLEDEs) *possesses a unique solution $X$ on $C_a$.*

**(2)** *$X$ is bounded, i.e.,*

$$\|X\|_\infty \le \xi_0 = \frac{\xi_2}{1 - \xi_1}, \quad \xi_2 = \|\varphi(a)\| + \|\bar{\Theta}_\alpha(\mathbb{I}_n - \bar{\Lambda}_\alpha)^{-1}\|G_D^0.$$

***Proof*** For $t \in [a, c_1]_{C_a}$, by Eq. (2.55) and considering the operator $\Psi_1 : \mathbb{X}_1 \to \mathbb{X}_1$ as noted below

$$(\Psi_1 X)(t) = \prod_{\iota=0}^{t^\circ-1} \Lambda_\alpha^{[\iota]} X(a) + \sum_{q=0}^{t^\circ-1} \prod_{\iota=q+1}^{t^\circ-1} \Lambda_\alpha^{[\iota]} \times \sum_{l=0}^{\nu_q} \Theta_\alpha^{[l,q]} G_D(c_{q-l}, X(c_{q-l})), \quad X \in \mathbb{X}_1. \tag{2.56}$$

For any $X, \tilde{X} \in \mathbb{X}_1$, it derives from $(H_2)$ and Lemma 2.1 that

$$\begin{aligned}
\|(\Psi_1 X)(t) - (\Psi_1 \tilde{X})(t)\| &\leq \left\| \sum_{q=0}^{t^\circ-1} \prod_{\iota=q+1}^{t^\circ-1} \Lambda_\alpha^{[\iota]} \sum_{l=0}^{\nu_q} \Theta_\alpha^{[l,q]} \right. \\
&\quad \left. \times \left[ G_D(c_{q-l}, X(c_{q-l})) - G_D(c_{q-l}, \tilde{X}(c_{q-l})) \right] \right\| \\
&\leq \sum_{q=0}^{t^\circ-1} \left\| \sum_{l=0}^{\nu_q} \Theta_\alpha^{[l,q]} \prod_{\iota=q+1}^{t^\circ-1} \Lambda_\alpha^{[\iota]} \right. \\
&\quad \left. \times \sup_{s \in [a,c_1]_{C_a}} \left\| G_D(s, X(s)) - G_D(s, \tilde{X}(s)) \right\| \right\| \\
&\leq \sum_{q=0}^{1-1} \|\bar{\Theta}_\alpha \bar{\Lambda}_\alpha^{1-q-1}\| \|L_{G_D}\| \|X - \tilde{X}\|_1 \\
&\leq \|\bar{\Theta}_\alpha\| \|L_{G_D}\| \|X - \tilde{X}\|_1 \\
&= \xi_1 \|X - \tilde{X}\|_1, \quad t \in [a, c_1]_{C_a}.
\end{aligned} \tag{2.57}$$

By (2.57), $\|\Psi_1 X - \Psi_1 \tilde{X}\|_1 \leq \xi_1 \|X - \tilde{X}\|_1, \forall X, \tilde{X} \in \mathbb{X}_1$. On account of $(H_3)$, $\Psi_1$ is a contractive mapping. Therefore, $\Psi_1$ possesses a unique fixed point $X_1 = \Psi_1 X_1 \in \mathbb{X}_1$ and $X_1$ solves (PMLEDEs) on interval $[a, c_1]_{C_a}$.

Similarly, thinking about the mapping $\Psi_2 : \mathbb{X}_2 \to \mathbb{X}_2$ as described below

$$(\Psi_2 X)(t) = \prod_{\iota=1}^{t^\circ-1} \Lambda_\alpha^{[\iota]} X(c_1) + \sum_{q=1}^{t^\circ-1} \prod_{\iota=q+1}^{t^\circ-1} \Lambda_\alpha^{[\iota]} \sum_{l=0}^{\nu_q} \Theta_\alpha^{[l,q]} G_D(c_{q-l}, X(c_{q-l})) \tag{2.58}$$

for $t \in (c_1, c_2]_{C_a}$; $(\Psi_2 X)(s) = X_1(s)$ for all $s \in [a, c_1]_{C_a}$ and arbitrary $X \in \mathbb{X}_2$. Supposing that $X, \tilde{X} \in \mathbb{X}_2$, it should be noted that $\|(\Psi_2 X)(s) - (\Psi_2 \tilde{X})(s)\| = 0$ for all $s \in [a, c_1]_{C_a}$. Similar to the derivation of (2.57), it yields from Eq. (2.58) that

$$\begin{aligned}\|(\Psi_2 X)(t)-(\Psi_2\tilde{X})(t)\| &\le \sum_{q=1}^{t^\circ-1}\left\|\sum_{l=0}^{\nu_q}\Theta_\alpha^{[l,q]}\prod_{\iota=q+1}^{t^\circ-1}\Lambda_\alpha^{[\iota]}\right\|\\ &\quad\times\left\|\sup_{s\in[a,c_2]_{C_a}}\left\|G_D(s,X(s))-G_D(s,\tilde{X}(s)))\right\|\right\|\\ &\le \sum_{q=1}^{2-1}\|\bar{\Theta}_\alpha\bar{\Lambda}_\alpha^{2-q-1}\|P_0\|L_{G_D}\|\|X-\tilde{X}\|_2\\ &\le \|\bar{\Theta}_\alpha\|\|L_{G_D}\|\|X-\tilde{X}\|_2\\ &= \xi_1\|X-\tilde{X}\|_2,\quad t\in(c_1,c_2]_{C_a}.\end{aligned}$$

In a consequence, $\|\Psi_2 X-\Psi_2\tilde{X}\|_2\le\xi_1\|X-\tilde{X}\|_2, \forall X,\tilde{X}\in\mathbb{X}_2$. By ($H_3$), $\Psi_2$ is contractive and (PMLEDEs) has a unique solution $X_2$ on interval $[a,c_2]_{C_a}$. By employing inductive method, (PMLEDEs) has a unique solution $X_m$ on interval $[a,c_m]_{C_a}$, $m\in\mathbb{N}_0$. Letting $m\to+\infty$, (PMLEDEs) has a unique solution $X$ on $C_a$.

Provided that $X\in\mathbb{X}_\infty$ is a solution of (PMLEDEs), then it follows from Eq. (2.55), ($H_2$), Lemma 2.1, ($H_1$) and ($H_3$) in turn that

$$\begin{aligned}\|X(t)\| &\le \prod_{\iota=0}^{t^\circ-1}\|\Lambda_\alpha^{[\iota]}\|\|\varphi(a)\|\\ &\quad+\left\|\sum_{q=0}^{t^\circ-1}\sum_{l=0}^{\nu_q}\left[\Theta_\alpha^{[l,q]}\prod_{\iota=q+1}^{t^\circ-1}\Lambda_\alpha^{[\iota]}\right]_{\text{abs}}\left\|G_D(c_{q-l},X(c_{q-l}))\right\|\right\|\\ &\le \|\varphi(a)\|+\left\|\bar{\Theta}_\alpha\sum_{q=0}^{t^\circ-1}\bar{\Lambda}_\alpha^{t^\circ-q-1}\right\|\big(\|L_{G_D}\|\|X\|_\infty+G_D^0\big)\\ &= \|\varphi(a)\|+\|\bar{\Theta}_\alpha(\mathbb{I}_n-\bar{\Lambda}_\alpha)^{-1}(\mathbb{I}_n-\bar{\Lambda}_\alpha^{t^\circ})\|\big(\|L_{G_D}\|\|X\|_\infty+G_D^0\big)\\ &\le \|\varphi(a)\|+\|\bar{\Theta}_\alpha(\mathbb{I}_n-\bar{\Lambda}_\alpha)^{-1}\|\big(\|L_{G_D}\|\|X\|_\infty+G_D^0\big)\\ &= \xi_1\|X\|_\infty+\xi_2,\quad t\in C_a.\end{aligned}$$

Consequently, $\|X\|_\infty\le\frac{\xi_2}{1-\xi_1}$. The proof is achieved. □

Let $\alpha=\mathbf{1}_n$ and $C_a=(h\mathbb{N}_0)_a$, then (PMLEDEs) is changed into EEDEs as below

$$\begin{cases}{}_D\Delta_{a,h}X(t)=G(t,X(t)), & \forall t\in(h\mathbb{N}_0)_a,\\ X(a)=\varphi(a)\in\mathbb{R}^n,\end{cases}\tag{EEDEs}$$

where $X=(x_1,\ldots,x_n)^T\in\mathbb{R}^n$, ${}_D\Delta_{a,h}X=\big({}_{d_1}\Delta_{a,h}x_1,\ldots,{}_{d_n}\Delta_{a,h}x_n\big)^T$, ${}_{d_i}\Delta_{a,h}$ stands for EED defined in $(\text{EEDO})_\Delta^\lambda$, $t\in(h\mathbb{N}_0)_a$, $i=1,2,\ldots,n$, $G:(h\mathbb{N}_0)_a\times\mathbb{R}^n\to\mathbb{R}^n$.

If $\alpha = \mathbf{1}_n$, then $\bar{\Lambda}_\alpha = e^{Dh}$ and $\bar{\Theta}_\alpha = (e^{Dh} - \mathbb{I}_n)D^{-1}$. Then (H$_3$) is turned into

(H$_4$) $[D]_{\text{dia}} < \mathbf{0}_n$ and $\acute{\xi}_1 = \|D^{-1}\|\|L_{G_D}\| < 1$.

Applying similar discussion of Theorem 2.2 into (EEDEs) induces

**Corollary 2.1** *Let* (H$_1$)*-*(H$_2$)*,* (H$_4$) *hold. Then* (EEDEs) *admits a unique solution* $X$ *on* $(h\mathbb{N}_0)_a$*. Further,* $X$ *is bounded, i.e.,* $\|X\|_\infty \le \frac{\acute{\xi}_2}{1-\acute{\xi}_1}$*, where* $\acute{\xi}_2 = \|\varphi(a)\| + \|D^{-1}\|G_D^0$.

## 2.6.2 Global $\bar{\lambda}_\alpha$-Exponential Stability

Assume that $X$ and $\tilde{X}$ are arbitrary two solutions of (PMLEDEs) with initial values $X(a)$ and $\tilde{X}(a)$, respectively. Define $\bar{\lambda}_\alpha = \|\bar{\Lambda}_\alpha\|$.

**Definition 2.5** Supposing that $0 < \bar{\lambda}_\alpha < 1$. (PMLEDEs) has the property of global $\bar{\lambda}$-exponential stability if there exist $M > 0$ and $0 < \kappa < 1$ such that

$$\|X(t) - \tilde{X}(t)\| \le M\|X(a) - \tilde{X}(a)\|\bar{\lambda}_\alpha^{\kappa t^\circ}, \quad \forall t \in C_a.$$

Here $\bar{\lambda}^\kappa$ depicts the convergent speed of the solutions of (PMLEDEs).

**Theorem 2.3** (PMLEDEs) *is globally* $\bar{\lambda}_\alpha$*-exponentially stable if* (H$_2$) *and* (H$_3$) *hold.*

***Proof*** Let $U(t) = \|X(t) - \tilde{X}(t)\|$, $\forall t \in C_a$. By Eq. (2.55), it yields from (H$_2$) and Remark 2.1 that

$$\begin{aligned} U(t) \le & \prod_{\iota=0}^{t^\circ-1} \|\Lambda_\alpha^{[\iota]}\|U(a) + \left\| \sum_{q=0}^{t^\circ-1} \prod_{\iota=q+1}^{t^\circ-1} \Lambda_\alpha^{[\iota]} \sum_{l=0}^{\nu_q} \Theta_\alpha^{[l,q]} \right. \\ & \left. \times \left[ G_D(c_{q-l}, X(c_{q-l})) - G_D(c_{q-l}, \tilde{X}(c_{q-l})) \right] \right\| \\ \le & \bar{\lambda}_\alpha^{t^\circ} U(a) + \|L_{G_D}\| \left\| \bar{\Theta}_\alpha \sum_{q=0}^{t^\circ-1} \bar{\Lambda}_\alpha^{t^\circ-q-1} \max_{l\in[0,N^*]_{\mathbb{Z}}} U(c_{q-l}) \right\| \end{aligned} \tag{2.59}$$

for $t \in C_a$.

By (H$_3$), it exists $M > 1$ large enough and $0 < \kappa < 1$ small enough such that

$$\frac{1}{M} + \|\bar{\Theta}_\alpha(\mathbb{I}_n - \bar{\Lambda}_\alpha^{1-\kappa})^{-1}\|\|L_{G_D}\|\bar{\lambda}_\alpha^{-\kappa N^*} < 1. \tag{2.60}$$

For any $t \in C_a$,

$$U(t) = \|X(t) - \tilde{X}(t)\| \le MU(a)\bar{\lambda}_\alpha^{\kappa t^\circ}$$

is supposed to be satisfied. If not, there exist $t_0 \in C_a \backslash \{a\}$ such that

$$U(t_0) > MU(a)\bar{\lambda}_\alpha^{\kappa t_0^\circ}; \quad U(s) \leq MU(a)\bar{\lambda}_\alpha^{\kappa s^\circ}, \quad \forall s \in [a, t_0)_{C_a}. \tag{2.61}$$

By utilizing (2.59), (2.61), Lemma 2.1 and (2.60) in turn, it results in

$$\begin{aligned}
U(t_0) \leq\ & \bar{\lambda}_\alpha^{t_0^\circ} U(a) + \|L_{G_D}\| \left\| \bar{\Theta}_\alpha \sum_{q=0}^{t_0^\circ - 1} \bar{\Lambda}_\alpha^{t_0^\circ - q - 1} \max_{l \in [0, N^*]_{\mathbb{Z}}} U(c_{q-l}) \right\| \\
\leq\ & \bar{\lambda}_\alpha^{t_0^\circ} U(a) + \|L_{G_D}\| \left\| \bar{\Theta}_\alpha \sum_{q=0}^{t_0^\circ - 1} \bar{\Lambda}_\alpha^{(1-\kappa)(t_0^\circ - q - 1)} \right. \\
& \left. \times \bar{\Lambda}_\alpha^{\kappa(t_0^\circ - q - 1)} \max_{l \in [0, N^*]_{\mathbb{Z}}} \bar{\lambda}_\alpha^{-\kappa l} MU(a)\bar{\lambda}_\alpha^{\kappa q} \right\| \\
\leq\ & \bar{\lambda}_\alpha^{t_0^\circ} U(a) + \|L_{G_D}\| \left\| \bar{\Theta}_\alpha \sum_{q=0}^{t_0^\circ - 1} \bar{\Lambda}_\alpha^{(1-\kappa)(t_0^\circ - q - 1)} \bar{\lambda}_\alpha^{-\kappa N^*} MU(a)\bar{\lambda}_\alpha^{\kappa t_0^\circ} \right\| \\
\leq\ & \bar{\lambda}_\alpha^{t_0^\circ} U(a) + \left\| \bar{\Theta}_\alpha (\mathbb{I}_n - \bar{\Lambda}_\alpha^{1-\kappa})^{-1} (\mathbb{I}_n - \bar{\Lambda}_\alpha^{t_0^\circ(1-\kappa)}) \right\| \|L_{G_D}\| \bar{\lambda}_\alpha^{-\kappa N^*} MU(a)\bar{\lambda}_\alpha^{\kappa t_0^\circ} \\
\leq\ & MU(a)\bar{\lambda}_\alpha^{\kappa t_0^\circ} \left\{ \frac{1}{M} \bar{\lambda}_\alpha^{(1-\kappa)t_0^\circ} + \|\bar{\Theta}_\alpha (\mathbb{I}_n - \bar{\Lambda}_\alpha^{1-\kappa})^{-1}\| \|L_{G_D}\| \bar{\lambda}_\alpha^{-\kappa N^*} \right\} \\
\leq\ & MU(a)\bar{\lambda}_\alpha^{\kappa t_0^\circ} \left\{ \frac{1}{M} + \|\bar{\Theta}_\alpha (\mathbb{I}_n - \bar{\Lambda}_\alpha^{1-\kappa})^{-1}\| \|L_{G_D}\| \bar{\lambda}_\alpha^{-\kappa N^*} \right\} \\
\leq\ & MU(a)\bar{\lambda}_\alpha^{\kappa t_0^\circ}.
\end{aligned}$$

This is a contradiction with the first part of (2.61). So

$$\|X(t) - \tilde{X}(t)\| \leq M\|X(a) - \tilde{X}(a)\|\bar{\lambda}_\alpha^{\kappa t^\circ}, \quad \forall t \in C_a.$$

Consequently, (PMLEDEs) is globally $\bar{\lambda}_\alpha$-exponentially stable. The proof is achieved. □

If $\alpha = \mathbf{1}_n$ in (PMLEDEs), then

$$\bar{\Theta}_\alpha = (\bar{\Lambda}_\alpha - \mathbb{I}_n)D^{-1}, \quad \|\bar{\Theta}_\alpha(\mathbb{I}_n - \bar{\Lambda}_\alpha)^{-1}\| = \|D^{-1}\|.$$

So $(H_3)$ is converted into $(H_4)$. By using Eq. (2.52) and similar to the proof of Theorem 2.3, it easily gains global exponential stability as below.

**Corollary 2.2** *Let* $(H_2)$ *and* $(H_4)$ *hold. Then* (EEDEs) *is globally exponentially stable, i.e., there exist* $M > 0$ *and* $0 < \kappa < 1$ *such that*

$$\|X(t) - \tilde{X}(t)\| \leq M\|X(a) - \tilde{X}(a)\|e^{\kappa d_0 t^\circ}, \quad \forall t \in (h\mathbb{N}_0)_a,$$

*where* $X, \tilde{X}$ *are arbitrary two solutions of* (EEDEs) *and* $d_0 = \max_{1\le i\le n} d_i$ *for* $D = \operatorname{diag}(d_1, d_2, \ldots, d_n)$. *Hereby,* $t^\circ = \frac{t-a}{h}$, $\forall t \in (h\mathbb{N}_0)_a$.

**Remark 2.13** This literature is mainly devoted to discussing MLEDEs with piecewise feature such as (PMLEDEs). Noting that the variation-of-constants formula of (PMLEDEs) is Eq. (2.55). Like the argument as that in Sect. 2.3, the trait of piecewise feature in (PMLEDEs) enables us to easily derive the upper bounds of the chief parameters $\Lambda_\alpha^{[k]}$ and $\Theta_\alpha^{[l,k]}$ of Eq. (2.55), $l, k \in \mathbb{N}_0$. It should be noted that the upper bounds of these parameters are crucial to the discussions of the existence of a unique global bounded solution and global exponential stability for (PMLEDEs). Observing the old investigations of FODCEs in [63–85], it is hard to calculate the upper bounds of these parameters and because of this, almost no literature considers the issues of semi-linear FODCEs such as the existence of a unique global bounded solution (e.g., periodic solution, almost periodic solution and almost automorphic solution) and global exponential stability, etc. However, (PMLEDEs) is capable of studying these problems. So the work of this chapter should create a new path to discuss the discrete-time models of fractional order in applied areas in the future.

**Remark 2.14** Since (EEDEs) has no memory property, in ($H_4$), memory term $\Theta_\alpha^{[l,k]}(l, k \in \mathbb{N})$ in (PMLEDEs) becomes $\mathbb{O}_n$. This future can be viewed from the definition of $\acute{\xi}_1$ in ($H_4$), which is more simple than $\xi$ in ($H_3$). The memory feature increases the complexity and difficulty of calculations of the existence of a unique bounded solution and global exponential stability to (PMLEDEs), see Theorems 2.2 and 2.3.

**Remark 2.15** By employing exponential Euler differences like $(\text{EEDO})_\Delta^\lambda$, literature [26–28] discussed the periodicity, almost periodicity and stability of discrete neural networks. The authors [27] considered mean dynamics to discrete stochastic neural networks with the same difference technique. This chapter studies the existence, uniqueness and boundedness of global solution, and global exponential stability to (PMLEDEs) with operator $(\text{PMLEDO})_\Delta^\lambda$. Thus, the work of this chapter supplements and extends the corresponding results in [26–28].

## 2.7 Numerical Experiments

**Example 2.1** Consider the FODEs below

$$\begin{cases} {}_{5p}^{\,C}D_t^\alpha X(t) = \begin{bmatrix} -0.3 & 0 \\ 0 & -0.3 \end{bmatrix} X(t) + \begin{bmatrix} 0 & 0.2 \\ 0.1 & 0 \end{bmatrix} F(X(t)) + \begin{bmatrix} 0.8 \\ 0.6 \end{bmatrix}, \\ X(0) = \xi \in \mathbb{R}^2, \end{cases} \tag{2.62}$$

where $t \in (5p, 5p+5]$,

$$ {}_{5p}^{C}D_t^{\alpha} X = ({}_{5p}^{C}D_t^{0.9} x_1, {}_{5p}^{C}D_t^{0.95} x_2)^T, \quad X(0) = \begin{bmatrix} x_1(0) \\ x_2(0) \end{bmatrix} \in \mathbb{R}^2, \quad p \in \mathbb{N}_0, $$

$$ F(X) = \begin{bmatrix} F_1(x_1) \\ F_2(x_2) \end{bmatrix} = \frac{1}{8} \begin{bmatrix} |x_1+1| - |x_1-1| \\ |x_2+1| - |x_2-1| \end{bmatrix}, \quad X = \begin{bmatrix} x_1 \\ x_2 \end{bmatrix} \in \mathbb{R}^2. $$

Let

$$ N \in \mathbb{N}, \quad h = 1/N, \quad \mathcal{T}_0 = \{t_p = 5p : p \in \mathbb{N}_0\}, $$

$$ \mathcal{N}_0 = \left\{N_p = 5p/h : p \in \mathbb{N}_0\right\}, \quad C_0 = \{c_k = kh : k \in \mathbb{N}_0\}. $$

Then $C_0$ is a $\mathcal{T}_0$-$\mathcal{N}_0$ sequence. By Property 2.1, numerical solution of FODEs (2.62) can be depicted by the difference equations involving $(\text{PMLEDO})_{\Delta}^{\lambda}$ as below

$$ \begin{cases} {}_{B}^{C}\Delta_{c_{\underline{t}},h}^{\alpha} X(t) = \begin{bmatrix} -0.3 & 0 \\ 0 & -0.3 \end{bmatrix} X(t) + \begin{bmatrix} 0 & 0.2 \\ 0.1 & 0 \end{bmatrix} F(X(t)) + \begin{bmatrix} 0.8 \\ 0.6 \end{bmatrix}, \\ X(0) = \xi \in \mathbb{R}^2, \end{cases} \tag{2.63} $$

where $t \in C_0$, $B = \begin{bmatrix} b_1 & 0 \\ 0 & b_2 \end{bmatrix}$ and ${}_{B}^{C}\Delta_{c_{\underline{t}},h}^{\alpha} = \left({}_{b_1}^{C}\Delta_{c_{\underline{t}},h}^{0.9}, {}_{b_2}^{C}\Delta_{c_{\underline{t}},h}^{0.95}\right)^T$, $t \in C_0$.

Taking $b_1 = b_2 = -0.3$ and $h = 0.01$. Corresponding to (PMLEDEs) and Theorem 2.2,

$$ N^* = 5, \quad \|L_{G_D}\| \leq 0.05, \quad \bar{\Lambda}_{\alpha} = \begin{bmatrix} 0.8138 & 0 \\ 0 & 0.7812 \end{bmatrix}, \quad \bar{\Theta}_{\alpha} = \begin{bmatrix} 1.1407 & 0 \\ 0 & 1.0169 \end{bmatrix}. $$

So $(H_1)$-$(H_3)$ in Theorems 2.2–2.3 are filled. Then MLEDEs (2.63) admits a unique bounded solution (it is an equilibrium point), which is globally 0.8138-exponentially stable with respect to an exponential function of base 0.8138. Taking some random initial values and $N = 100$, Figs. 2.20 and 2.21 describe the trajectories and exponential convergence of states $x_1$ and $x_2$ in MLEDEs (2.63), respectively.

Referring to MLEDEs (2.63), considering the corresponding MLEDEs with long-term memory (L-MLEDEs) as noted below

$$ {}_{B}^{C}\Delta_{a,h}^{\alpha} X(t) = DX(t) + \begin{bmatrix} 0 & 0.2 \\ 0.1 & 0 \end{bmatrix} F(X(t)) + \begin{bmatrix} 0.8 \\ 0.6 \end{bmatrix}, \quad X(0) = \xi \in \mathbb{R}^2, \tag{2.64} $$

where $t \in C_0 = (h\mathbb{N}_0)_0$, ${}_{B}^{C}\Delta_{a,h}^{\alpha} = \left({}_{b_1}^{C}\Delta_{a,h}^{0.9}, {}_{b_2}^{C}\Delta_{a,h}^{0.95}\right)^T$, ${}_{b_1}^{C}\Delta_{a,h}^{0.9}$ and ${}_{b_2}^{C}\Delta_{a,h}^{0.95}$ are defined as that in $(\text{MLEDO})_{\Delta}^{\lambda}$ with $b_1 = b_2 = -0.3$.

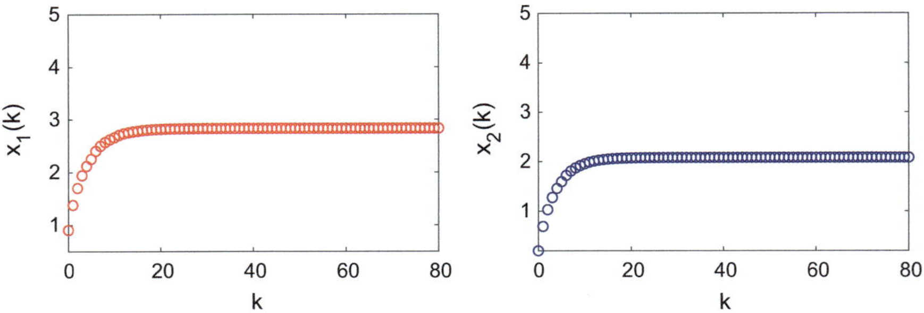


**Fig. 2.20** Trajectories of states $x_1$ and $x_2$ to MLEDEs (2.63)

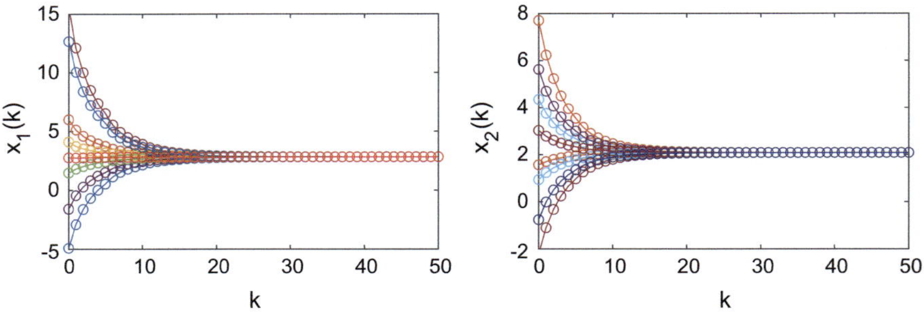


**Fig. 2.21** Exponential convergence of states $x_1$ and $x_2$ to MLEDEs (2.63)

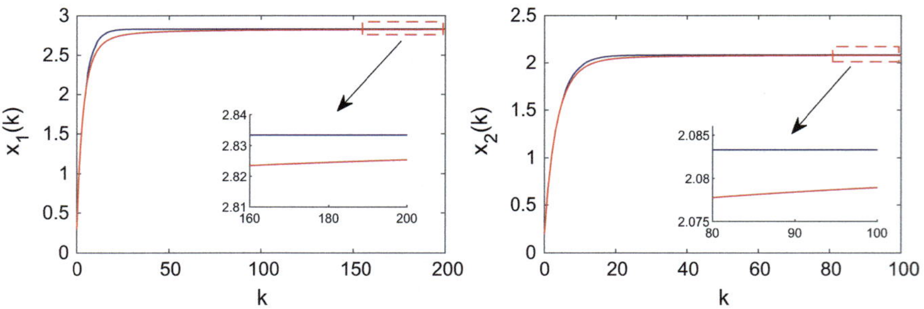


**Fig. 2.22** Comparisons of MLEDEs (2.63) (—) and L-MLEDEs (2.64) (—)

Figure 2.22 draws up the trajectories of states $x_1$ and $x_2$ of MLEDEs (2.63) and L-MLEDEs (2.64), respectively. The images display that the gap between MLEDEs (2.63) and L-MLEDEs (2.64) is very small. By careful observing, MLEDEs (2.63) reaches the equilibrium state more quickly than L-MLEDEs (2.64).

# Chapter 3
# A Fuzzy MLED-CGNNs with Time Lags

This chapter establishes a Mittag–Leffler Euler difference model for the fuzzy Cohen–Grossberg neural networks of fractional order in the sense of Caputo. Next, the existence of a unique global bounded solution and equilibrium point of the acquired difference model is studied by using the contractive mapping principle. Finally, global exponential stability and synchronization of the difference model are investigated by employing some inequality techniques and reduction.

## 3.1 Introduction

Cohen–Grossberg neural network was initially formulated by Cohen and Grossberg in 1983 [101]. This novel neural network had attracted a record amount of attention from all over the world for its prospective applicability in classification, associative memory, parallel computing and solving optimization problems. These operations rely heavily on the dynamics of the network, and consequently, the investigation of these dynamics is a necessary preliminary stage in the operational planning of neural networks. In this regard, stability is an important issue of both theoretical and practical significance. This is particularly true when neural networks are employed for finding optimization issues, where the existence or uniqueness of equilibrium points can guarantee the existence or uniqueness of the optimized solution to the optimization task. It is worth noting that in associative memory, the potential for information to be stored in the network also depends on the existence of equilibrium points, and the uniqueness of the recorded memories also lies in the uniqueness of the equilibrium points. Therefore, we ought to pay great attention to the existence and stability of the unique equilibrium point. Several investigators have researched the

 
T. Zhang et al., *Mittag-Leffler Euler Difference Techniques*,
https://doi.org/10.1007/978-981-96-6228-9_3

existence and stability of unique equilibrium point of Cohen–Grossberg type neural networks, as well as other dynamics in recent years, see [102–105].

Fuzzy logic was proposed in 1965 by Zadeh [106], which considers uncertainty and fuzziness and is one of the most popularly used and significant instruments for modeling realistic issues. Yang and Yang [107] in 1996 proposed a novel fuzzy cellular neural network, which incorporates fuzzy logic into the architecture of a cellular neural network. Fuzzy neural networks possess fuzzy logic for template inputs and/or outputs, in addition to summation of product operations. During recent years, fuzzy Cohen–Grossberg neural networks had received more and more attention because of their superiority in image processing and pattern recognition, see [108, 109]. On the other hand, the notion of synchronization was originally introduced by Picola and Carroll in [110] on behalf of driven-response systems. From then on, different synchronization problems were widely researched and a number of significant findings were achieved owing to the use of synchronization [111–114]. Therefore, studying the synchronization of Cohen–Grossberg neural networks is of great theoretical and practical importance, and numerous interesting achievements have been gained in [115–118].

Discrete-time neural networks are better fitted for implementations in real world applications. There are two advantages to working in a discrete-time framework. Firstly, appropriate technology can be used to implement digital controllers rather than analog ones. Secondly, the synthesized controller is directly implemented in a digital processor. Therefore, control methodologies developed for discrete-time nonlinear systems can be implemented in real systems more effectively. The main aim of this chapter is to set up the Mittag–Leffler Euler difference scheme of fractional-order fuzzy Cohen–Grossberg neural networks in the sense of Caputo. In addition, the existence of a unique bounded solution and equilibrium point, global exponential stability and synchronization of this type discrete-time models will be discussed, please see the publication [105].

The organization of this chapter is as follows. In Sect. 3.2, Mittag–Leffler discrete analogue is established for piecewise fractional-order fuzzy Cohen–Grossberg neural networks with time lags. The existence of a unique bounded solution and equilibrium point for the derived discrete-time models is discussed in Sect. 3.3. In Sects. 3.4–3.5, global exponential stability and synchronization are investigated. Some illustrative examples are employed to visually expound the current research findings in Sect. 3.6.

## 3.2 Discrete Analogue of CGNNs

In general, the form of fuzzy Cohen–Grossberg neural networks is described as

$$\begin{aligned}\frac{\mathrm{d}x_i(t)}{\mathrm{d}t} &= -d_i(x_i(t))\Big[a_i(x_i(t)) - \sum_{j=1}^{n} b_{ij} f_j(x_j(t-\tau_{ij}))\\&- \bigvee_{j=1}^{n} \varrho_{ij} f_j(x_j(t-\tau_{ij})) - \bigwedge_{j=1}^{n} \varsigma_{ij} f_j(x_j(t-\tau_{ij}))\\&- \bigvee_{j=1}^{n} T_{ij}\varpi_j - \bigwedge_{j=1}^{n} H_{ij}\varpi_j - I_i(t)\Big], \quad t \geq 0, \end{aligned} \tag{3.1}$$

where $x_i$ is the state of the $i$th unit, $d_i > 0$ represents an amplification function and $a_i$ is an appropriately behaved function with $a_i(0) = 0$, $f_j$ is the neuronal output signal function, $b_{ij}$ is the strength of the $j$th unit on the $i$th unit, $I_i$ corresponds to the external input, $\tau_{ij}$ denotes the transmission delay, $\varrho_{ij}$, $\varsigma_{ij}$, $T_{ij}$, $H_{ij}$ are the elements of fuzzy feedback MIN template and fuzzy feedback MAX template, fuzzy feed-forward MIN template and fuzzy feed-forward MAX template, respectively, $i, j = 1, 2, \ldots, n$. The fuzzy feed-forward MAX template and fuzzy feed-forward MIN template refer to the delay link parameters with the fuzzy OR and fuzzy AND operations, respectively; feedback MAX template and feedback MIN template refer to the link parameters with the fuzzy OR and fuzzy AND operations, respectively. $\bigwedge$ and $\bigvee$ denote the fuzzy AND and fuzzy OR operations, respectively. That is, $a \bigwedge b = \min\{a, b\}$ and $a \bigvee b = \max\{a, b\}$ for any $a, b \in \mathbb{R}$.

In view of Eq. (3.1), if $d_i > 0$, then Eq. (3.1) is equivalent to

$$\begin{aligned}\frac{\mathrm{d}z_i(t)}{\mathrm{d}t} &= -a_i\Big(h_{\diamond i}^{-1}(z_i(t))\Big) + \sum_{j=1}^{n} b_{ij} f_j\Big(h_{\diamond j}^{-1}(z_j(t-\tau_{ij}))\Big)\\&+ \bigvee_{j=1}^{n} \varrho_{ij} f_j\Big(h_{\diamond j}^{-1}(z_j(t-\tau_{ij}))\Big) + \bigwedge_{j=1}^{n} \varsigma_{ij} f_j\Big(h_{\diamond j}^{-1}(z_j(t-\tau_{ij}))\Big)\\&+ \bigvee_{j=1}^{n} T_{ij}\varpi_j + \bigwedge_{j=1}^{n} H_{ij}\varpi_j + I_i, \quad t \geq 0, \end{aligned} \tag{3.2}$$

where $z_i(t) = h_{\diamond i}(x_i(t))$, $h_{\diamond i}(x_i) = \int_0^{x_i} \frac{1}{d_i(s)}\,\mathrm{d}s$ with $h_{\diamond i}(0) = 0$, $h_{\diamond i}^{-1}$ stands for the inverse function of $h_{\diamond i}$, $i = 1, 2, \ldots, n$.

For convenience, let $z_j^{h_\diamond}(t) = h_{\diamond j}^{-1}(z_j(t))$ for $t \geq 0$, $j = 1, 2, \ldots, n$. Then the first equation of Eq. (3.2) is equivalent to

$$
\begin{aligned}
\dot{z}_i(t) = & -a_i\big(z_i^{h_\diamond}(t)\big) + \sum_{j=1}^{n} b_{ij} f_j\big(z_j^{h_\diamond}(t-\tau_{ij})\big) \\
& + \bigvee_{j=1}^{n} \varrho_{ij} f_j\big(z_j^{h_\diamond}(t-\tau_{ij})\big) + \bigwedge_{j=1}^{n} \varsigma_{ij} f_j\big(z_j^{h_\diamond}(t-\tau_{ij})\big) \\
& + \bigvee_{j=1}^{n} T_{ij}\varpi_j + \bigwedge_{j=1}^{n} H_{ij}\varpi_j + I_i, \quad t > 0, i = 1, 2, \ldots, n. \qquad (3.3)
\end{aligned}
$$

Let

$$
\mathcal{P} = \left\{ t_p \in \mathbb{R} : t_0 = 0, \lim_{p \to +\infty} t_p = +\infty, t_p \le t_{p+1}, p \in \mathbb{N}_0 \right\}
$$

and $\mathbb{R}_0 = \bigcup_{p=0}^{+\infty} [t_p, t_{p+1})$.

This chapter totally commits to discuss the following piecewise Caputo fractional-order fuzzy Cohen–Grossberg neural networks

$$
\begin{cases}
{}_{t_p}^{C}D_t^{\alpha_i} z_i(t) = -\lambda_i z_i(t) + \Big[ \Lambda_i^{h_\diamond}\big(z_i(t)\big) + \sum_{j=1}^{n} b_{ij} f_j\big(z_j^{h_\diamond}(t-\tau_{ij})\big) \\
\qquad + \bigvee_{j=1}^{n} \varrho_{ij} f_j\big(z_j^{h_\diamond}(t-\tau_{ij})\big) + \bigwedge_{j=1}^{n} \varsigma_{ij} f_j\big(z_j^{h_\diamond}(t-\tau_{ij})\big) \\
\qquad + \bigvee_{j=1}^{n} T_{ij}\varpi_j + \bigwedge_{j=1}^{n} H_{ij}\varpi_j + I_i \Big], \quad t \in (t_p, t_{p+1}], p \in \mathbb{N}_0, \\
z_i(s) = \varphi_i(s), \quad \forall s \in [-\tau_0, 0],
\end{cases} \qquad (3.4)
$$

where $\tau_0 = \max_{1 \le i,j \le n} \tau_{ij}$, $\lambda_i > 0$, $\Lambda_i^{h_\diamond}\big(z_i(t)\big) = \lambda_i z_i(t) - a_i\big(z_i^{h_\diamond}(t)\big)$, ${}_{t_p}^{C}D_t^{\alpha_i}$ denote Caputo fractional-order derivatives from initial point $t_p$, $p \in \mathbb{N}_0$, $\alpha_i \in (0, 1]$, $i = 1, 2, \ldots, n$.

Supposing that it exists

$$
\mathcal{N} = \left\{ N_p \in \mathbb{Z} : N_0 = 0, N_p < N_{p+1}, p \in \mathbb{N}_0 \right\}
$$

ensuring $\frac{t_{p+1}-t_p}{N_{p+1}-N_p} \equiv h$ for all $p \in \mathbb{N}_0$. Let

$$
\mu_k = \max_{p \in \mathbb{N}_0} \{N_p : N_p \le k\}, \quad \nu_k = k - \mu_k
$$

and

$$\begin{cases} \lambda_{\alpha_i}^{[k]} = \dfrac{E_{\alpha_i}[-\lambda_i(\nu_k h + h)^{\alpha_i}]}{E_{\alpha_i}[-\lambda_i(\nu_k h)^{\alpha_i}]}, \\ \theta_{\alpha_i}^{l,k} = \frac{1}{\lambda_i}\left[w_{\alpha_i}^{[l]} - \lambda_{\alpha_i}^{[k]} w_{\alpha_i}^{[l-1]}\right], \quad w_{\alpha_i}^{[-1]} = 0, \\ w_{\alpha_i}^{[l]} = E_{\alpha_i}[-\lambda_i(lh)^{\alpha_i}] - E_{\alpha_i}[-\lambda_i(lh+h)^{\alpha_i}], \end{cases} \tag{3.5}$$

where $l, k \in \mathbb{N}_0$, $i = 1, 2, \ldots, n$.

By Lemma 1.24, it easily verifies that the unique solution of Eq. (3.4) can be depicted by

$$\begin{aligned} z_i(t) = {} & E_{\alpha_i}[-\lambda_i(t-t_p)^{\alpha_i}] z_i(t_p) + \int_{t_p}^{t} (t-s)^{\alpha_i-1} E_{\alpha_i,\alpha_i}[-\lambda_i(t-s)^{\alpha_i}] \\ & \times \Bigg[ \Lambda_i^{h_\diamond}(z_i(s)) + \sum_{j=1}^{n} b_{ij} f_j\big(z_j^{h_\diamond}(s-\tau_{ij})\big) + \bigvee_{j=1}^{n} \varrho_{ij} f_j\big(z_j^{h_\diamond}(s-\tau_{ij})\big) \\ & + \bigwedge_{j=1}^{n} \varsigma_{ij} f_j\big(z_j^{h_\diamond}(s-\tau_{ij})\big) + \bigvee_{j=1}^{n} T_{ij}\varpi_j + \bigwedge_{j=1}^{n} H_{ij}\varpi_j + I_i \Bigg] \mathrm{d}s, \end{aligned} \tag{3.6}$$

where $t \in (t_p, t_{p+1}]$, $p \in \mathbb{N}_0$; $z_i(s) = \varphi_i(s)$, $\forall s \in [-\tau_0, 0]$, $i = 1, 2, \ldots, n$.

Considering Eq. (3.6) with piecewise constant arguments as follows:

$$\begin{aligned} z_i(t) = {} & E_{\alpha_i}[-\lambda_i(t-t_p)^{\alpha_i}] z_i(t_p) + \int_{t_p}^{t} (t-s)^{\alpha_i-1} E_{\alpha_i,\alpha_i}[-\lambda_i(t-s)^{\alpha_i}] \\ & \times \Bigg[ \Lambda_i^{h_\diamond}\big(z_i(\lfloor s/h \rfloor h)\big) + \sum_{j=1}^{n} b_{ij} f_j\big(z_j^{h_\diamond}(\lfloor s/h \rfloor h - \tau_{ij}^h h)\big) \\ & + \bigvee_{j=1}^{n} \varrho_{ij} f_j\big(z_j^{h_\diamond}(\lfloor s/h \rfloor h - \tau_{ij}^h h)\big) + \bigwedge_{j=1}^{n} \varsigma_{ij} f_j\big(z_j^{h_\diamond}(\lfloor s/h \rfloor h - \tau_{ij}^h h)\big) \\ & + \bigvee_{j=1}^{n} T_{ij}\varpi_j + \bigwedge_{j=1}^{n} H_{ij}\varpi_j + I_i \Bigg] \mathrm{d}s, \end{aligned} \tag{3.7}$$

where $\tau_{ij}^h = \lfloor \tau_{ij}/h \rfloor \in \mathbb{Z}$, $\lfloor s/h \rfloor$ stands for the integer part of $s/h$, $t \in (t_p, t_{p+1}]$, $p \in \mathbb{N}_0$, $i = 1, 2, \ldots, n$.

For $t \geq 0$, there exists $k \in \mathbb{N}_0$ ensuring that $t/h \in (k, k+1]$. For $t \in (kh, kh + h] \subseteq (t_p, t_{p+1}]$, $k, p \in \mathbb{N}_0$, by using Lemmas 1.20 and 1.23, then Eq. (3.7) becomes

$$
\begin{aligned}
z_i(t) = {} & E_{\alpha_i}[-\lambda_i(t-N_p h)^{\alpha_i}]z_i(N_p h) + \Big[\Lambda_i^{h_\diamond}\big(z_i(kh)\big) \\
& + \sum_{j=1}^{n} b_{ij} f_j\big(z_j^{h_\diamond}(kh-\tau_{ij}^h h)\big) + \bigvee_{j=1}^{n} \varrho_{ij} f_j\big(z_j^{h_\diamond}(kh-\tau_{ij}^h h)\big) \\
& + \bigwedge_{j=1}^{n} \varsigma_{ij} f_j\big(z_j^{h_\diamond}(kh-\tau_{ij}^h h)\big) + \bigvee_{j=1}^{n} T_{ij}\varpi_j + \bigwedge_{j=1}^{n} H_{ij}\varpi_j + I_i\Big] \\
& \times \int_{kh}^{t} (t-s)^{\alpha_i-1} E_{\alpha_i,\alpha_i}[-\lambda_i(t-s)^{\alpha_i}]\mathrm{d}s + \sum_{\iota=N_p}^{k-1} \Big[\Lambda_i^{h_\diamond}\big(z_i(\iota h)\big) \\
& + \sum_{j=1}^{n} b_{ij} f_j\big(z_j^{h_\diamond}(\iota h-\tau_{ij}^h h)\big) + \bigvee_{j=1}^{n} \varrho_{ij} f_j\big(z_j^{h_\diamond}(\iota h-\tau_{ij}^h h)\big) \\
& + \bigwedge_{j=1}^{n} \varsigma_{ij} f_j\big(z_j^{h_\diamond}(\iota h-\tau_{ij}^h h)\big) + \bigvee_{j=1}^{n} T_{ij}\varpi_j + \bigwedge_{j=1}^{n} H_{ij}\varpi_j + I_i\Big] \\
& \times \int_{\iota h}^{\iota h+h} (t-s)^{\alpha_i-1} E_{\alpha_i,\alpha_i}[-\lambda_i(t-s)^{\alpha_i}]\mathrm{d}s \\
= {} & E_{\alpha_i}[-\lambda_i(t-\mu_k h)^{\alpha_i}]z_i(\mu_k h) + \frac{1}{\lambda_i}\Big[1 - E_{\alpha_i}[-\lambda_i(t-kh)^{\alpha_i}]\Big] \\
& \times \Big[\Lambda_i^{h_\diamond}\big(z_i(kh)\big) + \sum_{j=1}^{n} b_{ij} f_j\big(z_j^{h_\diamond}(kh-\tau_{ij}^h h)\big) + \bigvee_{j=1}^{n} \varrho_{ij} f_j\big(z_j^{h_\diamond}(kh-\tau_{ij}^h h)\big) \\
& + \bigwedge_{j=1}^{n} \varsigma_{ij} f_j\big(z_j^{h_\diamond}(kh-\tau_{ij}^h h)\big) + \bigvee_{j=1}^{n} T_{ij}\varpi_j + \bigwedge_{j=1}^{n} H_{ij}\varpi_j + I_i\Big] \\
& + \sum_{l=1}^{k-\mu_k} \frac{1}{\lambda_i}\Big[E_{\alpha_i}[-\lambda_i(t-kh+lh-h)^{\alpha_i}] - E_{\alpha_i}[-\lambda_i(t-kh+lh)^{\alpha_i}]\Big] \\
& \times \Big[\Lambda_i^{h_\diamond}\big(z_i(kh-lh)\big) + \sum_{j=1}^{n} b_{ij} f_j\big(z_j^{h_\diamond}(kh-\tau_{ij}^h h-lh)\big) \\
& + \bigvee_{j=1}^{n} \varrho_{ij} f_j\big(z_j^{h_\diamond}(kh-\tau_{ij}^h h-lh)\big) + \bigwedge_{j=1}^{n} \varsigma_{ij} f_j\big(z_j^{h_\diamond}(kh-\tau_{ij}^h h-lh)\big) \\
& + \bigvee_{j=1}^{n} T_{ij}\varpi_j + \bigwedge_{j=1}^{n} H_{ij}\varpi_j + I_i\Big],
\end{aligned}
\tag{3.8}
$$

where $t \in (kh, kh+h] \subseteq (t_p, t_{p+1}]$, $k, p \in \mathbb{N}_0$, $i = 1, 2, \ldots, n$.

Letting $t \to (kh)^+$ and $t \to (kh+h)^-$ in Eq. (3.8) respectively, it deduces

$$
\begin{aligned}
z_i(kh) = {} & E_{\alpha_i}[-\lambda_i(\nu_k h)^{\alpha_i}] z_i(\mu_k h) + \frac{1}{\lambda_i}\sum_{l=1}^{\nu_k} w_{\alpha_i}^{[l-1]}\Big[\Lambda_i^{h_\diamond}\big(z_i(kh-lh)\big) \\
& + \sum_{j=1}^{n} b_{ij} f_j\big(z_j^{h_\diamond}(kh-\tau_{ij}^h h-lh)\big) + \bigvee_{j=1}^{n} \varrho_{ij} f_j\big(z_j^{h_\diamond}(kh-\tau_{ij}^h h-lh)\big) \\
& + \bigwedge_{j=1}^{n} \varsigma_{ij} f_j\big(z_j^{h_\diamond}(kh-\tau_{ij}^h h-lh)\big) + \bigvee_{j=1}^{n} T_{ij}\varpi_j \\
& + \bigwedge_{j=1}^{n} H_{ij}\varpi_j + I_i\Big],
\end{aligned} \tag{3.9}
$$

$$
\begin{aligned}
z_i(kh+h) = {} & E_{\alpha_i}[-\lambda_i(\nu_k h+h)^{\alpha_i}] z_i(\mu_k h) + \frac{1}{\lambda_i}\sum_{l=0}^{\nu_k} w_{\alpha_i}^{[l]}\Big[\Lambda_i^{h_\diamond}\big(z_i(kh-lh)\big) \\
& + \sum_{j=1}^{n} b_{ij} f_j\big(z_j^{h_\diamond}(kh-\tau_{ij}^h h-lh)\big) + \bigvee_{j=1}^{n} \varrho_{ij} f_j\big(z_j^{h_\diamond}(kh-\tau_{ij}^h h-lh)\big) \\
& + \bigwedge_{j=1}^{n} \varsigma_{ij} f_j\big(z_j^{h_\diamond}(kh-\tau_{ij}^h h-lh)\big) + \bigvee_{j=1}^{n} T_{ij}\varpi_j \\
& + \bigwedge_{j=1}^{n} H_{ij}\varpi_j + I_i\Big],
\end{aligned} \tag{3.10}
$$

where $k \in \mathbb{N}_0,\ i = 1, 2, \ldots, n$. From Eqs. (3.9)–(3.10), the Mittag–Leffler Euler discrete approximation of Eq. (3.4) is given by

$$
\begin{aligned}
z_i(kh+h) = {} & \lambda_{\alpha_i}^{[k]} z_i(kh) + \sum_{l=0}^{\nu_k} \theta_{\alpha_i}^{[l,k]}\Big[\Lambda_i^{h_\diamond}\big(z_i(kh-lh)\big) \\
& + \sum_{j=1}^{n} b_{ij} f_j\big(z_j^{h_\diamond}(kh-\tau_{ij}^h h-lh)\big) + \bigvee_{j=1}^{n} \varrho_{ij} f_j\big(z_j^{h_\diamond}(kh-\tau_{ij}^h h-lh)\big) \\
& + \bigwedge_{j=1}^{n} \varsigma_{ij} f_j\big(z_j^{h_\diamond}(kh-\tau_{ij}^h h-lh)\big) + \bigvee_{j=1}^{n} T_{ij}\varpi_j \\
& + \bigwedge_{j=1}^{n} H_{ij}\varpi_j + I_i\Big],
\end{aligned} \tag{3.11}
$$

where $k \in \mathbb{N}_0,\ i = 1, 2, \ldots, n$.

Let

$$
u_i(k) = z_i(kh), \quad u_j^{h_\diamond}(k) = z_j^{h_\diamond}(kh) = h_j^{-1}(u_j(k))
$$

in Eq. (3.11), then Eq. (3.11) is shifted into

$$u_i(k+1) = \lambda_{\alpha_i}^{[k]} u_i(k) + \sum_{l=0}^{\nu_k} \theta_{\alpha_i}^{[l,k]} \Bigg[ \Lambda_i^{h_\diamond}\big(u_i(k-l)\big) + \sum_{j=1}^{n} b_{ij} f_j\big(u_j^{h_\diamond}(k-l-\tau_{ij}^h)\big)$$
$$+ \bigvee_{j=1}^{n} \varrho_{ij} f_j\big(u_j^{h_\diamond}(k-l-\tau_{ij}^h)\big) + \bigwedge_{j=1}^{n} \varsigma_{ij} f_j\big(u_j^{h_\diamond}(k-l-\tau_{ij}^h)\big)$$
$$+ \bigvee_{j=1}^{n} T_{ij}\varpi_j + \bigwedge_{j=1}^{n} H_{ij}\varpi_j + I_i \Bigg] \tag{3.12}$$

with the following initial conditions

$$u_i(s) = \phi_i(s), \quad s \in [-\tau_0^h, 0]_{\mathbb{Z}}, \tag{3.13}$$

where $\tau_0^h = \max_{1\le i,j\le n} \lfloor \tau_{ij}/h \rfloor$, $\lambda_{\alpha_i}^{[k]}$ and $\theta_{\alpha_i}^{[l,k]}$ are defined as those in Eq. (3.5), $l, k \in \mathbb{N}_0$, $i, j = 1, 2, \ldots, n$.

**Remark 3.1** Similar to the argument as that in Sect. 2.4, it easily obtains the convergence of model Eq. (3.12). Precisely,

$$\|z(kh) - u(k)\| \le Ch,$$

where $C$ is a positive constant, $z = (z_1, \ldots, z_n)^T$ and $u = (u_1, \ldots, u_n)^T$ are the solutions of models (3.4) and (3.12), respectively.

**Remark 3.2** If $\alpha_i = 1$ in Eq. (3.12), then

$$\lambda_{\alpha_i}^{[k]} \equiv e^{-\lambda_i h}, \quad \theta_{\alpha_i}^{[0,k]} \equiv (1 - e^{-\lambda_i h})\lambda_i^{-1}, \quad \theta_{\alpha_i}^{[l,k]} \equiv 0$$

for all $l \in [1, \nu_k]_{\mathbb{Z}}$, $k \in \mathbb{N}$, $i = 1, 2, \ldots, n$. So Eq. (3.12) is turned into

$$u_i(k+1) = e^{-\lambda_i h} u_i(k) + (1 - e^{-\lambda_i h})\lambda_i^{-1} \Bigg[ \Lambda_i^{h_\diamond}\big(u_i(k)\big)$$
$$+ \sum_{j=1}^{n} b_{ij} f_j\big(u_j^{h_\diamond}(k-\tau_{ij}^h)\big) + \bigvee_{j=1}^{n} \varrho_{ij} f_j\big(u_j^{h_\diamond}(k-\tau_{ij}^h)\big)$$
$$+ \bigwedge_{j=1}^{n} \varsigma_{ij} f_j\big(u_j^{h_\diamond}(k-\tau_{ij}^h)\big) + \bigvee_{j=1}^{n} T_{ij}\varpi_j + \bigwedge_{j=1}^{n} H_{ij}\varpi_j + I_i \Bigg], \tag{3.14}$$

where $k \in \mathbb{N}_0$, $u_i(s) = \phi_i(s)$ for $s \in [-\tau_0^h, 0]_{\mathbb{Z}}$, $i = 1, 2, \ldots, n$, which corresponds to the numerical solutions of Eq. (3.2).

**Remark 3.3** According to Eq. (3.14), the authors [21, 26–30] studied the dynamic behaviours of different forms of discrete applied models, e.g., almost periodic dynamics [26, 27, 30], stochastic perturbations [21, 27] and stability [28, 29], etc. Obviously, Eq. (3.12) is an expansion of discrete analogues in literature [21, 26–29]. Thus the current work extends the works in literature [21, 26–30].

**Lemma 3.1** ([107]) *Let* $x = (x_1, x_2, \ldots, x_n)^T$, $y = (y_1, y_2, \ldots, y_n)^T \in \mathbb{R}^n$ *be the two states of Eq.* (3.12), *then*

$$\left|\bigvee_{j=1}^{n} \varrho_{ij} f_j(x_j) - \bigvee_{j=1}^{n} \varrho_{ij} f_j(y_j)\right| \le \sum_{j=1}^{n} |\varrho_{ij}||f_j(x_j) - f_j(y_j)|,$$

$$\left|\bigwedge_{j=1}^{n} \varsigma_{ij} f_j(x_j) - \bigwedge_{j=1}^{n} \varsigma_{ij} f_j(y_j)\right| \le \sum_{j=1}^{n} |\varsigma_{ij}||f_j(x_j) - f_j(y_j)|,$$

*where* $i = 1, 2, \ldots, n$.

The following variation-of-constants formula can be obtained directly from Lemma 2.3 in literature [27].

**Lemma 3.2** *If* $\alpha_i \in (0, 1]$ *and* $\lambda_i > 0$, *then the solution of Eq.* (3.12) *has the form of*

$$\begin{aligned} u_i(k) = {} & \prod_{s=0}^{k-1} \lambda_{\alpha_i}^{[s]} \phi_i(0) + \sum_{q=0}^{k-1} \prod_{s=q+1}^{k-1} \lambda_{\alpha_i}^{[s]} \sum_{l=0}^{\nu_q} \theta_{\alpha_i}^{[l,q]} \Big[ \Lambda_i^{h_\diamond}\big(u_i(q-l)\big) \\ & + \sum_{j=1}^{n} b_{ij} f_j\big(u_j^{h_\diamond}(q-l-\tau_{ij}^h)\big) + \bigvee_{j=1}^{n} \varrho_{ij} f_j\big(u_j^{h_\diamond}(q-l-\tau_{ij}^h)\big) \\ & + \bigwedge_{j=1}^{n} \varsigma_{ij} f_j\big(u_j^{h_\diamond}(q-l-\tau_{ij}^h)\big) + \bigvee_{j=1}^{n} T_{ij}\varpi_j + \bigwedge_{j=1}^{n} H_{ij}\varpi_j + I_i \Big], \end{aligned} \tag{3.15}$$

*where* $k \in \mathbb{Z}_0$, $u_i(s) = \phi_i(s)$ *for* $s \in [-\tau_0^h, 0]_{\mathbb{Z}}$, $i = 1, 2, \ldots, n$.

Let $N^* = \sup_{p\in\mathbb{N}_0}(N_{p+1} - N_p)$, $\bar{\lambda}_{\alpha_i} = \sup_{k\in\mathbb{N}_0} \lambda_{\alpha_i}^{[k]}$ and $\bar{\theta}_{\alpha_i} = \sup_{k\in\mathbb{N}_0} \sum_{l=0}^{N^*} |\theta_{\alpha_i}^{[l,k]}|$, $i = 1, 2, \ldots, n$.

**Lemma 3.3** *If* $N^* < +\infty$, $\alpha_i \in (0, 1]$ *and* $\lambda_i > 0$, *one has*

(1) $0 < \lambda_{\alpha_i}^{[k]} \le \bar{\lambda}_{\alpha_i} < 1$ *for* $k \in \mathbb{N}_0$, $i = 1, 2, \ldots, n$;

(2) $\sum_{l=0}^{\nu_k} |\theta_{\alpha_i}^{[l,k]}| \le \bar{\theta}_{\alpha_i} \le \dfrac{2}{\lambda_i}$ *for* $k \in \mathbb{N}_0$, $i = 1, 2, \ldots, n$;

(3) $\sum_{l=0}^{\nu_k} \theta_{\alpha_i}^{[l,k]} = \dfrac{1}{\lambda_i}[1 - \lambda_{\alpha_i}^{[k]}]$ *for* $k \in \mathbb{N}_0$, $i = 1, 2, \ldots, n$.

***Proof*** The proof of this lemma is similar to the proof in Lemma 2.1, so we omit it. The proof is achieved. □

**Remark 3.4** Since the studied model is piecewise, $\nu_k \in [0, N^*]_{\mathbb{Z}}$, $\forall k \in \mathbb{N}$. The boundedness of $\nu_k$ can ensure the boundedness of $\lambda_{\alpha_i}^{[k]}$ and $\sum_{l=0}^{\nu_k} \theta_{\alpha_i}^{[l,k]}$ for all $k \in \mathbb{N}_0$, $i = 1, 2, \ldots, n$. Apparently, non-piecewise model has no this trait. If $\alpha_i = 1$, then it yields

$$\bar{\lambda}_{\alpha_i} = e^{-\lambda_i h}, \quad \bar{\theta}_{\alpha_i} = (1 - e^{-\lambda_i h})\lambda_i^{-1}, \quad i = 1, 2, \ldots, n.$$

## 3.3 Global Solution and Equilibrium

Let

$$\mathbb{X} = \Big\{u = (u_1, \ldots, u_n)^T : \mathbb{N}_0 \to \mathbb{R}^n | u_i(s) = \phi_i(s) \text{ for } s \in [-\tau_0^h, 0]_{\mathbb{Z}}, i = 1, 2, \ldots, n\Big\}.$$

Then $\mathbb{X}$ is a Banach space with the norm $\|u\|_{\mathbb{X}} = \max_{1 \le i \le n} \sup_{k \in \mathbb{N}_0} |u_i(k)|$, $\forall u \in \mathbb{X}$. Set $\bar{d}_i = \sup_{s \in \mathbb{R}} d_i(s)$, $i = 1, 2, \ldots, n$.

**Theorem 3.1** *Let the assumptions below hold.*

(Q$_1$) *It exists* $L_i^\lambda > 0$ *such that*

$$|\Lambda_i^{h_\diamond}(x) - \Lambda_i^{h_\diamond}(y)| \le L_i^\lambda |x - y|, \quad \forall x, y \in \mathbb{R}, i = 1, 2, \ldots, n.$$

(Q$_2$) *It exists* $L_j^f > 0$ *such that*

$$|f_j(x) - f_j(y)| \le L_j^f |x - y|, \quad \forall x, y \in \mathbb{R}, \quad j = 1, 2, \ldots, n.$$

(Q$_3$) $\zeta_1 < 1$, *where*

$$\zeta_1 = \max_{1 \le i \le n} \frac{2}{\lambda_i (1 - \bar{\lambda}_{\alpha_i})} \Big[ L_i^\lambda + \sum_{j=1}^n L_j^f \bar{d}_j \big( |b_{ij}| + |\varrho_{ij}| + |\varsigma_{ij}| \big) \Big].$$

*Then*

(1) *Equation* (3.12) *with initial conditions* (3.13) *possesses a unique solution* $u^*$.

(2) $u^*$ *is bounded, i.e.,* $\|u^*\|_{\mathbb{X}} \le c_1 = \dfrac{\zeta_2}{1 - \zeta_1}$, *where*

$$\zeta_2 = \max_{1\le i\le n}\Big\{|\phi_i(0)| + \frac{2}{\lambda_i(1-\bar{\lambda}_{\alpha_i})}\Big[\sum_{j=1}^{n}|b_{ij}||f_j(0)| + \bigvee_{j=1}^{n}|\varrho_{ij}||f_j(0)| + \bigwedge_{j=1}^{n}|\varsigma_{ij}||f_j(0)| + \bigvee_{j=1}^{n}|T_{ij}\varpi_j| + \bigwedge_{j=1}^{n}|H_{ij}\varpi_j| + |I_i|\Big]\Big\}.$$

***Proof*** Define the operator $T:\mathbb{X}\to\mathbb{X}$ and

$$Tu = (\Phi_1 u,\ldots,\Phi_n u)^T,\quad \forall u=(u_1,\ldots,u_n)^T\in\mathbb{X},$$

where

$$\begin{aligned}(\Phi_i u)(k) = &\prod_{s=0}^{k-1}\lambda_{\alpha_i}^{[s]}\phi_i(0) + \sum_{q=0}^{k-1}\prod_{s=q+1}^{k-1}\lambda_{\alpha_i}^{[s]}\sum_{l=0}^{\nu_q}\theta_{\alpha_i}^{[l,q]}\Big[\Lambda_i^{h_\diamond}\big(u_i(q-l)\big)\\ &+\sum_{j=1}^{n} b_{ij} f_j\big(u_j^{h_\diamond}(q-l-\tau_{ij}^{h})\big) + \bigvee_{j=1}^{n}\varrho_{ij} f_j\big(u_j^{h_\diamond}(q-l-\tau_{ij}^{h})\big)\\ &+\bigwedge_{j=1}^{n}\varsigma_{ij} f_j\big(u_j^{h_\diamond}(q-l-\tau_{ij}^{h})\big) + \bigvee_{j=1}^{n}T_{ij}\varpi_j\\ &+\bigwedge_{j=1}^{n}H_{ij}\varpi_j + I_i\Big],\end{aligned} \tag{3.16}$$

where $k\in\mathbb{N}_0$, $i=1,2,\ldots,n$.

Set $\Omega=\{u\in\mathbb{X}: \|u\|_{\mathbb{X}}\le c_1\}$. If $u\in\Omega$, it gains from Eq. (3.16), Lemmas 3.1 and 3.3, ($Q_1$)–($Q_3$) that

$$\begin{aligned}|(\Phi_i u)(k)| \le &\prod_{s=0}^{k-1}\lambda_{\alpha_i}^{[s]}|\phi_i(0)| + \sum_{q=0}^{k-1}\prod_{s=q+1}^{k-1}\lambda_{\alpha_i}^{[s]}\sum_{l=0}^{\nu_q}|\theta_{\alpha_i}^{[l,q]}|\Big|\Lambda_i^{h_\diamond}\big(u_i(q-l)\big)\\ &+\sum_{j=1}^{n} b_{ij} f_j\big(u_j^{h_\diamond}(q-l-\tau_{ij}^{h})\big) + \bigvee_{j=1}^{n}\varrho_{ij} f_j\big(u_j^{h_\diamond}(q-l-\tau_{ij}^{h})\big)\\ &+\bigwedge_{j=1}^{n}\varsigma_{ij} f_j\big(u_j^{h_\diamond}(q-l-\tau_{ij}^{h})\big) + \bigvee_{j=1}^{n}T_{ij}\varpi_j + \bigwedge_{j=1}^{n}H_{ij}\varpi_j + I_i\Big|\\ = &\prod_{s=0}^{k-1}\lambda_{\alpha_i}^{[s]}|\phi_i(0)| + \sum_{q=0}^{k-1}\prod_{s=q+1}^{k-1}\lambda_{\alpha_i}^{[s]}\sum_{l=0}^{\nu_q}|\theta_{\alpha_i}^{[l,q]}|\\ &\times\Big|\Big[\Lambda_i^{h_\diamond}\big(u_i(q-l)\big) - \Lambda_i^{h_\diamond}(0)\Big] + \sum_{j=1}^{n} b_{ij}\Big[f_j\big(u_j^{h_\diamond}(q-l-\tau_{ij}^{h})\big)\end{aligned}$$

$$
\begin{aligned}
& - f_j(0)\Big] + \bigvee_{j=1}^{n} \varrho_{ij}\Big[f_j\big(u_j^{h_\diamond}(q-l-\tau_{ij}^h)\big) - f_j(0)\Big] \\
& + \bigwedge_{j=1}^{n} \varsigma_{ij}\Big[f_j\big(u_j^{h_\diamond}(q-l-\tau_{ij}^h)\big) - f_j(0)\Big] + \sum_{j=1}^{n} b_{ij} f_j(0) \\
& + \bigvee_{j=1}^{n} \varrho_{ij} f_j(0) + \bigwedge_{j=1}^{n} \varsigma_{ij} f_j(0) + \bigvee_{j=1}^{n} T_{ij}\varpi_j + \bigwedge_{j=1}^{n} H_{ij}\varpi_j + I_i\Bigg| \\
\le & \bar{\lambda}_{\alpha_i}^{k}|\phi_i(0)| + \frac{2}{\lambda_i}\sum_{q=0}^{k-1}\bar{\lambda}_{\alpha_i}^{k-q-1}\Bigg\{\Bigg[L_i^{\lambda} + \sum_{j=1}^{n} L_j^f \bar{d}_j(|b_{ij}| + |\varrho_{ij}| + |\varsigma_{ij}|)\Bigg]c_1 \\
& + \sum_{j=1}^{n}|b_{ij}||f_j(0)| + \bigvee_{j=1}^{n}|\varrho_{ij}||f_j(0)| + \bigwedge_{j=1}^{n}|\varsigma_{ij}||f_j(0)| \\
& + \bigvee_{j=1}^{n}|T_{ij}\varpi_j| + \bigwedge_{j=1}^{n}|H_{ij}\varpi_j| + |I_i|\Bigg\} \\
\le & |\phi_i(0)| + \frac{2}{\lambda_i(1-\bar{\lambda}_{\alpha_i})}\Bigg\{\Bigg[L_i^{\lambda} + \sum_{j=1}^{n} L_j^f \bar{d}_j(|b_{ij}| + |\varrho_{ij}| + |\varsigma_{ij}|)\Bigg]c_1 \\
& + \sum_{j=1}^{n}|b_{ij}||f_j(0)| + \bigvee_{j=1}^{n}|\varrho_{ij}||f_j(0)| + \bigwedge_{j=1}^{n}|\varsigma_{ij}||f_j(0)| \\
& + \bigvee_{j=1}^{n}|T_{ij}\varpi_j| + \bigwedge_{j=1}^{n}|H_{ij}\varpi_j| + |I_i|\Bigg\} \\
\le & \zeta_1 c_1 + \zeta_2 = c_1, \quad k \in \mathbb{N}_0, i = 1, 2, \ldots, n.
\end{aligned}
$$

Hence, $T : \Omega \to \Omega$.

In the end, for any $u, \tilde{u} \in \mathbb{X}$, it derives from Eq. (3.16) that

$$
\begin{aligned}
\big|(\Phi_i u)(k) - (\Phi_i \tilde{u})(k)\big| \le & \sum_{q=0}^{k-1}\prod_{s=q+1}^{k-1}\lambda_{\alpha_i}^{[s]}\sum_{l=0}^{\nu_q}|\theta_{\alpha_i}^{[l,q]}| \\
& \times \Bigg|\Big[\Lambda_i^{h_\diamond}\big(u_i(q-l)\big) - \Lambda_i^{h_\diamond}\big(\tilde{u}_i(q-l)\big)\Big] \\
& + \sum_{j=1}^{n} b_{ij}\Big[f_j\big(u_j^{h_\diamond}(q-l-\tau_{ij}^h)\big) - f_j\big(\tilde{u}_j^{h_\diamond}(q-l-\tau_{ij}^h)\big)\Big] \\
& + \bigvee_{j=1}^{n} \varrho_{ij}\Big[f_j\big(u_j^{h_\diamond}(q-l-\tau_{ij}^h)\big) - f_j\big(\tilde{u}_j^{h_\diamond}(q-l-\tau_{ij}^h)\big)\Big]
\end{aligned}
$$

$$
\begin{aligned}
&+\bigwedge_{j=1}^{n}\varsigma_{ij}\Big[f_j\big(u_j^{h_\diamond}(q-l-\tau_{ij}^h)\big)-f_j\big(\tilde{u}_j^{h_\diamond}(q-l-\tau_{ij}^h)\big)\Big]\Bigg|\\
\leq&\sum_{q=0}^{k-1}\prod_{s=q+1}^{k-1}\bar{\lambda}_{\alpha_i}\sum_{l=0}^{\nu_q}|\theta_{\alpha_i}^{[l,q]}|\\
&\times\Bigg[L_i^\lambda+\sum_{j=1}^{n}L_j^f\bar{d}_j\big(|b_{ij}|+|\varrho_{ij}|+|\varsigma_{ij}|\big)\Bigg]\|u-\tilde{u}\|_{\mathbb{X}}\\
\leq&\frac{2}{\lambda_i(1-\bar{\lambda}_{\alpha_i})}\Bigg[L_i^\lambda+\sum_{j=1}^{n}L_j^f\bar{d}_j\big(|b_{ij}|\\
&+|\varrho_{ij}|+|\varsigma_{ij}|\big)\Bigg]\|u-\tilde{u}\|_{\mathbb{X}}\\
\leq&\zeta_1\|u-\tilde{u}\|_{\mathbb{X}},
\end{aligned}
$$

where $k\in\mathbb{N}_0$, $i=1,2,\ldots,n$. So $\|Tu-T\tilde{u}\|_{\mathbb{X}}\leq\zeta_1\|u-\tilde{u}\|_{\mathbb{X}}$, $\forall u,\tilde{u}\in\mathbb{X}$. From ($Q_3$), $T$ is a contractive mapping. Therefore, $T$ possesses a unique fixed point $u^*=Tu^*\in\Omega$ and $u^*$ solves Eq. (3.12) with initial conditions (3.13). The proof is achieved. □

According to Eq. (3.12), define

$$
T^\star:\mathbb{R}^n\to\mathbb{R}^n,\quad T^\star u=(\Phi_1^\star u,\ldots,\Phi_n^\star u)^T,
$$

where

$$
\begin{aligned}
\Phi_i^\star u=&\frac{1}{\lambda_i}\Bigg[\Lambda_i^{h_\diamond}(u_i)+\sum_{j=1}^{n}b_{ij}f_j\big(u_j^{h_\diamond}\big)+\bigvee_{j=1}^{n}\varrho_{ij}f_j\big(u_j^{h_\diamond}\big)\\
&+\bigwedge_{j=1}^{n}\varsigma_{ij}f_j\big(u_j^{h_\diamond}\big)+\bigvee_{j=1}^{n}T_{ij}\varpi_j+\bigwedge_{j=1}^{n}H_{ij}\varpi_j+I_i\Bigg],
\end{aligned}
\tag{3.17}
$$

where $u=(u_1,\ldots,u_n)^T\in\mathbb{R}^n$, $u_j^{h_\diamond}=h_{\diamond j}^{-1}(u_j)$, $i,j=1,2,\ldots,n$.

**Remark 3.5** If $u^\star=(u_1^\star,\ldots,u_n^\star)^T\in\mathbb{R}^n$ is a fixed point of $T^\star$, then

$$
\begin{aligned}
u_i^\star=&\frac{1}{\lambda_i}\Bigg[\Lambda_i^{h_\diamond}(u_i^\star)+\sum_{j=1}^{n}b_{ij}f_j\big(h_{\diamond j}^{-1}(u_j^\star)\big)+\bigvee_{j=1}^{n}\varrho_{ij}f_j\big(h_{\diamond j}^{-1}(u_j^\star)\big)\\
&+\bigwedge_{j=1}^{n}\varsigma_{ij}f_j\big(h_{\diamond j}^{-1}(u_j^\star)\big)+\bigvee_{j=1}^{n}T_{ij}\varpi_j+\bigwedge_{j=1}^{n}H_{ij}\varpi_j+I_i\Bigg],
\end{aligned}
\tag{3.18}
$$

which implies from Lemma 3.3 that

$$\begin{aligned}(1-\lambda_{\alpha_i}^{[k]})u_i^\star=&\sum_{l=0}^{\nu_k}\theta_{\alpha_i}^{[l,k]}\Big[\Lambda_i^{h_\diamond}(u_i^\star)+\sum_{j=1}^{n}b_{ij}f_j\big(h_{\diamond_j}^{-1}(u_j^\star)\big)\\&+\bigvee_{j=1}^{n}\varrho_{ij}f_j\big(h_{\diamond_j}^{-1}(u_j^\star)\big)+\bigwedge_{j=1}^{n}\varsigma_{ij}f_j\big(h_{\diamond_j}^{-1}(u_j^\star)\big)\\&+\bigvee_{j=1}^{n}T_{ij}\varpi_j+\bigwedge_{j=1}^{n}H_{ij}\varpi_j+I_i\Big],\end{aligned}\tag{3.19}$$

where $k\in\mathbb{N}_0$, $i=1,2,\ldots,n$. Therefore, $u^\star\in\mathbb{R}^n$ is an equilibrium point of Eq. (3.12).

**Theorem 3.2** *Assume that* (Q$_1$)–(Q$_2$) *hold. Supposing further that*

(Q$_4$) $\zeta_3=\max\limits_{1\le i\le n}\dfrac{1}{\lambda_i}\Big[L_i^\lambda+\sum\limits_{j=1}^{n}L_j^f\bar{d}_j\big(|b_{ij}|+|\varrho_{ij}|+|\varsigma_{ij}|\big)\Big]<1.$

*Then Eq.* (3.12) *admits a unique equilibrium point* $u^\star$ *with* $\|u^\star\|\le c_2=\frac{\zeta_4}{1-\zeta_3}$, *where*

$$\begin{aligned}\zeta_4=\max_{1\le i\le n}\frac{1}{\lambda_i}\Big[&\sum_{j=1}^{n}|b_{ij}||f_j(0)|+\bigvee_{j=1}^{n}|\varrho_{ij}||f_j(0)|\\&+\bigwedge_{j=1}^{n}|\varsigma_{ij}||f_j(0)|+\bigvee_{j=1}^{n}|T_{ij}\varpi_j|+\bigwedge_{j=1}^{n}|H_{ij}\varpi_j|+|I_i|\Big].\end{aligned}$$

***Proof*** Define $\Omega_{eq}=\{u\in\mathbb{R}^n:\|u\|\le c_2\}$. If $u\in\Omega_{eq}$, it gets from Eq. (3.17), (Q$_1$)–(Q$_2$), (Q$_4$) that

$$\begin{aligned}|\Phi_i^\star u|=&\frac{1}{\lambda_i}\Big|\Lambda_i^{h_\diamond}(u_i)+\sum_{j=1}^{n}b_{ij}f_j\big(h_{\diamond_j}^{-1}(u_j)\big)+\bigvee_{j=1}^{n}\varrho_{ij}f_j\big(h_{\diamond_j}^{-1}(u_j)\big)\\&+\bigwedge_{j=1}^{n}\varsigma_{ij}f_j\big(h_{\diamond_j}^{-1}(u_j)\big)+\bigvee_{j=1}^{n}T_{ij}\varpi_j+\bigwedge_{j=1}^{n}H_{ij}\varpi_j+I_i\Big|\\=&\frac{1}{\lambda_i}\Big|\Big[\Lambda_i^{h_\diamond}(u_i)-\Lambda_i^{h_\diamond}(0)\Big]+\sum_{j=1}^{n}b_{ij}\Big[f_j\big(h_{\diamond_j}^{-1}(u_j)\big)-f_j(0)\Big]\\&+\bigvee_{j=1}^{n}\varrho_{ij}\Big[f_j\big(h_{\diamond_j}^{-1}(u_j)\big)-f_j(0)\Big]+\bigwedge_{j=1}^{n}\varsigma_{ij}\Big[f_j\big(h_{\diamond_j}^{-1}(u_j)\big)-f_j(0)\Big]\\&+\sum_{j=1}^{n}b_{ij}f_j(0)+\bigvee_{j=1}^{n}\varrho_{ij}f_j(0)+\bigwedge_{j=1}^{n}\varsigma_{ij}f_j(0)+\bigvee_{j=1}^{n}T_{ij}\varpi_j+\bigwedge_{j=1}^{n}H_{ij}\varpi_j+I_i\Big|\end{aligned}$$

$$\leq \frac{1}{\lambda_i}\Bigg\{\Bigg[L_i^{\lambda}+\sum_{j=1}^{n}L_j^f\bar{d}_j\big(|b_{ij}|+|\varrho_{ij}|+|\varsigma_{ij}|\big)\Bigg]\|u\|+\sum_{j=1}^{n}|b_{ij}||f_j(0)|$$
$$+\bigvee_{j=1}^{n}|\varrho_{ij}||f_j(0)|+\bigwedge_{j=1}^{n}|\varsigma_{ij}||f_j(0)|+\bigvee_{j=1}^{n}|T_{ij}\varpi_j|+\bigwedge_{j=1}^{n}|H_{ij}\varpi_j|+|I_i|\Bigg\}$$
$$\leq \zeta_3 c_2+\zeta_4=c_2,\quad i=1,2,\ldots,n.$$

This leads to $T^{\star}:\Omega_{eq}\to\Omega_{eq}$.

At last, $\forall u,\tilde{u}\in\mathbb{R}^n$, it acquires

$$|\Phi_i^{\star}u-\Phi_i^{\star}\tilde{u}| =\frac{1}{\lambda_i}\Bigg|\Big[\Lambda_i^{h_\diamond}(u_i)-\Lambda_i^{h_\diamond}(\tilde{u}_i)\Big]$$
$$+\sum_{j=1}^{n}b_{ij}\Big[f_j\big(h_{\diamond j}^{-1}(u_j)\big)-f_j\big(h_{\diamond j}^{-1}(\tilde{u}_j)\big)\Big]$$
$$+\bigvee_{j=1}^{n}\varrho_{ij}\Big[f_j\big(h_{\diamond j}^{-1}(u_j)\big)-f_j\big(h_{\diamond j}^{-1}(\tilde{u}_j)\big)\Big]$$
$$+\bigwedge_{j=1}^{n}\varsigma_{ij}\Big[f_j\big(h_{\diamond j}^{-1}(u_j)\big)-f_j\big(h_{\diamond j}^{-1}(\tilde{u}_j)\big)\Big]\Bigg|$$
$$\leq\frac{1}{\lambda_i}\Bigg[L_i^{\lambda}+\sum_{j=1}^{n}L_j^f\bar{d}_j\big(|b_{ij}|+|\varrho_{ij}|+|\varsigma_{ij}|\big)\Bigg]\|u-\tilde{u}\|$$
$$\leq\zeta_3\|u-\tilde{u}\|,\quad i=1,2,\ldots,n.$$

By employing ($Q_4$), $T^{\star}$ is contractive. Hence, $T^{\star}$ admits a unique fixed point $u^{\star}=T^{\star}u^{\star}\in\mathbb{R}^n$. That is, $u^{\star}$ solves Eq. (3.12) and it is a unique equilibrium point of Eq. (3.12). The proof is achieved. □

If $\alpha_i=1(i=1,2,\ldots,n)$, then ($Q_3$) is turned into ($Q_4$). Applying similar discussion of Theorem 3.1 into Eq. (3.14) induces

**Corollary 3.1** *Let* ($Q_1$), ($Q_2$) *and* ($Q_4$) *hold. Then Eq. (3.14) admits a unique equilibrium point.*

## 3.4 Global Exponential Stability

Let $u=(u_1,\ldots,u_n)^T\in\mathbb{R}^n$ and $\tilde{u}=(\tilde{u}_1,\ldots,\tilde{u}_n)^T\in\mathbb{R}^n$ be arbitrary two solutions of Eq. (3.12) with initial conditions $u_i(s)=\phi_i(s)$ and $\tilde{u}_i(s)=\tilde{\phi}_i(s)$ for $s\in[-\tau_0^h,0]_{\mathbb{Z}}$, respectively, $i=1,2,\ldots,n$.

**Theorem 3.3** *Equation* (3.12) *has the property of global exponential stability if* ($Q_1$)–($Q_3$) *hold. That is, there exist* $M > 0$ *and* $0 < \kappa < 1$ *such that*

$$\|u(k) - \tilde{u}(k)\| \leq M\gamma_0 e^{-\kappa k}, \quad \forall k \in \mathbb{N}_0,$$

*where* $\gamma_0 = \max_{1\leq i\leq n} \sup_{s\in[-\tau_0^h,0]_{\mathbb{Z}}} |\phi_i(s) - \tilde{\phi}_i(s)|$.

***Proof*** Let

$$U_i(k) = |u_i(k) - \tilde{u}_i(k)|, \quad \forall k \in \mathbb{N}_0, i = 1, 2, \ldots, n.$$

Based on Eq. (3.15), it yields

$$\begin{aligned}
U_i(k) \leq & \prod_{s=0}^{k-1} \lambda_{\alpha_i}^{[s]} U_i(0) \\
& + \sum_{q=0}^{k-1} \prod_{s=q+1}^{k-1} \lambda_{\alpha_i}^{[s]} \sum_{l=0}^{\nu_q} |\theta_{\alpha_i}^{[l,q]}| \Bigg[ \Big| \Lambda_i^{h_\diamond}\big(u_i(q-l)\big) - \Lambda_i^{h_\diamond}\big(\tilde{u}_i(q-l)\big) \Big| \\
& + \sum_{j=1}^{n} |b_{ij}| L_j^f \bar{d}_j U_j(q - l - \tau_{ij}^h) + \sum_{j=1}^{n} |\varrho_{ij}| L_j^f \bar{d}_j U_j(q - l - \tau_{ij}^h) \\
& + \sum_{j=1}^{n} |\varsigma_{ij}| L_j^f \bar{d}_j U_j(q - l - \tau_{ij}^h) \Bigg] \\
\leq & \bar{\lambda}_{\alpha_i}^k U_i(0) + \frac{2}{\lambda_i} \sum_{q=0}^{k-1} \bar{\lambda}_{\alpha_i}^{k-q-1} \Bigg[ L_i^\lambda \max_{l\in[0,\nu_q]_{\mathbb{Z}}} U_i(q-l) \\
& + \sum_{j=1}^{n} L_j^f \bar{d}_j \big(|b_{ij}| + |\varrho_{ij}| + |\varsigma_{ij}|\big) \max_{l\in[0,\nu_q+\tau_0^h]_{\mathbb{Z}}} U_j(q-l) \Bigg] \\
\leq & \bar{\lambda}_{\alpha_i}^k U_i(0) + \frac{2}{\lambda_i} \sum_{q=0}^{k-1} \bar{\lambda}_{\alpha_i}^{k-q-1} \Bigg[ L_i^\lambda \max_{l\in[0,\nu_q+\tau_0^h]_{\mathbb{Z}}} U_i(q-l) \\
& + \sum_{j=1}^{n} L_j^f \bar{d}_j \big(|b_{ij}| + |\varrho_{ij}| + |\varsigma_{ij}|\big) \max_{l\in[0,\nu_q+\tau_0^h]_{\mathbb{Z}}} U_j(q-l) \Bigg]
\end{aligned} \tag{3.20}$$

for $k \in \mathbb{N}_0, i = 1, 2, \ldots, n$.

Let $c = -\ln \min_{1\leq i\leq n} \bar{\lambda}_{\alpha_i}$. From ($Q_3$), it exists $M > 1$ large enough and $0 < \kappa \ll c$ small enough such that

$$\frac{1}{M} + \frac{2}{\lambda_i[1 - e^{-(c-\kappa)}]} \Bigg[ L_i^\lambda + \sum_{j=1}^{n} L_j^f \bar{d}_j \big(|b_{ij}| + |\varrho_{ij}| + |\varsigma_{ij}|\big) \Bigg] e^{\kappa(N^* + \tau_0^h + 1)} < 1,$$

where $i = 1, 2, \ldots, n$.

For any $k \in \mathbb{N}_0$, $\max_{1\le i\le n} U_i(k) \le M\gamma_0 e^{-\kappa k}$ is supposed to be satisfied. If not, then it exists $k_0 \in \mathbb{N}_0 \setminus \{0\}$ and $i_0 \in \{1, 2, \ldots, n\}$ such that

$$U_{i_0}(k_0) > M\gamma_0 e^{-\kappa k_0}; \quad \max_{1\le i\le n} U_i(k) \le M\gamma_0 e^{-\kappa k}, \quad \forall k \in [0, k_0)_{\mathbb{Z}}. \tag{3.21}$$

Noting that $\nu_k \in [0, N^*]_{\mathbb{Z}}$, $\forall k \in \mathbb{N}_0$. It has from (3.20) and (3.21) that

$$\begin{aligned}
U_{i_0}(k_0) \le& e^{-ck_0}\gamma_0 + \frac{2}{\lambda_{i_0}} \sum_{q=0}^{k_0-1} e^{-c(k_0-q-1)} \\
&\times \left[ L_{i_0}^{\lambda} + \sum_{j=1}^{n} L_j^f \bar{d}_j \big(|b_{i_0 j}| + |\varrho_{i_0 j}| + |\varsigma_{i_0 j}|\big) \right] \max_{l\in[0,\nu_q+\tau_0^h]_{\mathbb{Z}}} M\gamma_0 e^{-\kappa(q-l)} \\
\le& e^{-ck_0}\gamma_0 + \frac{2}{\lambda_{i_0}} \sum_{q=0}^{k_0-1} e^{-c(k_0-q-1)} \\
&\times \left[ L_{i_0}^{\lambda} + \sum_{j=1}^{n} L_j^f \bar{d}_j \big(|b_{i_0 j}| + |\varrho_{i_0 j}| + |\varsigma_{i_0 j}|\big) \right] e^{\kappa(N^*+\tau_0^h)} M\gamma_0 e^{-\kappa q} \\
\le& e^{-ck_0}\gamma_0 + \frac{2}{\lambda_{i_0}} \sum_{q=0}^{k_0-1} e^{-(c-\kappa)(k_0-q-1)} \\
&\times \left[ L_{i_0}^{\lambda} + \sum_{j=1}^{n} L_j^f \bar{d}_j \big(|b_{i_0 j}| + |\varrho_{i_0 j}| + |\varsigma_{i_0 j}|\big) \right] e^{\kappa(N^*+\tau_0^h+1)} M\gamma_0 e^{-\kappa k_0} \\
\le& M\gamma_0 e^{-\kappa k_0} \left\{ \frac{1}{M} e^{-(c-\kappa)k_0} + \frac{2}{\lambda_{i_0}[1 - e^{-(c-\kappa)}]} \right. \\
&\left. \times \left[ L_{i_0}^{\lambda} + \sum_{j=1}^{n} L_j^f \bar{d}_j \big(|b_{i_0 j}| + |\varrho_{i_0 j}| + |\varsigma_{i_0 j}|\big) \right] e^{\kappa(N^*+\tau_0^h+1)} \right\} \\
\le& M\gamma_0 e^{-\kappa k_0},
\end{aligned}$$

which is a contradiction with the first part of fjgs1. So

$$\max_{1\le i\le n} U_i(k) \le M\gamma_0 e^{-\kappa k}, \quad \forall k \in \mathbb{N}_0.$$

Consequently, Eq. (3.12) is globally exponentially stable. The proof is achieved. □

Together with Theorems 3.1, 3.2 and 3.3, it derives

**Theorem 3.4** *Let* $(Q_1)$–$(Q_3)$ *hold. Then Eq.* (3.12) *has a unique bounded solution, which is globally exponentially stable. If* $(Q_4)$ *is also satisfied, then Eq.* (3.12) *has a unique equilibrium point, which is globally exponentially stable.*

If $\alpha_i = 1 (i = 1, 2, \ldots, n)$ in Eq. (3.12), then it easily gains

**Corollary 3.2** *Let* $(\mathrm{Q}_1)$–$(\mathrm{Q}_2)$ *and* $(\mathrm{Q}_4)$ *hold. Then Eq. (3.14) has a unique equilibrium point, which is globally exponentially stable.*

**Remark 3.6** Since Eq. (3.14) has no memory property, in $(\mathrm{Q}_4)$, the memory term $\theta_{\alpha_i}^{[l,k]}(l, k \in \mathbb{N})$ in Eq. (3.12) becomes 0, $i = 1, 2, \ldots, n$. The memory feature increases the complexity and difficulty of calculations of the existence of a unique bounded solution and global exponential stability to Eq. (3.12), see the proofs of Theorems 3.1 and 3.3.

**Remark 3.7** From the proof of Theorem 3.3, the time delays have no influence on global exponential stability of Eq. (3.12). However, it should be noted that it affects the convergent speed $\kappa$.

**Remark 3.8** By employing exponential Euler differences like Eq. (3.14), literature [22, 26, 27] discussed the periodicity, almost periodicity and stability of discrete neural networks. The authors [27] considered mean dynamics to exponential Euler discrete-time Cohen–Grossberg neural networks with the same difference technique. Thus, the work of this chapter supplements and extends the corresponding results in [22, 26, 27].

**Remark 3.9** In allusion to continuous-time delayed fractional-order neural networks, it is difficulty to discuss the exponential stability and numerous literature focus on Mittag–Leffler stability, see [119–122]. By employing the Mittag–Leffler Euler difference technique, this chapter starts the study of global exponential stability for discrete-time fractional-order neural networks. So the works of this chapter are original and have important significance.

## 3.5 Exponential Synchronization

Considering Eq. (3.12) with initial conditions (3.13) as a drive system and taking the following Mittag–Leffler Euler discrete model as a response system

$$\begin{aligned} v_i(k+1) = \lambda_{\alpha_i}^{[k]} v_i(k) + \sum_{l=0}^{\nu_k} \theta_{\alpha_i}^{[l,k]} \Big[ \Lambda_i^{h_\diamond}\big(v_i(k-l)\big) + \sum_{j=1}^{n} b_{ij} f_j\big(v_j^{h_\diamond}(k-l-\tau_{ij}^{h})\big) \\ + \bigvee_{j=1}^{n} \varrho_{ij} f_j\big(v_j^{h_\diamond}(k-l-\tau_{ij}^{h})\big) + \bigwedge_{j=1}^{n} \varsigma_{ij} f_j\big(v_j^{h_\diamond}(k-l-\tau_{ij}^{h})\big) \\ + \bigvee_{j=1}^{n} T_{ij}\varpi_j + \bigwedge_{j=1}^{n} H_{ij}\varpi_j + I_i \Big] + \rho_i(k), \end{aligned} \tag{3.22}$$

where $v_i(s) = \psi_i(s)$ for $s \in [-\tau_0^h, 0]_{\mathbb{Z}}$, $v_j^{h_\diamond} = h_{\diamond j}^{-1}(v_j)$, $\rho_i$ is the control input, $k \in \mathbb{N}_0$, $i = 1, 2, \ldots, n$.

Let $e_i(k) = v_i(k) - u_i(k)$, $k \in \mathbb{N}_0$, $i = 1, 2, \ldots, n$. Then the error system of Eqs. (3.12) and (3.22) is described as below

$$
\begin{aligned}
e_i(k+1) = {} & \lambda_{\alpha_i}^{[k]} e_i(k) + \sum_{l=0}^{\nu_k} \theta_{\alpha_i}^{[l,k]} \Bigg\{ \Big[ \Lambda_i^{h_\diamond}\big(v_i(k-l)\big) - \Lambda_i^{h_\diamond}\big(u_i(k-l)\big) \Big] \\
& + \Big[ \sum_{j=1}^{n} b_{ij} f_j\big(v_j^{h_\diamond}(k-l-\tau_{ij}^h)\big) - \sum_{j=1}^{n} b_{ij} f_j\big(u_j^{h_\diamond}(k-l-\tau_{ij}^h)\big) \Big] \\
& + \Big[ \bigvee_{j=1}^{n} \varrho_{ij} f_j\big(v_j^{h_\diamond}(k-l-\tau_{ij}^h)\big) - \bigvee_{j=1}^{n} \varrho_{ij} f_j\big(u_j^{h_\diamond}(k-l-\tau_{ij}^h)\big) \Big] \\
& + \Big[ \bigwedge_{j=1}^{n} \varsigma_{ij} f_j\big(v_j^{h_\diamond}(k-l-\tau_{ij}^h)\big) - \bigwedge_{j=1}^{n} \varsigma_{ij} f_j\big(u_j^{h_\diamond}(k-l-\tau_{ij}^h)\big) \Big] \Bigg\} \\
& + \rho_i(k), \quad \forall k \in \mathbb{N}_0, i = 1, 2, \ldots, n. \qquad (3.23)
\end{aligned}
$$

This section will prove that the drive-response systems (3.12) and (3.22) remain synchronized by designing the control input $\rho = (\rho_1, \rho_2, \ldots, \rho_n)^T$ to make the error system (3.23) stable.

**Theorem 3.5** *Let* ($Q_1$)–($Q_2$) *and the assumption below hold.*

($Q_5$) *There exists* $\eta_i \in (0, 1)$ *ensuring that*

$$
\max_{1 \le i \le n} \frac{\bar{\theta}_{\alpha_i}}{1 - \eta_i} \Big[ L_i^{\lambda} + \sum_{j=1}^{n} L_j^f \bar{d}_j \big( |b_{ij}| + |\varrho_{ij}| + |\varsigma_{ij}| \big) \Big] < 1.
$$

*Then the drive-response systems* (3.12) *and* (3.22) *are globally exponentially synchronized under a controlling law designed by*

$$
\rho_i(k) = (\eta_i - \lambda_{\alpha_i}^{[k]}) e_i(k), \quad \forall k \in \mathbb{N}_0, i = 1, 2, \ldots, n. \qquad (3.24)
$$

***Proof*** By controlling law (3.24), the error system (3.23) is turned into

$$
\begin{aligned}
e_i(k+1) = {} & \eta_i e_i(k) + \sum_{l=0}^{\nu_k} \theta_{\alpha_i}^{[l,k]} \Bigg\{ \Big[ \Lambda_i^{h_\diamond}\big(v_i(k-l)\big) - \Lambda_i^{h_\diamond}\big(u_i(k-l)\big) \Big] \\
& + \Big[ \sum_{j=1}^{n} b_{ij} f_j\big(v_j^{h_\diamond}(k-l-\tau_{ij}^h)\big) - \sum_{j=1}^{n} b_{ij} f_j\big(u_j^{h_\diamond}(k-l-\tau_{ij}^h)\big) \Big]
\end{aligned}
$$

$$+\Big[\bigvee_{j=1}^{n}\varrho_{ij}f_j\big(v_j^{h_\diamond}(k-l-\tau_{ij}^h)\big)-\bigvee_{j=1}^{n}\varrho_{ij}f_j\big(u_j^{h_\diamond}(k-l-\tau_{ij}^h)\big)\Big]$$
$$+\Big[\bigwedge_{j=1}^{n}\varsigma_{ij}f_j\big(v_j^{h_\diamond}(k-l-\tau_{ij}^h)\big)-\bigwedge_{j=1}^{n}\varsigma_{ij}f_j\big(u_j^{h_\diamond}(k-l-\tau_{ij}^h)\big)\Big]\Big\},$$

where $k \in \mathbb{N}_0$, $i = 1, 2, \ldots, n$.

Similar to the derivation of Eq. (3.15), it follows

$$\begin{aligned} e_i(k) \le & \eta_i^k e_i(0) + \sum_{q=0}^{k-1}\eta_i^{k-q-1}\sum_{l=0}^{\nu_q}\theta_{\alpha_i}^{[l,q]}\Big\{\Big[\Lambda_i^{h_\diamond}\big(v_i(q-l)\big)-\Lambda_i^{h_\diamond}\big(u_i(q-l)\big)\Big] \\ & +\Big[\sum_{j=1}^{n}b_{ij}f_j\big(v_j^{h_\diamond}(q-l-\tau_{ij}^h)\big)-\sum_{j=1}^{n}b_{ij}f_j\big(u_j^{h_\diamond}(q-l-\tau_{ij}^h)\big)\Big] \\ & +\Big[\bigvee_{j=1}^{n}\varrho_{ij}f_j\big(v_j^{h_\diamond}(q-l-\tau_{ij}^h)\big)-\bigvee_{j=1}^{n}\varrho_{ij}f_j\big(u_j^{h_\diamond}(q-l-\tau_{ij}^h)\big)\Big] \\ & +\Big[\bigwedge_{j=1}^{n}\varsigma_{ij}f_j\big(v_j^{h_\diamond}(q-l-\tau_{ij}^h)\big)-\bigwedge_{j=1}^{n}\varsigma_{ij}f_j\big(u_j^{h_\diamond}(q-l-\tau_{ij}^h)\big)\Big]\Big\}, \end{aligned} \tag{3.25}$$

where $k \in \mathbb{N}_0$, $i = 1, 2, \ldots, n$.

Set

$$\tilde{\gamma}_0 = \max_{1\le i\le n}\sup_{s\in[-\tau_0^h,0]_{\mathbb{Z}}}|e_i(s)| = \max_{1\le i\le n}\sup_{s\in[-\tau_0^h,0]_{\mathbb{Z}}}|\phi_i(s)-\psi_i(s)|$$

and $\tilde{c} = -\ln\min_{1\le i\le n}\eta_i$. In view of ($Q_5$), it exists $\tilde{M} > 1$ large enough and $0 < \tilde{\kappa} \ll \tilde{c}$ small enough such that

$$\frac{1}{\tilde{M}} + \frac{\bar{\theta}_{\alpha_i}}{1-e^{-(\tilde{c}-\tilde{\kappa})}}\Big[L_i^\lambda + \sum_{j=1}^{n}L_j^f\bar{d}_j\big(|b_{ij}|+|\varrho_{ij}|+|\varsigma_{ij}|\big)\Big]e^{\tilde{\kappa}(N^*+\tau_0^h+1)} < 1$$

for $i = 1, 2, \ldots, n$.

Supposing that $\max_{1\le i\le n}|e_i(k)| \le \tilde{M}\tilde{\gamma}_0 e^{-\tilde{\kappa}k}$, $\forall k \in \mathbb{N}_0$. If not, then it exists $\tilde{k}_0 \in \mathbb{N}_0 \setminus \{0\}$ and $\tilde{i}_0 \in \{1, 2, \ldots, n\}$ such that

$$|e_{\tilde{i}_0}(\tilde{k}_0)| > \tilde{M}\tilde{\gamma}_0 e^{-\tilde{\kappa}\tilde{k}_0};\quad \max_{1\le i\le n}|e_i(k)| \le \tilde{M}\tilde{\gamma}_0 e^{-\tilde{\kappa}k},\quad \forall k \in [-\tau_0^h, \tilde{k}_0)_{\mathbb{Z}}. \tag{3.26}$$

In case that (3.26) is valid, it has from (3.25) that

$$
\begin{aligned}
|e_{\tilde{i}_0}(\tilde{k}_0)| \leq & e^{-\tilde{c}\tilde{k}_0}\tilde{\gamma}_0 + \bar{\theta}_{\alpha_{\tilde{i}_0}} \sum_{q=0}^{\tilde{k}_0-1} e^{-\tilde{c}(\tilde{k}_0-q-1)} \\
& \times \left[ L_{\tilde{i}_0}^{\lambda} + \sum_{j=1}^{n} L_j^f \bar{d}_j \big( |b_{\tilde{i}_0 j}| + |\varrho_{\tilde{i}_0 j}| + |\varsigma_{\tilde{i}_0 j}| \big) \right] \max_{l \in [0, \nu_q + \tau_0^h]_{\mathbb{Z}}} \tilde{M}\tilde{\gamma}_0 e^{-\tilde{\kappa}(q-l)} \\
\leq & e^{-\tilde{c}\tilde{k}_0}\tilde{\gamma}_0 + \bar{\theta}_{\alpha_{\tilde{i}_0}} \sum_{q=0}^{\tilde{k}_0-1} e^{-\tilde{c}(\tilde{k}_0-q-1)} \\
& \times \left[ L_{\tilde{i}_0}^{\lambda} + \sum_{j=1}^{n} L_j^f \bar{d}_j \big( |b_{\tilde{i}_0 j}| + |\varrho_{\tilde{i}_0 j}| + |\varsigma_{\tilde{i}_0 j}| \big) \right] e^{\tilde{\kappa}(N^* + \tau_0^h)} \tilde{M}\tilde{\gamma}_0 e^{-\tilde{\kappa}q} \\
\leq & e^{-\tilde{c}\tilde{k}_0}\tilde{\gamma}_0 + \bar{\theta}_{\alpha_{\tilde{i}_0}} \sum_{q=0}^{\tilde{k}_0-1} e^{-(\tilde{c}-\tilde{\kappa})(\tilde{k}_0-q-1)} \\
& \times \left[ L_{\tilde{i}_0}^{\lambda} + \sum_{j=1}^{n} L_j^f \bar{d}_j \big( |b_{\tilde{i}_0 j}| + |\varrho_{\tilde{i}_0 j}| + |\varsigma_{\tilde{i}_0 j}| \big) \right] e^{\tilde{\kappa}(N^* + \tau_0^h + 1)} \tilde{M}\tilde{\gamma}_0 e^{-\tilde{\kappa}\tilde{k}_0} \\
\leq & \tilde{M}\tilde{\gamma}_0 e^{-\tilde{\kappa}\tilde{k}_0} \left\{ \frac{1}{\tilde{M}} e^{-(\tilde{c}-\tilde{\kappa})\tilde{k}_0} + \frac{\bar{\theta}_{\alpha_{\tilde{i}_0}}}{1 - e^{-(\tilde{c}-\tilde{\kappa})}} \right. \\
& \left. \times \left[ L_{\tilde{i}_0}^{\lambda} + \sum_{j=1}^{n} L_j^f \bar{d}_j \big( |b_{\tilde{i}_0 j}| + |\varrho_{\tilde{i}_0 j}| + |\varsigma_{\tilde{i}_0 j}| \big) \right] e^{\tilde{\kappa}(N^* + \tau_0^h + 1)} \right\} \\
\leq & \tilde{M}\tilde{\gamma}_0 e^{-\tilde{\kappa}\tilde{k}_0},
\end{aligned}
$$

which is a contradiction with the first part of (3.26). So

$$
\max_{1 \leq i \leq n} |e_i(k)| \leq \tilde{M}\tilde{\gamma}_0 e^{-\tilde{\kappa}k}
$$

is valid for all $k \in \mathbb{N}_0$. So the error system (3.23) is globally exponentially stable under controlling law (3.24). The proof is achieved. □

If $\alpha_i = 1 (i = 1, 2, \ldots, n)$ in the drive system (3.12), then the response system (3.22) and the error system (3.23) are changed to

$$
\begin{aligned}
v_i(k+1) = & e^{-\lambda_i h} v_i(k) + (1 - e^{-\lambda_i h}) \lambda_i^{-1} \bigg[ \Lambda_i^{h_\diamond}\big(v_i(k)\big) \\
& + \sum_{j=1}^{n} b_{ij} f_j\big(v_j^{h_\diamond}(k - \tau_{ij}^h)\big) + \bigvee_{j=1}^{n} \varrho_{ij} f_j\big(v_j^{h_\diamond}(k - \tau_{ij}^h)\big) \\
& + \bigwedge_{j=1}^{n} \varsigma_{ij} f_j\big(v_j^{h_\diamond}(k - \tau_{ij}^h)\big) + \bigvee_{j=1}^{n} T_{ij}\varpi_j + \bigwedge_{j=1}^{n} H_{ij}\varpi_j + I_i \bigg] \\
& + \rho_i(k),
\end{aligned}
\tag{3.27}
$$

and

$$e_i(k+1) = e^{-\lambda_i h} e_i(k) + (1 - e^{-\lambda_i h})\lambda_i^{-1} \Big\{ \Big[ \Lambda_i^{h_\diamond}\big(v_i(k)\big) - \Lambda_i^{h_\diamond}\big(u_i(k)\big) \Big]$$
$$+ \Big[ \sum_{j=1}^{n} b_{ij} f_j\big(v_j^{h_\diamond}(k - \tau_{ij}^h)\big) - \sum_{j=1}^{n} b_{ij} f_j\big(u_j^{h_\diamond}(k - \tau_{ij}^h)\big) \Big]$$
$$+ \Big[ \bigvee_{j=1}^{n} \varrho_{ij} f_j\big(v_j^{h_\diamond}(k - \tau_{ij}^h)\big) - \bigvee_{j=1}^{n} \varrho_{ij} f_j\big(u_j^{h_\diamond}(k - \tau_{ij}^h)\big) \Big]$$
$$+ \Big[ \bigwedge_{j=1}^{n} \varsigma_{ij} f_j\big(v_j^{h_\diamond}(k - \tau_{ij}^h)\big) - \bigwedge_{j=1}^{n} \varsigma_{ij} f_j\big(u_j^{h_\diamond}(k - \tau_{ij}^h)\big) \Big] \Big\} + \rho_i(k),$$

where $k \in \mathbb{N}_0$, $i = 1, 2, \ldots, n$. Hereby, the control input $\rho = (\rho_1, \rho_2, \ldots, \rho_n)^T$ is designed by

$$\rho_i(k) = (\eta_i - e^{-\lambda_i h}) e_i(k), \quad \forall k \in \mathbb{N}_0, i = 1, 2, \ldots, n. \tag{3.28}$$

Similar to the derivation of Theorem 3.5, it gains.

**Corollary 3.3** *The drive-response systems (3.14) and* (3.27) *are globally exponentially synchronized under controlling law* (3.28) *in case* ($Q_1$)–($Q_2$) *and the assumption below hold.*

($Q_6$) *There exists* $\eta_i \in (0, 1)$ *ensuring that*

$$\max_{1 \le i \le n} \frac{1 - e^{-\lambda_i h}}{\lambda_i (1 - \eta_i)} \Big[ L_i^{\lambda} + \sum_{j=1}^{n} L_j^f \bar{d}_j \big( |b_{ij}| + |\varrho_{ij}| + |\varsigma_{ij}| \big) \Big] < 1.$$

**Remark 3.10** In Sects. 3.3–3.4, $\lambda_i > 0$ is necessary, $i = 1, 2, \ldots, n$. However, this condition is not indispensable in Theorem 3.5 and Corollary 3.3.

## 3.6 Numerical Experiments

**Example 3.1** Considering the following Caputo fractional-order fuzzy CGNNs

$${}_{t_p}^{C}D_t^{\alpha} \begin{bmatrix} z_1(t) \\ z_2(t) \end{bmatrix} = - \begin{bmatrix} 5 z_1^{h_\diamond}(t) \\ 6 z_2^{h_\diamond}(t) \end{bmatrix} + 0.001 \begin{bmatrix} |z_2^{h_\diamond}(t - 0.1)| \\ |z_1^{h_\diamond}(t - 0.1)| \end{bmatrix}$$
$$+ 0.1 \bigvee_{j=1}^{2} f_j\big(z_j^{h_\diamond}(t - 0.1)\big) + 0.2 \bigwedge_{j=1}^{2} f_j\big(z_j^{h_\diamond}(t - 0.1)\big) + 3, \tag{3.29}$$

where $t \in (t_p, t_{p+1}]$, $p \in \mathbb{N}_0$, $z_i(s) = \varphi_i(s)$ for $s \in [-0.1, 0]$, $f_i(x) = 0.01|x|$ for $x \in \mathbb{R}$,

$$z_i^{h_\diamond}(t) = h_{\diamond i}^{-1}(z_i(t)), \quad h_{\diamond i}(z_i) = \int_0^{z_i} \frac{1}{d_i(s)} \mathrm{d}s,$$

$d_i(s) = \frac{1}{1+0.01\cos s}$ for $s \in \mathbb{R}$, $i = 1, 2$.

Taking $h = 0.1$, $\mathcal{P} = \{t_p = 3p : p \in \mathbb{N}_0\}$ and $\mathcal{N} = \{N_p = 30p : p \in \mathbb{N}_0\}$. By using Mittag–Leffler Euler difference technique in Eq. (3.12), Mittag–Leffler discrete analogue of Eq. (3.29) is depicted as

$$\begin{aligned}
\begin{bmatrix} z_1(kh+h) \\ z_2(kh+h) \end{bmatrix} &= \begin{bmatrix} \lambda_{\alpha_1}^{[k]} z_1(kh) \\ \lambda_{\alpha_2}^{[k]} z_2(kh) \end{bmatrix} + \sum_{l=0}^{\nu_k} \begin{bmatrix} \theta_{\alpha_1}^{[l,k]} \\ \theta_{\alpha_2}^{[l,k]} \end{bmatrix} \Bigg\{ \begin{bmatrix} 5z_1(kh-lh) \\ 6z_2(kh-lh) \end{bmatrix} \\
&\quad - \begin{bmatrix} 5z_1^{h_\diamond}(kh-lh) \\ 6z_2^{h_\diamond}(kh-lh) \end{bmatrix} + 0.001 \begin{bmatrix} |z_2^{h_\diamond}(kh-lh-0.1)| \\ |z_1^{h_\diamond}(kh-lh-0.1)| \end{bmatrix} \\
&\quad + 0.1 \bigvee_{j=1}^{2} f_j\big(z_j^{h_\diamond}(kh-lh-0.1)\big) + 0.2 \bigwedge_{j=1}^{2} f_j\big(z_j^{h_\diamond}(kh-lh-0.1)\big) \\
&\quad + 3 \Bigg\}, \quad \forall k \in \mathbb{N}_0,
\end{aligned} \tag{3.30}$$

where $z_i(sh) = \varphi_i(sh)$ for $s = -1, 0$, $\nu_k = k - \mu_k$, $\mu_k = \max_{p\in\mathbb{N}_0} \{30p : 30p \le k\}$,

$$\lambda_{\alpha_1}^{[k]} = \frac{E_\alpha[-5(\nu_k h + h)^\alpha]}{E_\alpha[-5(\nu_k h)^\alpha]}, \quad \lambda_{\alpha_2}^{[k]} = \frac{E_\alpha[-6(\nu_k h + h)^\alpha]}{E_\alpha[-6(\nu_k h)^\alpha]},$$

$$\theta_{\alpha_1}^{[l,k]} = -\frac{1}{5}\Big[w_{\alpha_1}^{[l]} - \lambda_{\alpha_1}^{[k]} w_{\alpha_1}^{[l-1]}\Big], \quad \theta_{\alpha_2}^{[l,k]} = -\frac{1}{6}\Big[w_{\alpha_2}^{[l]} - \lambda_{\alpha_2}^{[k]} w_{\alpha_2}^{[l-1]}\Big],$$

$$w_{\alpha_1}^{[-1]} = w_{\alpha_2}^{[-1]} = 0, \quad w_{\alpha_1}^{[l]} = E_\alpha[-5(lh+h)^\alpha] - E_\alpha[-5(lh)^\alpha],$$

$$w_{\alpha_2}^{[l]} = E_\alpha[-6(lh+h)^\alpha] - E_\alpha[-6(lh)^\alpha],$$

where $l \in [0, \nu_k]_{\mathbb{Z}}$, $k \in \mathbb{N}_0$.

Through the use of fractional Euler differences, it obtains another type of numerical solution of Eq. (3.29) as below

$$\begin{bmatrix} z_{*1}(kh+h) \\ z_{*2}(kh+h) \end{bmatrix} = \begin{bmatrix} z_{*1}(kh) \\ z_{*2}(kh) \end{bmatrix} + \sum_{l=0}^{\nu_k} \hat{\theta}_{\alpha}^{[l]} \Bigg\{ - \begin{bmatrix} 5z_{*1}^{h_{\diamond}}(kh-lh) \\ 6z_{*2}^{h_{\diamond}}(kh-lh) \end{bmatrix}$$
$$+0.001 \begin{bmatrix} |z_{*2}^{h_{\diamond}}(kh-lh-0.1)| \\ |z_{*1}^{h_{\diamond}}(kh-lh-0.1)| \end{bmatrix} + 0.1 \bigvee_{j=1}^{2} f_j\big(z_{*j}^{h_{\diamond}}(kh-lh-0.1)\big)$$
$$+0.2 \bigwedge_{j=1}^{2} f_j\big(z_{*j}^{h_{\diamond}}(kh-lh-0.1)\big) + 3 \Bigg\}, \quad \forall k \in \mathbb{N}_0, \tag{3.31}$$

where $z_{i*}(sh) = \varphi_{i*}(sh)$ for $s = -1, 0$,

$$\hat{\theta}_{\alpha}^{[l]} = \frac{1}{\Gamma(\alpha+1)} \big[\hat{w}_{\alpha}^{[l]} - \hat{w}_{\alpha}^{[l-1]}\big], \quad \hat{w}_{\alpha}^{[-1]} = 0,$$

$$\hat{w}_{\alpha}^{[l]} = \big[(l+1)^{\alpha} - l^{\alpha}\big] h^{\alpha}, \quad \forall l \in [0, \nu_k]_{\mathbb{Z}}, k \in \mathbb{N}_0.$$

The solutions of Eqs. (3.30) and (3.31) with step length $h = 0.1$ and different orders $\alpha \in \{0.3, 0.5, 0.7, 0.9\}$ are drawn in Figs. 3.1, 3.2, 3.3 and 3.4.

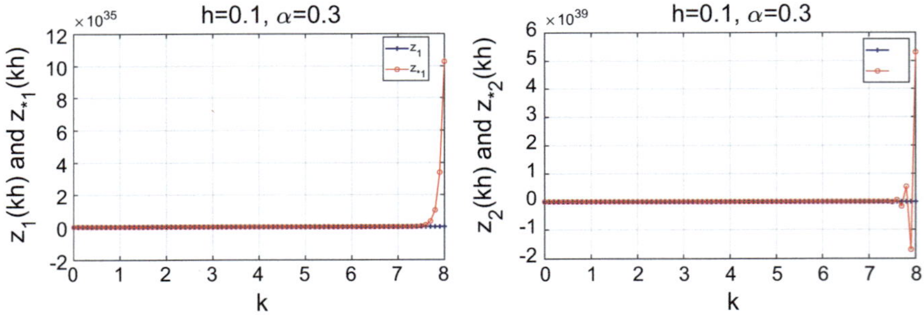


**Fig. 3.1** $(z_1, z_2)^T$ and $(z_{*1}, z_{*2})^T$ of Eqs. (3.30) and (3.31) with $h = 0.1$ and $\alpha = 0.3$

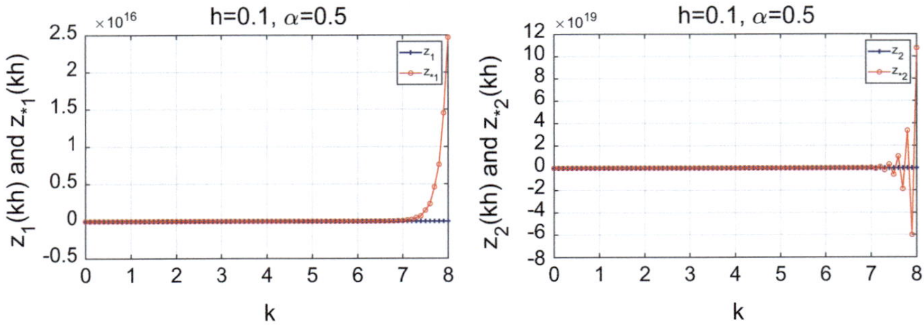


**Fig. 3.2** $(z_1, z_2)^T$ and $(z_{*1}, z_{*2})^T$ of Eqs. (3.30) and (3.31) with $h = 0.1$ and $\alpha = 0.5$

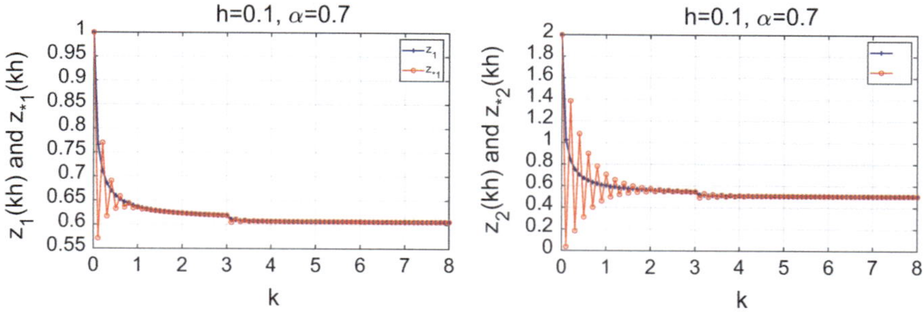


**Fig. 3.3** $(z_1, z_2)^T$ and $(z_{*1}, z_{*2})^T$ of Eqs. (3.30) and (3.31) with $h = 0.1$ and $\alpha = 0.7$

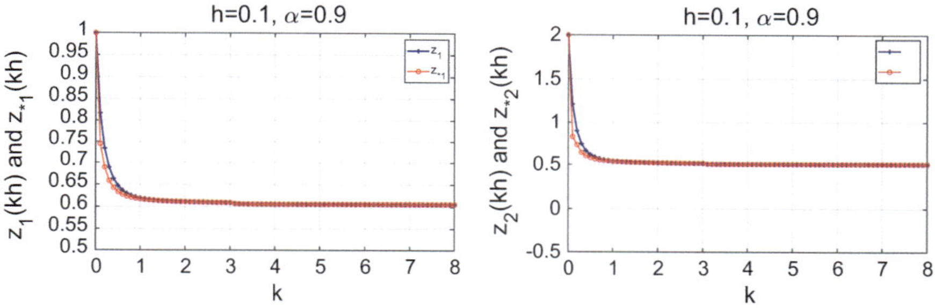


**Fig. 3.4** $(z_1, z_2)^T$ and $(z_{*1}, z_{*2})^T$ of Eqs. (3.30) and (3.31) with $h = 0.1$ and $\alpha = 0.9$

Corresponding to Eq. (3.12) and Theorem 3.1, it easily verifies that all conditions of Theorems 3.1 and 3.3 are filled. Then Eq. (3.30) admits a unique bounded solution, which is globally exponentially stable. Taking some random initial values, Fig. 3.5 describes the trajectories and exponential convergence of states $z_1$ and $z_2$ in Eq. (3.30), respectively.

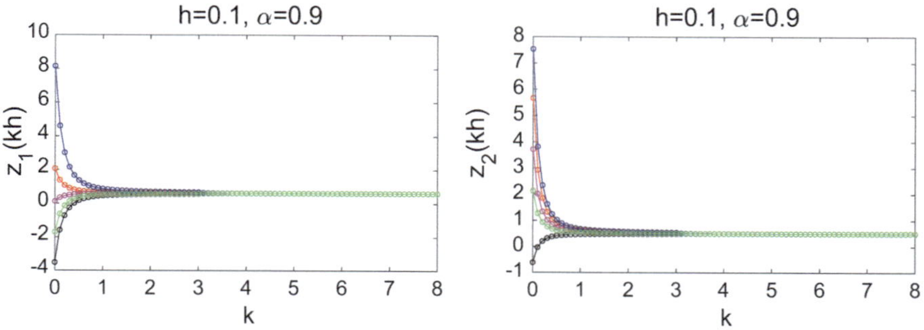


**Fig. 3.5** Global exponential stability of $(z_1, z_2)^T$ of Eq. (3.30) with $h = 0.1$ and $\alpha = 0.9$

**Example 3.2** Taking

$$h = 0.1, \quad \mathcal{P} = \{t_p = 3p : p \in \mathbb{N}_0\}, \quad \mathcal{N} = \left\{N_p = 30p : p \in \mathbb{N}_0\right\}.$$

Considering the following fuzzy Mittag–Leffler discrete-time Cohen–Grossberg neural networks as a drive system

$$\begin{aligned}\begin{bmatrix} u_1(k+1) \\ u_2(k+1) \end{bmatrix} &= \begin{bmatrix} \lambda_{\alpha_1}^{[k]} u_1(k) \\ \lambda_{\alpha_2}^{[k]} u_2(k) \end{bmatrix} + \sum_{l=0}^{\nu_k} \begin{bmatrix} \theta_{\alpha_1}^{[l,k]} \\ \theta_{\alpha_2}^{[l,k]} \end{bmatrix} \left\{ \begin{bmatrix} 5u_1(k-l) \\ 6u_2(k-l) \end{bmatrix} - \begin{bmatrix} 5u_1^{h_\diamond}(k-l) \\ 6u_2^{h_\diamond}(k-l) \end{bmatrix} \right. \\ &\quad +0.001 \begin{bmatrix} |u_2^{h_\diamond}(k-l-1)| \\ |u_1^{h_\diamond}(k-l-1)| \end{bmatrix} + 0.1 \bigvee_{j=1}^{2} f_j\big(u_j^{h_\diamond}(k-l-1)\big) \\ &\quad \left. +0.2 \bigwedge_{j=1}^{2} f_j\big(u_j^{h_\diamond}(k-l-1)\big) + 3 \right\}, \quad \forall k \in \mathbb{N}_0, \end{aligned} \tag{3.32}$$

where $u_i(s) = \phi_i(s)$ for $s = -1, 0$, $f_i(x) = 0.01|x|$ for $x \in \mathbb{R}$, $u_i^{h_\diamond} = h_{\diamond i}^{-1}(u_i)$,

$$h_{\diamond i}(u_i) = \int_0^{u_i} \frac{1}{d_i(s)} \mathrm{d}s, \quad d_i(s) = \frac{1}{1 + 0.01 \cos s}$$

for $s \in \mathbb{R}$, $\nu_k = k - \mu_k$, $\mu_k = \max_{p \in \mathbb{N}_0} \left\{30p : 30p \leq k\right\}$, $\lambda_{\alpha_i}^{[k]}$ and $\theta_{\alpha_i}^{[l,k]}$ are defined as those in Example 3.1, $l \in [0, \nu_k]_{\mathbb{Z}}$, $k \in \mathbb{N}_0$, $i = 1, 2$.

Taking the following Mittag–Leffler Euler discrete model as a response system

$$\begin{aligned}\begin{bmatrix} v_1(k+1) \\ v_2(k+1) \end{bmatrix} &= \begin{bmatrix} \lambda_{\alpha_1}^{[k]} v_1(k) \\ \lambda_{\alpha_2}^{[k]} v_2(k) \end{bmatrix} + \sum_{l=0}^{\nu_k} \begin{bmatrix} \theta_{\alpha_1}^{[l,k]} \\ \theta_{\alpha_2}^{[l,k]} \end{bmatrix} \left\{ \begin{bmatrix} 5v_1(k-l) \\ 6v_2(k-l) \end{bmatrix} - \begin{bmatrix} 5v_1^{h_\diamond}(k-l) \\ 6v_2^{h_\diamond}(k-l) \end{bmatrix} \right. \\ &\quad +0.001 \begin{bmatrix} |v_2^{h_\diamond}(k-l-1)| \\ |v_1^{h_\diamond}(k-l-1)| \end{bmatrix} + 0.1 \bigvee_{j=1}^{2} f_j\big(v_j^{h_\diamond}(k-l-1)\big) \\ &\quad \left. +0.2 \bigwedge_{j=1}^{2} f_j\big(v_j^{h_\diamond}(k-l-1)\big) + 3 \right\} + \begin{bmatrix} \rho_1(k) \\ \rho_2(k) \end{bmatrix}, \quad \forall k \in \mathbb{N}_0, \end{aligned} \tag{3.33}$$

where $v_i(s) = \psi_i(s)$ for $s = -1, 0$, $v_i^{h_\diamond} = h_{\diamond i}^{-1}(v_i)$, $\rho_i$ is the control input, $i = 1, 2$.

Design a controlling law as

$$\rho_i(k) = (0.1 - \lambda_{\alpha_i}^{[k]})(v_i(k) - u_i(k)), \quad \forall k \in \mathbb{N}_0, i = 1, 2. \tag{3.34}$$

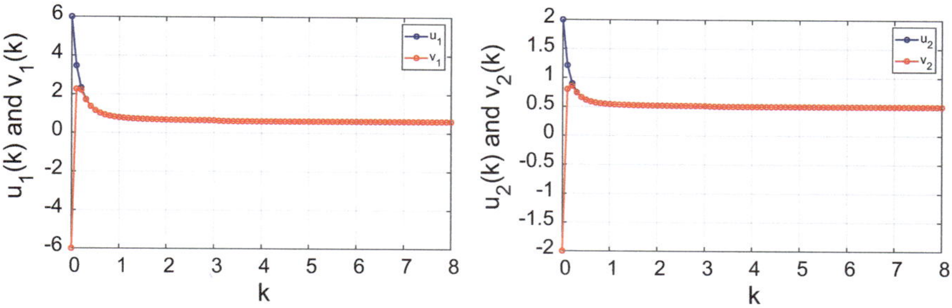


**Fig. 3.6** Global exponential synchronization of the drive-response systems (3.32) and (3.33) with $h = 0.1$ and $\alpha = 0.9$

**Fig. 3.7** Image pattern "I"

It easily verifies that all conditions of Theorem 3.5 are fulfilled. Then the drive-response systems (3.32) and (3.33) are globally exponentially synchronized under a controlling law designed by (3.34), see Fig. 3.6.

**Example 3.3** This example should structure a Mittag–Leffler Euler discrete model of four-neuron neural networks to memorize the image pattern "I" displayed in Fig. 3.7.

Considering the following Mittag–Leffler discrete-time CGNNs

$$\begin{bmatrix} u_1(k+1) \\ u_2(k+1) \\ u_3(k+1) \\ u_4(k+1) \end{bmatrix} = \begin{bmatrix} \lambda_{\alpha_1}^{[k]} u_1(k) \\ \lambda_{\alpha_2}^{[k]} u_2(k) \\ \lambda_{\alpha_3}^{[k]} u_3(k) \\ \lambda_{\alpha_4}^{[k]} u_4(k) \end{bmatrix} + \sum_{l=0}^{\nu_k} \begin{bmatrix} \theta_{\alpha_1}^{[l,k]} \\ \theta_{\alpha_2}^{[l,k]} \\ \theta_{\alpha_3}^{[l,k]} \\ \theta_{\alpha_4}^{[l,k]} \end{bmatrix} \left\{ \begin{bmatrix} 0.3u_1(k-l) \\ 0.4u_2(k-l) \\ 0.4u_3(k-l) \\ 0.2u_4(k-l) \end{bmatrix} - \begin{bmatrix} 0.3u_1^{h_\diamond}(k-l) \\ 0.4u_2^{h_\diamond}(k-l) \\ 0.4u_3^{h_\diamond}(k-l) \\ 0.2u_4^{h_\diamond}(k-l) \end{bmatrix} \right.$$
$$\left. + \begin{bmatrix} 0 & 0.2 & 0 & 0 \\ 0 & 0 & 0.1 & 0 \\ 0 & 0 & 0 & 0.3 \\ 0 & 0 & 0 & 0 \end{bmatrix} \begin{bmatrix} f_1(u_1(k)) \\ f_2(u_2(k)) \\ f_3(u_3(k)) \\ f_4(u_4(k)) \end{bmatrix} + I \right\}, \quad \forall k \in \mathbb{N}_0, \quad (3.35)$$

where $\alpha_i = 0.95$,

$$f_j(x) = \frac{1}{8}(|x+1| - |x-1|),$$

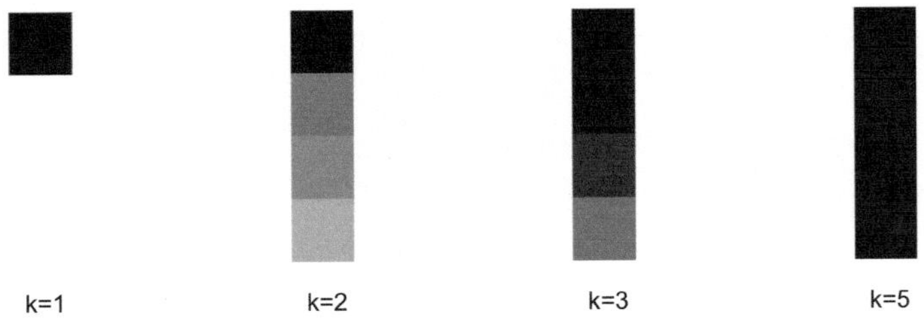


**Fig. 3.8** The memory processes with initial value $u_0 = (1, -1, -1, -1)^T$

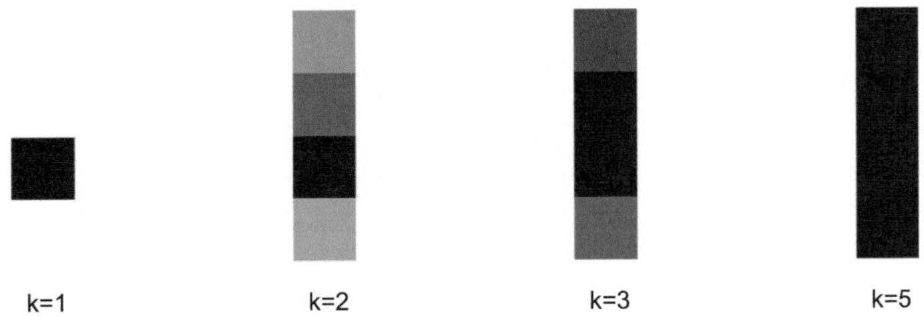


**Fig. 3.9** The memory processes with initial value $u_0 = (-1, -1, 1, -1)^T$

$d_i(s) \equiv 1$ for $s \in \mathbb{R}$, $\mu_k = \max_{p \in \mathbb{N}_0} \{5p : 5p \leq k\}$, $\lambda_{\alpha_i}^{[k]}$ and $\theta_{\alpha_i}^{[l,k]}$ are defined as those in (3.5) with $\lambda_i = 0.3, 0.4, 0.4, 0.2$, respectively, $l \in [0, \nu_k]_{\mathbb{Z}}$, $k \in \mathbb{N}_0$, $i, j = 1, 2, 3, 4$; $I = (I_1, I_2, I_3, I_4)^T$ needs to be confirmed.

Selecting four points from the values of gray-scale image in as the equilibrium point $u_*$ of model (3.35), i.e.,

$$u_* = (-1, -1, -1, -1)^T.$$

Then the external inputs of model (3.35) can be computed as

$$I = (-0.25, -0.375, -0.325, -0.2)^T.$$

By Theorem 3.3, the equilibrium point $u_*$ of model (3.35) is stable. Figures 3.8 and 3.9 describe the memory processes of the designed discrete model with two different initial values (i.e., $u_0 = (1, -1, -1, -1)^T$ and $u_0 = (-1, -1, 1, -1)^T$), respectively. These images show that model (3.35) could remember the image pattern "I" very well.

# Chapter 4
# The Theory of Nabla MLEDs

This chapter outlines the fundamental structure of the Nabla Mittag–Leffler Euler difference scheme for FODEs in the sense of Caputo. Additionally, the fractional PECE algorithm is proposed as an effective solution for this type of implicit difference in numerical calculations. Furthermore, the corresponding nonlocal Nabla difference equations are modelled and the well-posed problem is investigated.

## 4.1 Introduction

This chapter is devoted to studying the Nabla Mittag–Leffler Euler differences of FODEs from the viewpoints of both numerical computations and the theory of difference equations, which the considered FODEs are described as

$$\begin{cases} {}_a^C D_t^{\alpha} x(t) = \lambda x(t) + f(t, x(t)), & \forall t \in (a, +\infty), \\ x(a) = \xi_0 \in \mathbb{R}, \end{cases} \tag{4.1}$$

where $a \in \mathbb{R}$, $x \in C([a, +\infty), \mathbb{R})$, $\alpha \in (0, 1]$, $\lambda \in \mathbb{R}\backslash\{0\}$ and $f \in C([a, +\infty) \times \mathbb{R}, \mathbb{R})$. It exists a constant $L_f > 0$ such that

$$|f(t, x) - f(t, x)| \leq L_f|x - y|, \quad \forall x, y \in \mathbb{R}, t \in [a, +\infty).$$

During the last two decades, many practical models were depicted by FODEs (4.1) and some excellent results were obtained in literature [123–126]. For instance, the authors in [123] discussed the existence and uniqueness of $S$-asymptotically periodic solution to a kind of BAM neural networks. Paper [124] designed the impulsive controllers to a fractional gene regulatory networks modelled by FODEs (4.1). Therefore, the study of FODEs (4.1) has very important practical significance. The main

 
T. Zhang et al., *Mittag-Leffler Euler Difference Techniques*,
https://doi.org/10.1007/978-981-96-6228-9_4

aim of this chapter is to establish the basic structure of Nabla Mittag–Leffler Euler difference scheme to FODEs (4.1), which extends the traditional exponential Euler differences [19–29], please see publication [127].

The principal contributions of this chapter are concluded as below. **(1)** Nabla Mittag–Leffler Euler difference scheme (with short-term memory) to FODEs (4.1) is set up. **(2)** Fractional PECE algorithm is introduced to solve the acquired implicit difference in numerical calculations effectively. **(3)** A novel Nabla Mittag–Leffler Euler difference operator and equations are proposed from the viewpoint of the theory of difference equations and some relative properties are presented. **(4)** The existence and boundedness of a unique globally exponentially stable solution of Nabla Mittag–Leffler Euler difference equations with short-term memory are discussed. **(5)** A set of theories and methods in studying Nabla Mittag–Leffler Euler difference equations is built for the further investigations in the future.

The organization of this chapter is as follows. In Sect. 4.2, the basic structure of Nabla Mittag–Leffler Euler difference is established. In Sect. 4.3, the fractional PECE algorithm is introduced to solve the acquired implicit difference. In Sect. 4.4, Nabla Mittag–Leffler Euler difference operator and equations are structured. Section 4.5 discusses the well-posed problem to the nonlocal Nabla difference model. Two explicative example are employed to visually elaborate the current findings in Sect. 4.6.

## 4.2 Nabla Mittag–Leffler Euler Difference

Let $h > 0$,

$$\begin{cases} \lambda_{\alpha,h}^{[k]} = \dfrac{\mathbb{E}_\alpha[\lambda(kh)^\alpha]}{\mathbb{E}_\alpha[\lambda(kh-h)^\alpha]}, \\ \theta_{\alpha,h}^{[l,k]} = \dfrac{w_{\alpha,h}^{[l,k]} - \lambda_{\alpha,h}^{[k]} w_{\alpha,h}^{[l,k-1]}}{\lambda}, \\ w_{\alpha,h}^{[l,k]} = \mathbb{E}_\alpha[\lambda(kh-lh+h)^\alpha] - \mathbb{E}_\alpha[\lambda(kh-lh)^\alpha], \\ w_{\alpha,h}^{[l,k]} \equiv 0 \text{ if } l > k, \end{cases} \tag{4.2}$$

where $l \in [1, k]_{\mathbb{Z}}$, $k \in \mathbb{N}$.

By Lemma 1.24, Eq. (4.1) admits a unique solution described by Eq. (1.19). Consider the approximate model of Eq. (1.19) as noted below

$$\begin{aligned} x(t) = {} & \mathbb{E}_\alpha\big[\lambda(t-a)^\alpha\big]\xi_0 \\ & + \int_a^t (t-s)^{\alpha-1}\mathbb{E}_{\alpha,\alpha}[\lambda(t-s)^\alpha] f(\lceil s\rceil, x(\lceil s\rceil))\mathrm{d}s, \end{aligned} \tag{4.3}$$

where $\lceil s \rceil = \min\{a + kh : s \leq a + kh, k \in \mathbb{N}\}$, $t \in \mathbb{R}_a$. For arbitrary $t > a$, there exists $k \in \mathbb{N}$ ensuring that $t \in (a + kh - h, a + kh] \subseteq \mathbb{R}_a$, then Eq. (4.3) is turned into

$$\begin{aligned}
x(t) &= \mathbb{E}_\alpha\big[\lambda(t-a)^\alpha\big]\xi_0 \\
&\quad + \int_{a+kh-h}^{t} (t-s)^{\alpha-1}\mathbb{E}_{\alpha,\alpha}[\lambda(t-s)^\alpha]\mathrm{d}s\, f(a+kh, x(a+kh)) \\
&\quad + \sum_{l=1}^{k-1}\int_{a+lh-h}^{a+lh} (t-s)^{\alpha-1}\mathbb{E}_{\alpha,\alpha}[\lambda(t-s)^\alpha]\mathrm{d}s\, f(a+lh, x(a+lh)) \\
&= \mathbb{E}_\alpha\big[\lambda(t-a)^\alpha\big]\xi_0 \\
&\quad + \big[\mathbb{E}_\alpha(\lambda(t-a-kh+h)^\alpha) - 1\big]\lambda^{-1} f(a+kh, x(a+kh)) \\
&\quad + \sum_{l=1}^{k-1}\Big\{\mathbb{E}_\alpha\big[\lambda(t-a-lh+h)^\alpha\big] - \mathbb{E}_\alpha\big[\lambda(t-a-lh)^\alpha\big]\Big\} \\
&\quad \times \lambda^{-1} f(a+lh, x(a+lh)),
\end{aligned} \tag{4.4}$$

where $t \in (a + kh - h, a + kh] \subseteq \mathbb{R}_a$, $k \in \mathbb{N}$. In the calculation of Eq. (4.4), Lemma 1.23 is employed. Letting $t \to (a + kh)^-$ and $t \to (a + kh - h)^+$ in Eq. (4.4), it derives

$$\begin{cases}
x(a+kh) &= \mathbb{E}_\alpha\big[\lambda(kh)^\alpha\big]\xi_0 \\
&\quad + \sum_{l=1}^{k} w_{\alpha,h}^{[l,k]}\lambda^{-1} f(a+lh, x(a+lh)), \\
x(a+kh-h) &= \mathbb{E}_\alpha\big[\lambda(kh-h)^\alpha\big]\xi_0 \\
&\quad + \sum_{l=1}^{k-1} w_{\alpha,h}^{[l,k-1]}\lambda^{-1} f(a+lh, x(a+lh)),
\end{cases} \tag{4.5}$$

where $k \in \mathbb{N}$.

By using cancellation law in Eq. (4.5), Nabla Mittag–Leffler Euler difference form of FODEs (4.1) is given by

$$\begin{aligned}
x^{[k]} &= \lambda_{\alpha,h}^{[k]} x^{[k-1]} + \sum_{l=1}^{k} \theta_{\alpha,h}^{[l,k]} f^{[l]}(x^{[l]}) \qquad (\mathrm{MLED})_\nabla \\
&= \lambda_{\alpha,h}^{[k]} x^{[k-1]} + \sum_{l=0}^{k-1} \theta_{\alpha,h}^{[k-l,k]} f^{[k-l]}(x^{[k-l]}), \quad \forall k \in \mathbb{N},
\end{aligned}$$

where $x^{[0]} = \xi_0$, $x^{[k]} = x(a + kh)$, $h > 0$ denotes the step length, $f^{[k]}(x) = f(a + kh, x)$ is an abbreviation for arbitrary $x \in \mathbb{R}$, $\lambda_\alpha^{[k]}$ and $\theta_\alpha^{[l,k]}$ are defined as those in (4.2) for $l \in [1, k]_{\mathbb{Z}}$ and $k \in \mathbb{N}$.

**Property 4.1** *The solutions of* $(\text{MLED})_\nabla$ *denote the numerical solutions of FODEs* (4.1).

**Remark 4.1** The authors in [42–45] achieved Mittag–Leffler Euler schemes for some *(*stochastic*)* FODEs, which are approximately depicted by the first equation of Eq. (4.5). From the viewpoint of the theory of difference equations, difference form $(\text{MLED})_\nabla$ clearly presents the relationships of the state variable $x$ in $k - 1$ and $k$.

**Implicit Exponential Euler Difference**

If $\alpha = 1$ in $(\text{MLED})_\nabla$, then

$$\lambda_{\alpha,h}^{[k]} \equiv e^{\lambda h}, \quad \theta_{\alpha,h}^{[k,k]} \equiv (e^{\lambda h} - 1)\lambda^{-1}, \quad \theta_{\alpha,h}^{[l,k]} \equiv 0$$

for all $l \in [1, k)_{\mathbb{Z}}$, $k \in \mathbb{N}$. So $(\text{MLED})_\nabla$ is turned into

$$\begin{cases} x(t) = e^{\lambda h} x(t - h) + (e^{\lambda h} - 1)\lambda^{-1} f(t, x(t)), & \forall t \in (h\mathbb{N})_a, \\ x(a) = \xi_0 \in \mathbb{R}. \end{cases} \tag{4.6}$$

Obviously, Eq. (4.6) is the exponential Euler difference equation corresponding to semi-linear problem of first order differential equation as below

$$\begin{cases} \dfrac{\mathrm{d}x(t)}{\mathrm{d}t} = \lambda x(t) + f(t, x(t)), & \forall t \in \mathbb{R}_a, \\ x(a) = \xi_0 \in \mathbb{R}. \end{cases}$$

**Remark 4.2** In Eq. (4.6), the memory terms $f^{[k-l]}(x^{[k-l]})$ $(l = 1, 2, \ldots, k - 1)$ in $(\text{MLED})_\nabla$ have become zero. On the basis of exponential Euler difference form such as Eq. (4.6), the authors in the old literature [19–22, 26–32] studied the various dynamics of different forms of discrete-time applied models, e.g., periodicity [28], almost periodicity [26, 27, 29], almost automorphy [31, 32], stability [21, 26, 27, 31, 32], stochastic disturbances [20, 21, 27], etc. Obviously, the proposed model $(\text{MLED})_\nabla$ complements the discrete analogues in literature [19–22, 26–32].

Let $\mathcal{T}_a = \{t_p \in \mathbb{R} : t_0 = a, \lim_{p\to+\infty} t_p = +\infty, t_p \le t_{p+1}, p \in \mathbb{N}_0\}$ for some $a \in \mathbb{R}$. $(h\mathbb{N}_0)_a = \{a + kh : k \in \mathbb{N}_0\}$ is called a $\mathcal{T}_a$-$\mathcal{N}_0$ sequence if it exists $\mathcal{T}_a$ and $\mathcal{N}_0 = \{N_p \in \mathbb{Z} : N_0 = 0, N_p < N_{p+1}, p \in \mathbb{N}_0\}$ satisfying $\frac{t_{p+1}-t_p}{N_{p+1}-N_p} \equiv h$ and $t_p = a + N_p h$, $\forall p \in \mathbb{N}_0$. Throughout this article, suppose that $(h\mathbb{N}_0)_a$ is a $\mathcal{T}_a$-$\mathcal{N}_0$ sequence.

Now let us focus on the following FODEs with piecewise characteristics (i.e., short-term memory)

$$\begin{cases} {}_{t_p}^{C}D_t^{\alpha}x(t) = \lambda x(t) + f(t, x(t)), \quad \forall t \in (t_p, t_{p+1}], \quad p = 0, 1, \ldots, \\ x(a) = \xi_0 \in \mathbb{R}, \end{cases} \tag{4.7}$$

where $x \in C(\mathbb{R}_a, \mathbb{R})$, $\alpha \in (0, 1]$, $\lambda \in \mathbb{R}\backslash\{0\}$ and $f$ is defined in before.

Similar to the argument as that in $(\text{MLED})_{\nabla}$, piecewise difference form of FODEs (4.7) is depicted by

$$\begin{cases} x^{[k]} = \tilde{\lambda}_{\alpha,h}^{[k]} x^{[k-1]} + \sum_{l=0}^{\nu_k - 1} \tilde{\theta}_{\alpha,h}^{[l,k]} f^{[k-l]}(x^{[k-l]}), \quad \forall k \in \mathbb{N}, \\ x^{[0]} = \xi_0, \end{cases} \tag{PMLED$_{\nabla}$}$$

where $x^{[k]} = x(a + kh)$, $\nu_k = k - \mu_k$, $\mu_k = \max\{N_p \in \mathcal{N}_0 : N_p < k,\ p \in \mathbb{N}_0\}$,

$$\begin{cases} \tilde{\lambda}_{\alpha,h}^{[k]} = \dfrac{\mathbb{E}_{\alpha}[\lambda(\nu_k h)^{\alpha}]}{\mathbb{E}_{\alpha}[\lambda(\nu_k h - h)^{\alpha}]}, \\ \tilde{\theta}_{\alpha,h}^{[l,k]} = \dfrac{\tilde{w}_{\alpha,h}^{[l]} - \tilde{\lambda}_{\alpha,h}^{[k]} \tilde{w}_{\alpha,h}^{[l-1]}}{\lambda}, \\ \tilde{w}_{\alpha,h}^{[l]} = \mathbb{E}_{\alpha}[\lambda(lh + h)^{\alpha}] - \mathbb{E}_{\alpha}[\lambda(lh)^{\alpha}], \end{cases} \tag{4.8}$$

where $l \in [0, \nu_k)_{\mathbb{Z}}$, $\tilde{w}_{\alpha,h}^{[-1]} = 0$, $k \in \mathbb{N}$.

**Property 4.2** *The solutions of* $(\text{PMLED})_{\nabla}$ *denote the numerical solutions of FODEs* (4.7).

Let $N^* = \sup_{p \in \mathbb{N}_0}(N_{p+1} - N_p) < +\infty$ and $t_{\diamond} = N^* h$.

**Remark 4.3** In $(\text{PMLED})_{\nabla}$, it exists some indetectable facts as noted below.

(1) $\nu_k \in (0, N_{p+1} - N_p]_{\mathbb{Z}} \subseteq (0, N^*]_{\mathbb{Z}}$ for $k \in (N_p, N_{p+1}]_{\mathbb{Z}}$, $p \in \mathbb{N}_0$.
(2) $\nu_k h \in (0, t_{\diamond}]$ for all $k \in \mathbb{N}$.
(3) $\tilde{\lambda}_{\alpha,h}^{[k]} \in (0, 1]$ for $\lambda \leq 0$ and $\tilde{\lambda}_{\alpha,h}^{[k]} \in [1, +\infty)$ for $\lambda \geq 0$, $k \in \mathbb{N}$.
(4) $\dfrac{\tilde{w}_{\alpha,h}^{[l]}}{\lambda} > 0$ for $\lambda \neq 0$, $l \in \mathbb{N}_0$.
(5) If $\alpha = 1$, then $\tilde{\theta}_{\alpha,h}^{[l,k]} \equiv 0$ for all $l \in [0, \nu_k)_{\mathbb{Z}}$, $k \in \mathbb{N}$.

**Remark 4.4** Clearly, $\nu_k \in [0, N^*]_{\mathbb{Z}}$, $\forall k \in \mathbb{N}$. The boundedness of $\nu_k$ can ensure the boundedness of $\tilde{\lambda}_{\alpha,h}^{[k]}$ and $\sum_{l=0}^{\nu_k - 1} \tilde{\theta}_{\alpha,h}^{[l,k]}$ for all $k \in \mathbb{N}$. Apparently, non-piecewise model $(\text{MLED})_{\nabla}$ has no this trait. This is the most important reason to discuss piecewise Mittag–Leffler Euler difference systems in this literature.

**Remark 4.5** By Fig. 4.1 and the relative definitions as described above, $t_p = a + N_p h$ and $t_{p+1} - t_p = (N_{p+1} - N_p)h$ portray the starting point and the length of each piece (i.e., $(t_p, t_{p+1}]$) of short-term memory in $(\text{PMLED})_\nabla$, respectively, $p \in \mathbb{N}_0$. As a consequence, $(\text{PMLED})_\nabla$ is a fractional-order difference equations with short-term memory and $\sum_{l=0}^{\nu_k-1} \tilde{\theta}_{\alpha,h}^{[l,k]} f^{[k-l]}(x^{[k-l]})$ represents the trait of short-term memory of $(\text{PMLED})_\nabla$, $k \in \mathbb{N}_0$.

**Remark 4.6** In the practical modelling of $(\text{PMLED})_\nabla$, it only predetermines some appropriate sets $\mathcal{T}_a$, $\mathcal{N}_0$ and $(h\mathbb{N}_0)_a$ according to the actual situations. If it needs to model $(\text{PMLED})_\nabla$ in each interval $[t_p, t_{p+1}]$ with step length $h$, $\forall p \in \mathbb{N}_0$, then $(\text{PMLED})_\nabla$ could be modelled by the procedure as noted in Fig. 4.2.

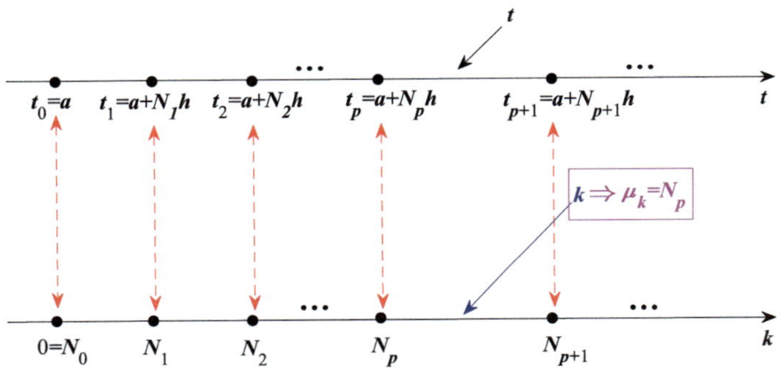


**Fig. 4.1** Graphic illustrations of sequences $\{t_p\}_{p\in\mathbb{N}_0}$ and $\{N_p\}_{p\in\mathbb{N}_0}$

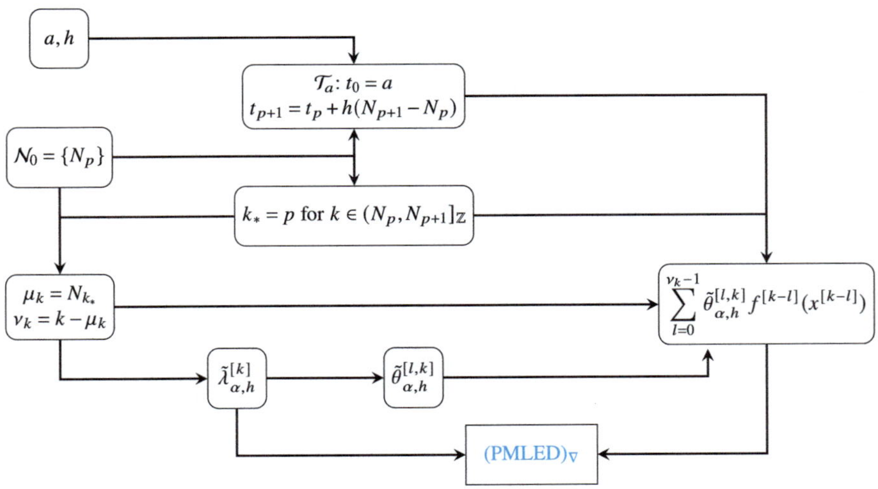


**Fig. 4.2** Modelling procedure of $(\text{PMLED})_\nabla$

**Nabla Piecewise Mittag–Leffler Euler Difference in Matrix Form**

Consider the vector-valued FODEs with piecewise feature as noted below

$$\begin{cases} {}_{t_p}^{C}D_t^{\alpha} X(t) = AX(t) + F(t, X(t)), \quad t \in (t_p, t_{p+1}], \quad p = 0, 1, \dots, \\ X(a) = \xi \in \mathbb{R}^n, \end{cases} \tag{4.9}$$

where $X = (x_1, \dots, x_n)^T \in C(\mathbb{R}_a, \mathbb{R}^n)$, ${}^cD_{t_p}^{\alpha} X = ({}^cD_{t_p}^{\alpha} x_1, \dots, {}^cD_{t_p}^{\alpha} x_n)^T$, $0 < \alpha \le 1$, $A \in \mathbb{R}^{n\times n}$, $F \in C(\mathbb{R}_a \times \mathbb{R}^n, \mathbb{R}^n)$ meets the Lipschitz condition.

Let $(h\mathbb{N}_0)_a$, $\mu_k$ and $\nu_k$ be defined as those in (PMLED)$_\nabla$ for all $k \in \mathbb{N}$. By the same derivation as that in (PMLED)$_\nabla$, it easily gets piecewise Mittag–Leffler Euler implicit difference form of FODEs (4.9) described by

$$\begin{cases} X^{[k]} = \tilde{\Lambda}_{\alpha,h}^{[k]} X^{[k-1]} + \sum_{l=0}^{\nu_k - 1} \tilde{\Theta}_{\alpha,h}^{[l,k]} F^{[k-l]}(X^{[k-l]}), \\ X^{[0]} = \xi \in \mathbb{R}^n, \end{cases} \tag{4.10}$$

where $X^{[k]} = X(a + kh)$,

$$\tilde{\Lambda}_{\alpha,h}^{[k]} = \mathbb{E}_\alpha[A(\nu_k h)^\alpha]\Big\{\mathbb{E}_\alpha[A(\nu_k h - h)^\alpha]\Big\}^{-1}, \quad \tilde{\Theta}_{\alpha,h}^{[l,k]} = \Big[\tilde{W}_{\alpha,h}^{[l]} - \tilde{\Lambda}_{\alpha,h}^{[k]} \tilde{W}_{\alpha,h}^{[l-1]}\Big] A^{-1},$$

$$\tilde{W}_{\alpha,h}^{[-1]} = 0, \quad \tilde{W}_{\alpha,h}^{[l]} = \mathbb{E}_\alpha[A(lh + h)^\alpha] - \mathbb{E}_\alpha[A(lh)^\alpha], \quad \forall l \in [0, \nu_k)_{\mathbb{Z}}, k \in \mathbb{N}.$$

**Property 4.3** *The solutions of Eq.* (4.10) *correspond to the numerical solutions of vector-valued FODEs* (4.9).

**Nabla Mittag–Leffler Euler Difference in Matrix Form**

Let $\nu_k = k$ in Eq. (4.10), then it is converted into

$$\begin{cases} X^{[k]} = \Lambda_{\alpha,h}^{[k]} X^{[k-1]} + \sum_{l=0}^{k-1} \Theta_{\alpha,h}^{[l,k]} F^{[k-l]}(X^{[k-l]}), \\ X^{[0]} = \xi \in \mathbb{R}^n, \end{cases} \tag{4.11}$$

where $X^{[k]} = X(a + kh)$,

$$\Lambda_{\alpha,h}^{[k]} = \mathbb{E}_\alpha[A(kh)^\alpha]\left\{\mathbb{E}_\alpha[A(kh-h)^\alpha]\right\}^{-1}, \quad \Theta_{\alpha,h}^{[l,k]} = \left[W_{\alpha,h}^{[l]} - \Lambda_{\alpha,h}^{[k]} W_{\alpha,h}^{[l-1]}\right]A^{-1},$$

$$W_{\alpha,h}^{[-1]} = 0, \quad W_{\alpha,h}^{[l]} = \mathbb{E}_\alpha[A(lh+h)^\alpha] - \mathbb{E}_\alpha[A(lh)^\alpha], \quad \forall k \in \mathbb{N}.$$

**Property 4.4** *The solutions of Eq.* (4.11) *correspond to the numerical solutions of vector-valued FODEs without piecewise feature as noted below*

$${}_a^C D_t^\alpha X(t) = AX(t) + F(t, X(t)), \quad t \in \mathbb{R}_a, \quad X(a) = \xi \in \mathbb{R}^n.$$

**Remark 4.7** $(\text{MLED})_\nabla$, $(\text{PMLED})_\nabla$, Eqs. (4.10) and (4.11) are implicit single step Euler forms. Based on $(\text{MLED})_\nabla$ or $(\text{PMLED})_\nabla$, the highly accurate forms can be discussed, e.g., two steps Euler form [86], Runge-Kutta form [22], linear multi-step form [23], etc. So the work of this chapter lays a certain theoretical foundation for the future researches of exponential Euler methods of FODEs.

## 4.3 Fractional PECE Algorithm

The both sides of implicit Euler form $(\text{MLED})_\nabla$ include the term $x^{[k]}$, $k \in \mathbb{N}$. In order to avoid solving nonlinear differential equations, the term $x^{[k]}$ in the right side of $(\text{MLED})_\nabla$ should be replaced by a predicted value $x_p^{[k]}$, $k \in \mathbb{N}$. Consequently, $(\text{MLED})_\nabla$ can be rewritten to the corrected formula described by

$$\begin{aligned} x^{[k]} &= \lambda_{\alpha,h}^{[k]} x^{[k-1]} + \sum_{l=1}^{k-1} \theta_{\alpha,h}^{[l,k]} f^{[l]}(x^{[l]}) + \theta_{\alpha,h}^{[k,k]} f^{[k]}(x_p^{[k]}) \\ &= \lambda_{\alpha,h}^{[k]} x^{[k-1]} + \sum_{l=1}^{k-1} \theta_{\alpha,h}^{[k-l,k]} f^{[k-l]}(x^{[k-l]}) + \theta_{\alpha,h}^{[k,k]} f^{[k]}(x_p^{[k]}), \end{aligned} \tag{4.12}$$

where $k \in \mathbb{N}$, $x^{[0]} = \xi_0$.

The predicted value $x_p^{[k]}$ can be computed by the fractional Euler differences (also known as Adams–Bashforth differences) in literature [37, 91–93] as below

$$x_p^{[k]} = x^{[k-1]} + \sum_{l=0}^{k-1} \hat{\theta}_\alpha^{[l]} \left[\lambda x^{[k-l-1]} + f^{[k-l-1]}(x^{[k-l-1]})\right], \tag{4.13}$$

where

$$\hat{\theta}_{\alpha}^{[l]} = \frac{1}{\Gamma(\alpha+1)}\left[\hat{w}_{\alpha}^{[l]} - \hat{w}_{\alpha}^{[l-1]}\right], \quad \hat{w}_{\alpha}^{[-1]} = 0, \quad \hat{w}_{\alpha}^{[l]} = \left[(l+1)^{\alpha} - l^{\alpha}\right]h^{\alpha},$$

where $l \in [0, k-1]_{\mathbb{Z}}$, $k \in \mathbb{N}$. Equation (4.13) is called predicted formula.

Similarly, the corrected and predicted formulas of $(\text{PMLED})_{\nabla}$ are described as

$$x^{[k]} = \tilde{\lambda}_{\alpha,h}^{[k]} x^{[k-1]} + \sum_{l=1}^{\nu_k - 1} \tilde{\theta}_{\alpha,h}^{[l,k]} f^{[k-l]}(x^{[k-l]}) + \tilde{\theta}_{\alpha,h}^{[0,k]} f^{[k]}(x_p^{[k]}), \tag{4.14}$$

where $x^{[0]} = \xi_0$,

$$x_p^{[k]} = x^{[k-1]} + \sum_{l=0}^{\nu_{k-1}} \hat{\theta}_{\alpha}^{[l]} \left[\lambda x^{[k-l-1]} + f^{[k-l-1]}(x^{[k-l-1]})\right], \tag{4.15}$$

in which $k \in \mathbb{N}$.

---

**Algorithm 1** PECE algorithm for $(\text{MLED})_{\nabla}$ or $(\text{PMLED})_{\nabla}$

(1) Predict: predict $x_p^{[k]}$ by Eq. (6.20) or (4.15).
(2) Evaluate: compute $f^{[k]}(x_p^{[k]})$.
(3) Correct: correct $x^{[k]}$ by Eq. 6.17 or (4.14).
(4) Evaluate: compute $f^{[k]}(x^{[k]})$ and start next iteration.

---

**Remark 4.8** In literature [37, 91–93], PECE algorithm had been proposed for Adams-Moulton difference form corresponding to FODEs. That is, the corrected formula is depicted by Adams-Moulton difference and the predicted formula is achieved by Adams–Bashforth difference. Nevertheless, the corrected formula in the proposed PECE algorithm of this chapter is described by Mittag–Leffler Euler difference $(\text{MLED})_{\nabla}$ or $(\text{PMLED})_{\nabla}$. So the work of this chapter complements the results in literature [37, 91–93].

## 4.4 Nabla Mittag–Leffler Euler Difference Equations

In the light of Mittag–Leffler Euler difference forms in before, the Nabla difference equations corresponding to FODEs (4.1) will be proposed in this section.

Define $t_a = \frac{t-a}{h}, \forall t \in (h\mathbb{N}_0)_a$. Let $\mathrm{S}((h\mathbb{N})_a, \mathbb{R})$ denote the totality of all mappings $x : (h\mathbb{N})_a \to \mathbb{R}$. Define an operator ${}_a\mathbf{M}_{\lambda}^{-\alpha} : \mathrm{S}((h\mathbb{N})_a, \mathbb{R}) \to \mathrm{S}((h\mathbb{N})_a, \mathbb{R})$ by

$${}_a\mathbf{M}_{\lambda}^{-\alpha} y(t) = \sum_{l=0}^{t_a - 1} \theta_{\alpha,h}^{[t_a - l, t_a]} y(t - lh), \quad \forall y \in \mathrm{S}((h\mathbb{N})_a, \mathbb{R}), \forall t \in (h\mathbb{N})_a.$$

Obviously, ${}_a\mathbf{M}_\lambda^{-\alpha} : \mathrm{S}((h\mathbb{N})_a, \mathbb{R}) \to \mathrm{S}((h\mathbb{N})_a, \mathbb{R})$ is a linear operator. By Cramer's rule in the theory of linear algebra, ${}_a\mathbf{M}_\lambda^{-\alpha} : \mathrm{S}((h\mathbb{N})_a, \mathbb{R}) \to \mathrm{S}((h\mathbb{N})_a, \mathbb{R})$ has a unique inverse operator denoted by ${}_a\mathbf{M}_\lambda^{\alpha} : \mathrm{S}((h\mathbb{N})_a, \mathbb{R}) \to \mathrm{S}((h\mathbb{N})_a, \mathbb{R})$, which can be calculated by a sequence of iterations as noted below

$$ {}_a\mathbf{M}_\lambda^{\alpha} z(t) = \frac{\lambda}{\mathbb{E}_\alpha(\lambda h^\alpha) - 1}\left[ z(t) - \sum_{j=1}^{t_a - 1} \theta_{\alpha,h}^{[t_a - j, t_a]}\, {}_a\mathbf{M}_\lambda^{\alpha} z(t - jh) \right], \tag{4.16} $$

where $z \in \mathrm{S}((h\mathbb{N})_a, \mathbb{R})$, $t$ is successively equal to $a + h, a + 2h, \ldots$.

By employing operators ${}_a\mathbf{M}_\lambda^{-\alpha}$ and ${}_a\mathbf{M}_\lambda^{\alpha}$, $(\mathrm{MLED})_\nabla$ can be reformulated by

$$ x(t) = \lambda_{\alpha,h}^{[t_a]} x(t - h) + {}_a\mathbf{M}_\lambda^{-\alpha} f(t, x(t)), \quad \forall t \in (h\mathbb{N})_a $$

is equivalent to

$$ {}_a\mathbf{M}_\lambda^{\alpha} \left[ x(t) - \lambda_{\alpha,h}^{[t_a]} x(t - h) \right] = f(t, x(t)), \quad \forall t \in (h\mathbb{N})_a. \tag{4.17} $$

It should be noted that the solution $x$ of Eq. (4.17) is a solution of difference form for FODEs (4.1) by an approach of Mittag–Leffler Euler difference. Motivated by Eq. (4.17), the Caputo Nabla Mittag–Leffler Euler difference operator with parameter $\lambda \in \mathbb{R}\backslash\{0\}$ can be defined as below.

**Definition 4.1** Let $\alpha \in (0, 1]$, $\lambda \in \mathbb{R}\backslash\{0\}$ and ${}_a\mathbf{M}_\lambda^{\alpha}$ be defined as that in iterative formula (4.16). The $\alpha$-order Caputo Nabla Mittag–Leffler Euler difference operator of $x : (h\mathbb{N}_0)_a \to \mathbb{R}$ with step length $h$ and parameter $\lambda$ is defined by

$$ {}_\lambda^C\nabla_{a,h}^{\alpha} x(t) = {}_a\mathbf{M}_\lambda^{\alpha} \left[ x(t) - \lambda_{\alpha,h}^{[t_a]} x(t - h) \right] + \lambda x(t), \quad \forall t \in (h\mathbb{N})_a. \tag{MLEDO$)_\nabla^\lambda$} $$

Noting that $(\mathrm{MLEDO})_\nabla^\lambda$ is a difference operator from initial point $a \in (h\mathbb{N}_0)_a$.

In $(\mathrm{MLEDO})_\nabla^\lambda$, if $t = a + kh \in (h\mathbb{N})_a$, then

$$ {}_\lambda^C\nabla_{a,h}^{\alpha} x(t) = y_k = z_k + \lambda x(a + kh), $$

where

$$ z_k = {}_a\mathbf{M}_\lambda^{\alpha} z(a + kh), \quad z(a + kh) = x(a + kh) - \lambda_{\alpha,h}^{[k]} x(a + kh - h), \quad \forall k \in \mathbb{N}. $$

More clearly, $z_k$ can be calculated by

$$ z_k = \frac{\lambda}{\mathbb{E}_\alpha(\lambda h^\alpha) - 1}\left[ z\,(a + kh) - \sum_{j=1}^{k-1} \theta_{\alpha,h}^{[k-j,k]} z_{k-j} \right], \quad \forall k \in \mathbb{N}. $$

Apparently, the following properties hold.

**Property 4.5** *Let* $\alpha \in (0, 1]$, $\lambda \in \mathbb{R}\backslash\{0\}$ *and* $K \in \mathbb{R}$ *be a constant. Then*

$$ {}_{\lambda}^{C}\nabla_{a,h}^{\alpha} K = 0, \quad \forall t \in (h\mathbb{N})_a. $$

*Further,*

$$ {}_{\lambda}^{C}\nabla_{a,h}^{\alpha} [m_1 x(t) + m_2 y(t)] = m_1\, {}_{\lambda}^{C}\nabla_{a,h}^{\alpha} x(t) + m_2\, {}_{\lambda}^{C}\nabla_{a,h}^{\alpha} y(t), \quad \forall t \in (h\mathbb{N})_a, $$

*where* $m_1, m_2 \in \mathbb{R}$ *and* $x, y : (h\mathbb{N}_0)_a \to \mathbb{R}$.

**Property 4.6** *(MLED)*$_{\nabla}$ *can be rewritten to*

$$ {}_{\lambda}^{C}\nabla_{a,h}^{\alpha} x(t) = \lambda x(t) + f(t, x(t)), \quad \forall t \in (h\mathbb{N})_a, \quad x(a) = \xi_0. \tag{4.18} $$

*The solutions of Eq.* (4.18) *denote the numerical solutions of FODEs* (4.1).

**Property 4.7** *(PMLED)*$_{\nabla}$ *can be rewritten to*

$$ {}_{\lambda}^{C}\nabla_{t_p,h}^{\alpha} x(t) = \lambda x(t) + f(t, x(t)), \quad \forall t \in (t_p, t_{p+1}] \cap (h\mathbb{N}_0)_a, \tag{4.19} $$

*where* $x(a) = \xi_0$. *The solutions of Eq.* (4.19) *denote the numerical solutions of piecewise FODEs* (4.7).

Hereby, Eqs. (4.18) and (4.19) include (MLEDO)$_{\nabla}^{\lambda}$, which are called (piecewise) Mittag–Leffler Euler difference equations (i.e., MLEDEs or PMLEDEs in short), respectively.

**Nabla Exponential and Euler Difference Operators**

If $\alpha = 1$ in (MLEDO)$_{\nabla}^{\lambda}$, then it becomes Nabla exponential Euler difference operator denoted by

$$ {}_{\lambda}\nabla_h x(t) = {}_{\lambda}^{C}\nabla_{a,h}^{1} x(t) = \frac{\lambda[x(t) - x(t-h)]}{e^{\lambda h} - 1}, \quad \forall t \in (h\mathbb{N})_a, \qquad \text{(EEDO)}_{\nabla}^{\lambda} $$

where $x : (h\mathbb{N}_0)_a \to \mathbb{R}$.

Letting $\lambda \to 0$ in (EEDO)$_{\nabla}^{\lambda}$, it approaches the famous Nabla Euler difference operator as below

$$ \nabla_h x(t) = \frac{x(t) - x(t-h)}{h}, \quad \forall t \in (h\mathbb{N})_a. \qquad \text{(EDO)}_{\nabla} $$

## 4.5 Qualitative Analysis

Let $(h\mathbb{N}_0)_a$, $\mu_k$ and $\nu_k$ be defined as those in (PMLED)$_\nabla$ for all $k \in \mathbb{N}$. Consider a Cauchy problem of PMLEDEs as described below

$$\begin{cases} {}_D^C\nabla_{t_p,h}^{\alpha} X(t) = F(t, X(t)), & \forall t \in (t_p, t_{p+1}] \cap (h\mathbb{N})_a, \\ X(a) = \varphi(a) \in \mathbb{R}^n, \end{cases} \tag{4.20}$$

where

$$X = (x_1, \ldots, x_n)^T \in \mathbb{R}^n, \quad \alpha = (\alpha_1, \ldots, \alpha_n)^T \in \mathbb{R}^n, \quad 0 < \alpha_i \le 1,$$

$D = \operatorname{diag}(d_1, \ldots, d_n)$ is $n$-order real diagonal matrix,

$${}_D^C\nabla_{t_p,h}^{\alpha} X = \left( {}_{d_1}^C\nabla_{t_p,h}^{\alpha_1} x_1, \ldots, {}_{d_n}^C\nabla_{t_p,h}^{\alpha_n} x_n \right)^T,$$

${}_{d_i}^C\nabla_{t_p,h}^{\alpha_i}$ is defined as that in Definition 4.1 with parameter $\lambda = d_i$ and $a = t_p$, $p \in \mathbb{N}_0$, $i = 1, 2, \ldots, n$, $F : (h\mathbb{N})_a \times \mathbb{R}^n \to \mathbb{R}^n$.

In view of Property 4.7, Eq. (4.20) can be rewritten to

$$X(t) = d_{\alpha,h}^{[t_a]} X(t-h) + \sum_{l=0}^{\nu_{t_a}-1} \hat{\theta}_{\alpha,h}^{[l,t_a]} F_D\left(a + (t_a - l)h, X(a + (t_a - l)h)\right), \tag{4.21}$$

where $X(a) = \varphi(a)$,

$$t_a = \frac{t-a}{h}, \quad \forall t \in (h\mathbb{N}_0)_a; \quad F_D(t, X) = F(t, X) - DX$$

for arbitrary $(t, X) \in (h\mathbb{N})_a \times \mathbb{R}^n$,

$$d_{\alpha,h}^{[k]} = \operatorname{diag}\left(d_{\alpha_1,h}^{[k]}, \ldots, d_{\alpha_n,h}^{[k]}\right), \quad d_{\alpha_i,h}^{[k]} = \frac{\mathbb{E}_{\alpha_i}[d_i(\nu_k h)^{\alpha_i}]}{\mathbb{E}_{\alpha_i}[d_i(\nu_k h - h)^{\alpha_i}]},$$

$$\hat{\theta}_{\alpha,h}^{[l,k]} = \operatorname{diag}\left(\hat{\theta}_{\alpha_1,h}^{[l,k]}, \ldots, \hat{\theta}_{\alpha_n,h}^{[l,k]}\right),$$

where $\hat{\theta}_{\alpha_i,h}^{[l,k]}$ is equal to $\tilde{\theta}_{\alpha_i,h}^{[l,k]}$ in (4.8) with $\lambda = d_i$ and $\alpha = \alpha_i$, $l \in [0, k)_{\mathbb{Z}}$, $k \in \mathbb{N}$, $i = 1, 2, \ldots, n$.

By employing the variation-of-constants formula in Theorem 3.39 of literature [128], the solution of Eq. (4.21) has the form of

$$X(t) = \prod_{\iota=1}^{t_a} d_{\alpha,h}^{[\iota]} \varphi(a) + \sum_{q=1}^{t_a} \prod_{\iota=q+1}^{t_a} d_{\alpha,h}^{[\iota]} \times \sum_{l=0}^{\nu_q - 1} \hat{\theta}_{\alpha,h}^{[l,q]} F_D(a + (q-l)h, X(a + (q-l)h)), \tag{4.22}$$

where $t \in (h\mathbb{N})_a$.

Since Eq. (4.20) is equal to Eq. (4.21), then it obtains

**Lemma 4.1** *The solutions of Eq. (4.20) possess the form of Eq. (4.22). If $\alpha_i \equiv 1$ and $d_i < 0$ for $i = 1, 2, \ldots, n$, then Eq. (4.22) becomes*

$$X(t) = e^{Dt_a h} \varphi(a) + D^{-1}(e^{Dh} - \mathbb{I}_n) \times \sum_{q=1}^{t_a} e^{D(t_a - q)h} F_D(a + qh, X(a + qh)), \quad \forall t \in (h\mathbb{N})_a, \tag{4.23}$$

*which is the solution of exponential Euler difference equations described by*

$$\begin{cases} {}_D\nabla_h X(t) = F(t, X(t)), & \forall t \in (h\mathbb{N})_a, \\ X(a) = \varphi(a) \in \mathbb{R}^n, \end{cases} \tag{4.24}$$

*where ${}_D\nabla_h X = \left({}_{d_1}\nabla_h x_1, \ldots, {}_{d_n}\nabla_h x_n\right)^T$, ${}_{d_i}\nabla_h$ is defined as that in $(EEDO)_\nabla^\lambda$ with parameter $\lambda = d_i$, $i = 1, 2, \ldots, n$.*

Let

$$\bar{d}_{\alpha,h} = \operatorname{diag}\left(\bar{d}_{\alpha_1,h}, \ldots, \bar{d}_{\alpha_n,h}\right), \quad \bar{d}_{\alpha_i,h} = \sup_{k \in \mathbb{N}} d_{\alpha_i,h}^{[k]},$$

$$\bar{\hat{\theta}}_{\alpha,h} = \sup_{k \in \mathbb{N}} \sum_{l=0}^{\nu_k - 1} \operatorname{diag}(|\hat{\theta}_{\alpha_1,h}^{[l,k]}|, \ldots, |\hat{\theta}_{\alpha_n,h}^{[l,k]}|),$$

where $k \in \mathbb{N}$, $i = 1, 2, \ldots, n$. Set $\mathbf{d}_{\alpha,h} = \|\bar{d}_{\alpha,h}\|$.

**Remark 4.9** The boundedness of $\nu_k$ $(k \in \mathbb{N})$ can ensure the boundedness of $\|\bar{d}_{\alpha,h}\|$ and $\|\bar{\hat{\theta}}_{\alpha,h}\|$. Apparently, non-piecewise model has no this trait. If $d_i < 0$ $(i = 1, 2, \ldots, n)$, then $\mathbf{d}_{\alpha,h} \in (0, 1)$ and $\|\bar{\hat{\theta}}_{\alpha,h}\| \leq 2\|D^{-1}\|$.

Define $\mathbb{X} = \{X : (h\mathbb{N}_0)_a \to \mathbb{R}^n | X(a) = \varphi(a)\}$, which is a Banach space with the norm $\|X\|_\infty = \sup_{t \in (h\mathbb{N}_0)_a} \|X(t)\|$, $\forall X \in \mathbb{X}$.

**Theorem 4.1** *Equation* (4.20) *owns a unique bounded globally* $\mathbf{d}_{\alpha,h}$*-exponentially stable solution* $X \in \mathbb{X}$ *with*

$$\|X\|_\infty < c = \frac{\|\varphi(a)\| + \|\bar{\hat{\theta}}_{\alpha,h}(\mathbb{I}_n - \bar{d}_{\alpha,h})^{-1}\|F_D^0}{1 - \|\bar{\hat{\theta}}_{\alpha,h}(\mathbb{I}_n - \bar{d}_{\alpha,h})^{-1}\|\|L_{F_D}\|}$$

*if the conditions below hold.*

($K_1$) $F_D^0 = \sup_{s\in(h\mathbb{N}_0)_a} \|F_D(s,0))\| < +\infty$ *and it exists a n-order constant matrix* $L_{F_D}$ *such that*

$$\|F_D(t,X) - F_D(t,Y)\| \le \|L_{F_D}\|\|X-Y\|, \quad \forall t \in (h\mathbb{N})_a, X, Y \in \mathbb{R}^n.$$

($K_2$) $N^* < +\infty$, $d_i < 0\,(i = 1, 2, \ldots, n)$ *and*

$$\xi = \|\bar{\hat{\theta}}_{\alpha,h}(\mathbb{I}_n - \bar{d}_{\alpha,h})^{-1}\|\|L_{F_D}\| < 1.$$

***Proof*** By Lemma 4.1, considering the operator $\Psi : \mathbb{X} \to \mathbb{X}$ as noted below

$$\begin{aligned}(\Psi X)(t) = {} & \prod_{\iota=1}^{t_a} d_{\alpha,h}^{[\iota]}\varphi(a) + \sum_{q=1}^{t_a}\prod_{\iota=q+1}^{t_a} d_{\alpha,h}^{[\iota]} \\ & \times \sum_{l=0}^{\nu_q-1} \hat{\theta}_{\alpha,h}^{[l,q]} F_D(a+(q-l)h, X(a+(q-l)h)),\end{aligned} \tag{4.25}$$

where $t \in (h\mathbb{N}_0)_a$. Clearly, the unique fixed point of $\Psi$ in $\mathbb{X}$ is a unique solution of Eq. (4.20).

For any $X, \tilde{X} \in \mathbb{X}$, it derives from Eq. (4.25) and ($K_1$) that

$$\begin{aligned}\|(\Psi X)(t) - (\Psi\tilde{X})(t)\| \le & \sum_{q=1}^{t_a}\left\|\sum_{l=0}^{\nu_q-1}\hat{\theta}_{\alpha,h}^{[l,q]}\prod_{\iota=q+1}^{t_a} d_{\alpha,h}^{[\iota]}\right\| \\ & \times \left\|\sup_{s\in(h\mathbb{N}_0)_a}\left\|F_D(s,X(s)) - F_D(s,\tilde{X}(s)))\right\|\right\| \\ \le & \sum_{q=1}^{t_a}\|\bar{\hat{\theta}}_{\alpha,h}\bar{d}_{\alpha,h}^{t_a-q}\|\|L_{F_D}\|\|X-\tilde{X}\|_\infty \\ \le & \|\bar{\hat{\theta}}_{\alpha,h}(\mathbb{I}_n - \bar{d}_{\alpha,h})^{-1}\|\|L_{F_D}\|\|X-\tilde{X}\|_\infty \\ = & \,\xi\|X-\tilde{X}\|_\infty, \quad t \in (h\mathbb{N}_0)_a.\end{aligned} \tag{4.26}$$

By (4.26), $\|\Psi X - \Psi \tilde{X}\|_\infty \le \xi \|X - \tilde{X}\|_\infty$, $\forall X, \tilde{X} \in \mathbb{X}$. On account of ($K_2$), $\Psi$ is a contractive mapping. Therefore, $\Psi$ possesses a unique fixed point $X^* = \Psi X^* \in \mathbb{X}$ and $X^*$ solves Eq. (4.20).

Let $X \in \mathbb{X}$ be a solution of Eq. (4.20). Then it follows from ($K_1$)–($K_2$) that

$$\begin{aligned}
\|X(t)\| \le& \prod_{\iota=1}^{t_a} \|d_{\alpha,h}^{[\iota]}\| \|\varphi(a)\| \\
&+ \sum_{q=1}^{t_a} \left\| \sum_{l=0}^{\nu_q - 1} \hat{\theta}_{\alpha,h}^{[l,q]} \prod_{\iota=q+1}^{t_a} d_{\alpha,h}^{[\iota]} \right\| \sup_{s \in (h\mathbb{N}_0)_a} \|F_D(s, X(s)) - F_D(s, 0))\| \\
&+ \left\| \sum_{l=0}^{\nu_q - 1} \hat{\theta}_{\alpha,h}^{[l,q]} \prod_{\iota=q+1}^{t_a} d_{\alpha,h}^{[\iota]} \right\| \sup_{s \in (h\mathbb{N}_0)_a} \|F_D(s, 0))\| \\
\le& \|\varphi(a)\| + \|\bar{\hat{\theta}}_{\alpha,h} (\mathbb{I}_n - \bar{d}_{\alpha,h})^{-1}\| \big( \|L_{F_D}\| \|X\|_\infty + F_D^0 \big), \quad t \in (h\mathbb{N}_0)_a,
\end{aligned}$$

which deduces $\|X\|_\infty \le c$.

Let $U(t) = \|X(t) - \tilde{X}(t)\|$, $\forall t \in (h\mathbb{N}_0)_a$. By Lemma 4.1, it yields from ($K_1$) that

$$U(t) \le \mathbf{d}_{\alpha,h}^{t_a} U(a) + \|L_{F_D}\| \left\| \bar{\hat{\theta}}_{\alpha,h} \sum_{q=1}^{t_a} \bar{d}_{\alpha,h}^{t_a - q} \max_{l \in [0, N^* - 1]_{\mathbb{Z}}} U(a + (q - l)h) \right\| \quad (4.27)$$

for $t \in (h\mathbb{N}_0)_a$.

From ($K_2$), it exists $M > 1$ large enough and $0 < \kappa < 1$ small enough such that

$$\frac{1}{M} + \|\bar{\hat{\theta}}_{\alpha,h} (\mathbb{I}_n - \bar{d}_{\alpha,h}^{1-\kappa})^{-1}\| \|L_{F_D}\| \mathbf{d}_{\alpha,h}^{-\kappa(N^* - 1)} < 1. \quad (4.28)$$

Suppose that $U(t) = \|X(t) - \tilde{X}(t)\| \le MU(a) \mathbf{d}_{\alpha,h}^{\kappa t_a}$ holds for any $t \in (h\mathbb{N}_0)_a$. If not, it exists a point $\tilde{t} \in (h\mathbb{N})_a$ such that

$$U(\tilde{t}) > MU(a) \mathbf{d}_{\alpha,h}^{\kappa \tilde{t}_a}; \quad U(s) \le MU(a) \mathbf{d}_{\alpha,h}^{\kappa s^\circ}, \quad \forall s \in [a, \tilde{t}) \cap (h\mathbb{N}_0)_a. \quad (4.29)$$

By utilizing (4.27), (4.29) and (4.28) in turn, it results in

$$\begin{aligned}
U(\tilde{t}) \le& \mathbf{d}_{\alpha,h}^{\tilde{t}_a} U(a) \\
&+ \|L_{F_D}\| \left\| \bar{\hat{\theta}}_{\alpha,h} \sum_{q=1}^{\tilde{t}_a} \bar{d}_{\alpha,h}^{(1-\kappa)(\tilde{t}_a - q)} \bar{d}_{\alpha,h}^{\kappa(\tilde{t}_a - q)} \max_{l \in [0, N^* - 1]_{\mathbb{Z}}} \mathbf{d}_{\alpha,h}^{-\kappa l} MU(a) \mathbf{d}_{\alpha,h}^{\kappa q} \right\| \\
\le& \mathbf{d}_{\alpha,h}^{\tilde{t}_a} U(a) + \|L_{F_D}\| \left\| \bar{\hat{\theta}}_{\alpha,h} \sum_{q=1}^{\tilde{t}_a} \bar{d}_{\alpha,h}^{(1-\kappa)(\tilde{t}_a - q)} \mathbf{d}_{\alpha,h}^{-\kappa(N^* - 1)} MU(a) \mathbf{d}_{\alpha,h}^{\kappa \tilde{t}_a} \right\|
\end{aligned}$$

$$
\begin{aligned}
\leq & MU(a)\mathbf{d}_{\alpha,h}^{\kappa\tilde{t}_a}\left\{\frac{1}{M}\mathbf{d}_{\alpha,h}^{(1-\kappa)\tilde{t}_a} + \|\bar{\hat{\theta}}_{\alpha,h}(\mathbb{I}_n - \bar{d}_{\alpha,h}^{1-\kappa})^{-1}\|\|L_{F_D}\|\mathbf{d}_{\alpha,h}^{-\kappa(N^*-1)}\right\} \\
\leq & MU(a)\bar{\lambda}_{\alpha,h}^{\kappa\tilde{t}_a}\left\{\frac{1}{M} + \|\bar{\hat{\theta}}_{\alpha,h}(\mathbb{I}_n - \bar{d}_{\alpha,h}^{1-\kappa})^{-1}\|\|L_{F_D}\|\mathbf{d}_{\alpha,h}^{-\kappa(N^*-1)}\right\} \\
\leq & MU(a)\mathbf{d}_{\alpha,h}^{\kappa\tilde{t}_a}.
\end{aligned}
$$

This is a contradiction with the first part of (4.29). So

$$
\|X(t) - \tilde{X}(t)\| \leq M\|X(a) - \tilde{X}(a)\|\mathbf{d}_{\alpha,h}^{\kappa t_a}, \quad \forall t \in (h\mathbb{N}_0)_a.
$$

Consequently, Eq. (4.20) is globally $\mathbf{d}_{\alpha,h}$-exponentially stable. The proof is achieved. □

**Remark 4.10** Taking advantage of variation-of-constants formula (4.23), literature [21, 22, 27–32] studied the existence, uniqueness and global exponential stability of various solutions for different forms of applied difference models depicted by exponential Euler difference equations, e.g., periodicity [28], almost periodicity [26, 27, 29, 30], almost automorphy [31, 32], etc. Up to now, almost no literature focuses on exponential Euler difference models of (piecewise) FODEs. Consequently, the current work is novel and complements the works in literature [21, 22, 27–32].

**Remark 4.11** Literature [42–44] mainly discussed the Mittag–Leffler Euler schemes of FODEs from the viewpoint of numerical computations. As a result, the relative qualitative analyses (e.g., the existence of a unique solution, global exponential stability, etc) of Eq. (4.20) are not considered until now. The work of this chapter fills these gaps from the viewpoint of the theory of difference equations.

## 4.6 Numerical Experiments

**Example 4.1** Considering a FODEs described by

$$
{}_{5p}^{C}D_t^{\alpha}x(t) = -7x(t) + |\cos(\sqrt{5}t)|\,[|x(t)+1| - |x(t)-1|] + 6\sin^2(\sqrt{7}t), \tag{4.30}
$$

where $t \in (5p, 5p+5]$, $p \in \mathbb{N}_0$, $x(0) = \xi_0 \in \mathbb{R}$.

Taking $N \in \mathbb{N}$, $h = 1/N$, $\mathcal{T}_0 = \{t_p = 5p : p \in \mathbb{N}_0\}$, $\mathcal{N}_0 = \{N_p = 5pN : p \in \mathbb{N}_0\}$ and $(h\mathbb{N}_0)_0 = \{kh : k \in \mathbb{N}_0\}$. Then $(h\mathbb{N}_0)_0$ is a $\mathcal{T}_0$-$\mathcal{N}_0$ sequence. By utilizing the predictor-corrector method (i.e., PECE method), the numerical solution of FODEs (4.30) is given by

$$x^{[k]} = \tilde{\lambda}_{\alpha,h}^{[k]} x^{[k-1]} + \sum_{l=1}^{\nu_k - 1} \tilde{\theta}_{\alpha,h}^{[l,k]} \Big\{ |\cos(\sqrt{5}(k-l)h)| \Big[ |x^{[k-l]} + 1| - |x^{[k-l]} - 1| \Big]$$
$$+ 6\sin^2(\sqrt{7}(k-l)h) \Big\} + \tilde{\theta}_{\alpha,h}^{[0,k]} \Big\{ |\cos(\sqrt{5}(k)h)| \Big[ |x_p^{[k]} + 1| - |x_p^{[k]} - 1| \Big]$$
$$+ 6\sin^2(\sqrt{7}(k)h) \Big\}, \quad \forall k \in \mathbb{N}, \tag{4.31}$$

where $x^{[0]} = \xi_0$, $x^{[k]} = x(kh)$, $\nu_k = k - \mu_k$, $\mu_k = \max_{p \in \mathbb{N}_0} \{5pN : 5pN < k\}$,

$$\tilde{\lambda}_{\alpha,h}^{[k]} = \frac{\mathbb{E}_\alpha[-7(\nu_k h)^\alpha]}{\mathbb{E}_\alpha[-7(\nu_k h - h)^\alpha]}, \quad \tilde{\theta}_{\alpha,h}^{[l,k]} = -\frac{1}{7} \Big[ \tilde{w}_{\alpha,h}^{[l]} - \tilde{\lambda}_{\alpha,h}^{[k]} \tilde{w}_{\alpha,h}^{[l-1]} \Big],$$

$$\tilde{w}_{\alpha,h}^{[-1]} = 0, \quad \tilde{w}_{\alpha,h}^{[l]} = \mathbb{E}_\alpha[-7(lh+h)^\alpha] - \mathbb{E}_\alpha[-7(lh)^\alpha], \quad \forall l \in [0, \nu_k)_{\mathbb{Z}}, k \in \mathbb{N},$$

and the predicted value

$$x_p^{[k]} = x^{[k-1]} + \sum_{l=0}^{\nu_{k-1}} \hat{\theta}_\alpha^{[l]} \Big\{ -7x^{[k-l-1]} + |\cos(\sqrt{5}(k-l-1)h)|$$
$$\times \Big[ |x^{[k-l-1]} + 1| - |x^{[k-l-1]} - 1| \Big] + 6\sin^2(\sqrt{7}(k-l-1)h) \Big\}, \quad \forall k \in \mathbb{N},$$

in which $\hat{\theta}_\alpha$ is defined as that in Eq. (4.13).

Through the use of fractional Euler differences (i.e., Adams–Bashforth method), it has another type of numerical solution of FODEs (4.30) denoted by

$$x_*^{[k]} = x_*^{[k-1]} + \sum_{l=1}^{\nu_k - 1} \hat{\theta}_{\alpha,h}^{[l]} \Big\{ -7x_*^{[k-l]} + |\cos(\sqrt{5}(k-l)h)| \Big[ |x_*^{[k-l]} + 1| - |x_*^{[k-l]} - 1| \Big]$$
$$+ 6\sin^2(\sqrt{7}(k-l)h) \Big\} + \hat{\theta}_{\alpha,h}^{[0]} \Big\{ -7x_{*p}^{[k]} + |\cos(\sqrt{5}(k)h)| \Big[ |x_{*p}^{[k]} + 1|$$
$$- |x_{*p}^{[k]} - 1| \Big] + 6\sin^2(\sqrt{7}(k)h) \Big\}, \quad \forall k \in \mathbb{N}, \tag{4.32}$$

where $x_*^{[0]} = \xi_0$, $x_*^{[k]} = x_*(kh)$, and the predicted value

$$x_{*p}^{[k]} = x_*^{[k-1]} + \sum_{l=0}^{\nu_{k-1}} \hat{\theta}_\alpha^{[l]} \Big\{ -7x_*^{[k-l-1]} + |\cos(\sqrt{5}(k-l-1)h)|$$
$$\times \Big[ |x_*^{[k-l-1]} + 1| - |x_*^{[k-l-1]} - 1| \Big] + 6\sin^2(\sqrt{7}(k-l-1)h) \Big\}, \quad \forall k \in \mathbb{N}.$$

The solutions of Eqs. (4.31) and (4.32) with different step lengths $h \in \{1, 0.1, 0.01, 0.001\}$ and orders $\alpha \in \{0.3, 0.6, 0.9\}$ are drawn in Figs. 4.3, 4.4, 4.5, 4.6, 4.7 and 4.8 by using the tool of Matlab.

Figures 4.3, 4.4, 4.5, 4.6, 4.7 and 4.8 give some implications as below.

**(a)** *The solutions produced by Adams–Bashforth method (i.e.,* Eq. (4.32)*) have non-periodic oscillations (divergence) for smaller orders $\alpha$ or bigger step lengths $h$, please see* Figs. 4.3, 4.4, 4.5 *and* 4.7. *Inversely, the solutions produced by PECE method (i.e.,* Eq. 4.31*) always keep the almost periodic oscillations (convergence) for arbitrary step lengths $h \in \{0.001, 0.01, 0.1, 1\}$ and orders $\alpha \in \{0.3, 0.6, 0.9\}$. Consequently, Mittag–Leffler Euler differences can depict and retain the dynamics of FODEs better than fractional Euler differences (i.e., Adams–Bashforth differences).*

**(b)** *In view of the right figures in* Figs. 4.4, 4.6 *and* 4.8, *the solutions produced by Adams–Bashforth method are almost consistent with the solutions produced by PECE method only for very small $h$. This is a result of the error caused by the predicted formula* (4.13). *Consequently, improving the precision of the predicted formula such as* Eq. (4.13) *can available modify the execution efficiency of PECE and Adams–Bashforth methods. This work will be considered in the future.*

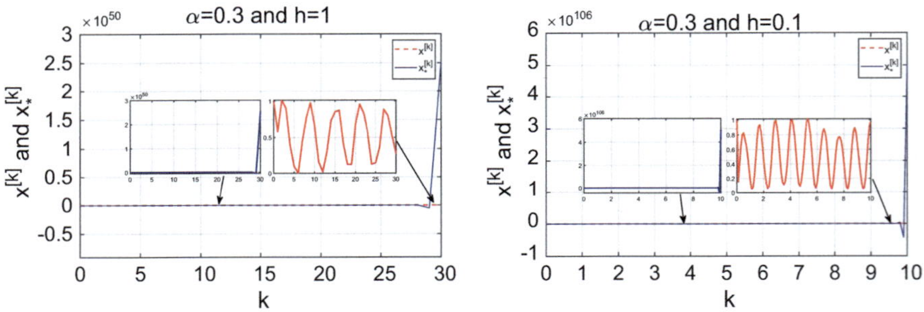


**Fig. 4.3** PECE and Adams–Bashforth methods for Eq. (4.30) with $\alpha = 0.3$ and $h = 1, 0.1$

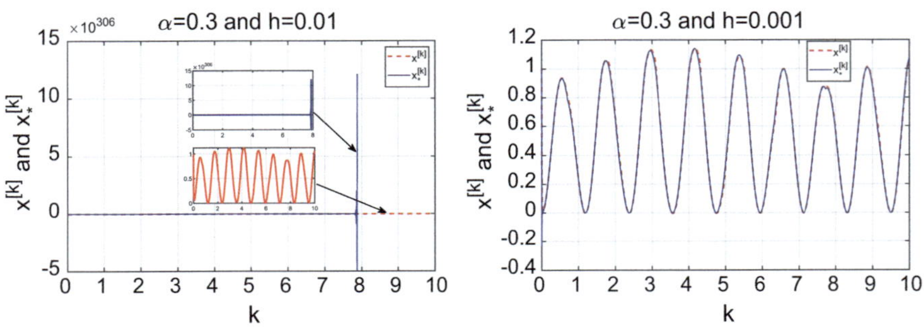


**Fig. 4.4** PECE and Adams–Bashforth methods for Eq. (4.30) with $\alpha = 0.3$ and $h = 0.01, 0.001$

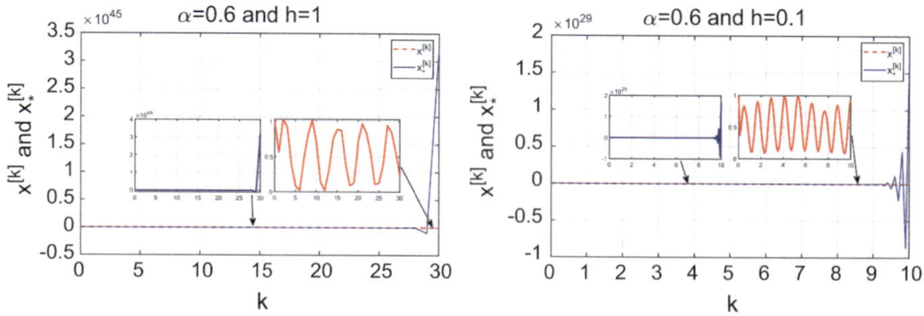


**Fig. 4.5** PECE and Adams–Bashforth methods for Eq. (4.30) with $\alpha = 0.6$ and $h = 1, 0.1$

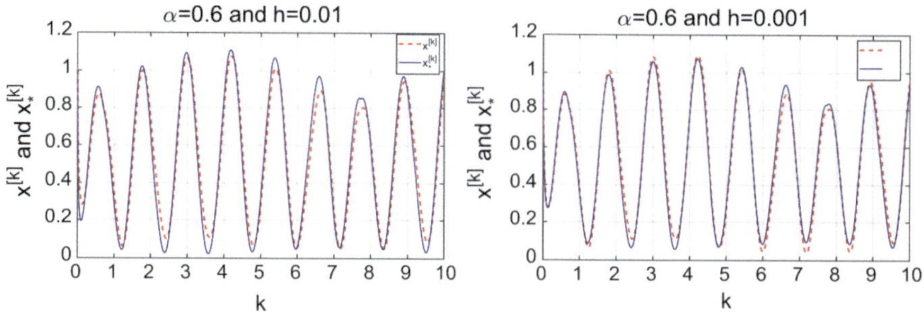


**Fig. 4.6** PECE and Adams–Bashforth methods for Eq. (4.30) with $\alpha = 0.6$ and $h = 0.01, 0.001$

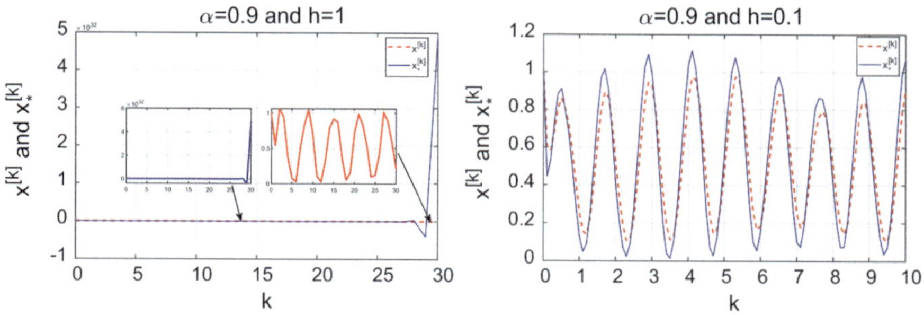


**Fig. 4.7** PECE and Adams–Bashforth methods for Eq. (4.30) with $\alpha = 0.9$ and $h = 1, 0.1$

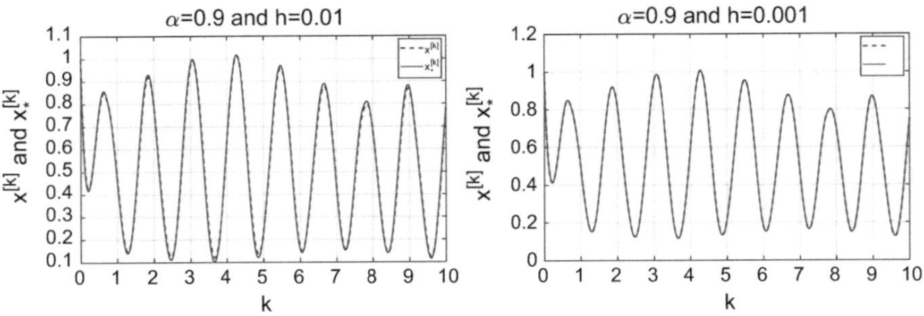


**Fig. 4.8** PECE and Adams–Bashforth methods for Eq. (4.30) with $\alpha = 0.9$ and $h = 0.01, 0.001$

**(c)** *For the same step length h, the bigger the order $\alpha$, the higher the goodness of fit of PECE and Adams–Bashforth methods.*

**Example 4.2** Let $h = 0.01$, $\mathcal{T}_0 = \{t_p = 5p : p \in \mathbb{N}_0\}$, $\mathcal{N}_0 = \{N_p = 5p/h : p \in \mathbb{N}_0\}$ and $(0.01\mathbb{N}_0)_0 = \{0.01k : k \in \mathbb{N}_0\}$. Then $(0.01\mathbb{N}_0)_0$ is a $\mathcal{T}_0$-$\mathcal{N}_0$ sequence. Considering a PMLEDEs as noted below

$$
\begin{aligned}
{}^{C}_{D}\nabla^{\alpha}_{0,0.01} X(t) = & \begin{bmatrix} -0.5 & 0 \\ 0 & -0.5 \end{bmatrix} X(t) \\
& + \frac{1}{8} \begin{bmatrix} 0 & 0.2 \\ 0.1 & 0 \end{bmatrix} \begin{bmatrix} |x_1 + 1| - |x_1 - 1| \\ |x_2 + 1| - |x_2 - 1| \end{bmatrix}, \ \forall t \in (0.01\mathbb{N})_0, \quad (4.33)
\end{aligned}
$$

where $X(0) = \xi \in \mathbb{R}^2$, $X = \begin{bmatrix} x_1 \\ x_2 \end{bmatrix}$,

$$
D = \begin{bmatrix} -0.5 & 0 \\ 0 & -0.5 \end{bmatrix}, \quad {}^{C}_{D}\nabla^{\alpha}_{0,0.01} = \left( {}^{C}_{-0.5}\nabla^{0.9}_{0,0.01}, {}^{C}_{-0.5}\nabla^{0.95}_{0,0.01} \right)^T.
$$

Corresponding to Eq. (4.20) and Theorem 4.1, $N^* = 500$, $\|L_{F_D}\| \leq 0.045$,

$$
\bar{d}_{\alpha,h} = \begin{bmatrix} 0.9972 & 0 \\ 0 & 0.9964 \end{bmatrix}, \quad \bar{\hat{\theta}}_{\alpha,h} = \begin{bmatrix} 0.0056 & 0 \\ 0 & 0.0072 \end{bmatrix}.
$$

So $(K_1)$–$(K_2)$ in Theorem 4.1 hold. Then Eq. (4.33) admits a unique solution, which is globally 0.9972-exponentially stable with respect to exponential function of base 0.9972, see Fig. 4.9.

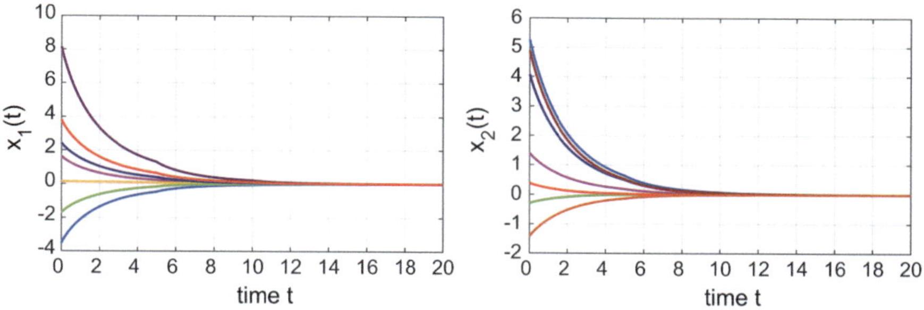


**Fig. 4.9** Global 0.9972-exponential stability of states $x_1$ and $x_2$ of Eq. (4.33) with random initial values

**Remark 4.12** By Examples 4.1 and 4.2, the applied models in literature [123–126] could be directly considered by using the methods in this chapter.

# Chapter 5
# The Theory of CF-EEDs

This chapter presents the exponential Euler difference scheme for Caputo-Fabrizio operator equations with multiple lags. The research demonstrates that the acquired difference falls within the scope of implicit Euler differences. Consequently, the fractional PECE algorithm is proposed to solve it. Finally, the global exponential synchronization of the difference model is discussed from the viewpoint of control theory.

## 5.1 Introduction

In order to escape the singular kernels existing in the fractional derivative, Caputo and Fabrizio [51] proposed the Caputo-Fabrizio operator, which is involved with the convolution of an exponential function and first order derivative. This new concept is receiving more and more attention in recent years, see [52–55, 129–133]. In [55], Mozyrska et al. considered the solution of a Caputo-Fabrizio operator equations (CFOEs) on time scales. Atanacković et al. [129] discussed several properties of Caputo-Fabrizio operators and presented the distributional framework of CFOEs. Paper [130] investigated the Cauchy problems of impulsive CFOEs by employing fixed point theorems. In applications, literature [131, 132] studied the models of COVID-19 with Caputo-Fabrizio operator. Because Caputo-Fabrizio operator is a new concept in recent years, a number of meaningful research topics for CFOEs have not been concerned by the scholars, e.g., almost periodicity, global stability, chaos, bifurcation, etc.

This chapter studies the discrete-time analogue for multi-delay CFOEs

$$\begin{cases} {}^{CF}_{\ 0}D^{\alpha}_{t}X(t) = DX(t) + F\left(t, X(t), X_{\tau}(t)\right) + H_{\alpha}(t), & t \in (0, +\infty), \\ X(s) = \phi(s), \quad s \in [-\tau, 0], \end{cases} \tag{5.1}$$



T. Zhang et al., *Mittag-Leffler Euler Difference Techniques*,
https://doi.org/10.1007/978-981-96-6228-9_5

where $X = (x_1, \ldots, x_n)^T$, $\alpha = \mathrm{diag}(\alpha_1, \ldots, \alpha_n) \in \mathbb{R}^{n\times n}$, $\alpha_i \in [0, 1)$,

$$ {}^{CF}_{0}D^{\alpha}_{t} X = \left({}^{CF}_{0}D^{\alpha_1}_{t} x_1, \ldots, {}^{CF}_{0}D^{\alpha_n}_{t} x_n\right)^T, \quad D = \mathrm{diag}(d_1, \ldots, d_n) \in \mathbb{R}^{n\times n}, $$

$$ X_\tau(t) = (x_1(t - \tau_1(t)), \ldots, x_n(t - \tau_n(t)))^T, $$

$\tau_i(t) \geq 0$ for all $t \geq 0$, $i = 1, 2, \ldots, n$, $\tau = \max_{1\leq i\leq n} \sup_{t\geq 0} \tau_i(t)$ and $F \in C([0, +\infty) \times \mathbb{R}^n \times \mathbb{R}^n, \mathbb{R}^n)$ meets the Lipschitz condition, i.e., it exists a $n$-order nonnegative constant matrix $L_F$ such that

$$ \|F(t, X, Y) - F(t, \tilde{X}, \tilde{Y})\| \leq \|L_F\|(\|X - \tilde{X}\| + \|Y - \tilde{Y}\|), \quad \forall X, Y, \tilde{X}, \tilde{Y} \in \mathbb{R}^n; $$

$$ H_\alpha(t) = -\big[DX(0) + F(0, X(0), X_\tau(0))\big] e^{\frac{\alpha t}{\alpha - 1}}, \quad \forall t \in [0, +\infty). $$

The contents of this chapter are concluded as below. **(1)** Implicit exponential Euler difference for CFOEs (5.1) is established. **(2)** Fractional PECE algorithm is built to solve the proposed implicit Euler difference. **(3)** From the viewpoint of control theory, global exponential synchronization for the discrete-time CFOEs is discussed. **(4)** The proposed difference form can depict and retain the dynamics of CFOEs better than the traditional Euler differences. **(5)** What's more important is to construct a set of theories and methods in studying discrete-time CFOEs and start a new avenue to further discuss discrete-time CFOEs.

This chapter are organized as below. In Sect. 5.2, the implicit exponential Euler difference of CFOEs (5.1) is set up. In Sect. 5.3, the fractional PECE algorithm is proposed. Global exponential synchronization for discrete-time CFOEs is studied in Sect. 5.4. Numerical illustrations are given in Sect. 5.5.

## 5.2 Implicit Exponential Euler Difference

Let

$$ D_\alpha := \mathbb{I}_n - D1_\alpha, \quad 1_\alpha := \mathrm{diag}\{1 - \alpha_1, \ldots, 1 - \alpha_n\}. $$

**Lemma 5.1** *If $M(\alpha) = 1$ in Eq. (1.20), $|D_\alpha|_{\det} \neq 0$ and $\|D_\alpha^{-1} 1_\alpha\| \|L_F\| < 1$ hold, then Eq.* (5.1) *has a unique solution described by*

$$ \begin{aligned} X(t) = {} & e^{P_\alpha t}\phi(0) + D_\alpha^{-1} 1_\alpha \left[F(t, X(t), X_\tau(t)) - e^{P_\alpha t} F(0, \phi(0), \phi_\tau(0))\right] \\ & + \alpha(D_\alpha^2)^{-1} \int_0^t e^{P_\alpha (t-s)} F(s, X(s), X_\tau(s))\, \mathrm{d}s, \quad t \in \mathbb{R}_0, \end{aligned} \tag{5.2} $$

*where $P_\alpha := D_\alpha^{-1} D\alpha$.*

***Proof*** By employing Lemma 1.26, Banach fixed point theorem and the extension method, this lemma is easily obtained and the detailed proof is omitted. The proof is achieved. □

For simplicity, let

$$t_h = \lfloor t/h \rfloor = \max\{k : t/h \geq k, k \in \mathbb{N}_0\}$$

for some $h > 0$,

$$F^*_{X_k} = F\left(kh, X(kh), X_{\tau_h}(kh)\right),$$

where

$$X_{\tau_h}(kh) = (x_1(kh - \lfloor \tau_1(kh)/h \rfloor h), \ldots, x_n(kh - \lfloor \tau_n(kh)/h \rfloor h))^T$$

for $k \in \mathbb{N}_0$.

Referring to Eq. (5.2), consider the following approximate model of CFOEs (5.1) in the form of

$$\begin{aligned} X(t) = {} & e^{P_\alpha t}\phi(0) + D_\alpha^{-1} 1_\alpha \left(F^*_{X_{t_h}} - e^{P_\alpha t} F^*_{X_0}\right) \\ & + \alpha (D_\alpha^2)^{-1} \int_0^t e^{P_\alpha (t-s)} F^*_{X_{s_h}} \mathrm{d}s, \end{aligned} \tag{5.3}$$

where $t \in \mathbb{R}_0$. If $t \in [kh, kh + h)$ for some $k \in \mathbb{N}_0$, then Eq. (5.3) is turned into

$$\begin{aligned} X(t) = {} & e^{P_\alpha t}\phi(0) + D_\alpha^{-1} 1_\alpha \left[F^*_{X_k} - e^{P_\alpha t_h} F^*_{X_0}\right] \\ & + \alpha (D_\alpha^2)^{-1} \left[\int_{kh}^t e^{P_\alpha (t-s)} F^*_{X_k} \mathrm{d}s + \sum_{l=0}^{k-1} \int_{lh}^{lh+h} e^{P_\alpha (t-s)} F^*_{X_l} \mathrm{d}s\right] \\ = {} & e^{P_\alpha t}\phi(0) + D_\alpha^{-1} 1_\alpha \left[F^*_{X_k} - e^{P_\alpha kh} F^*_{X_0}\right] \\ & + \alpha (D_\alpha^2)^{-1} P_\alpha^{-1} \Bigg\{ \left[e^{P_\alpha (t-kh)} - \mathbb{I}_n\right] F^*_{X_k} \\ & + \sum_{l=0}^{k-1} \left[e^{P_\alpha (t-lh)} - e^{P_\alpha (t-lh-h)}\right] F^*_{X_l} \Bigg\}, \end{aligned}$$

which derives the equation below by letting $t \to kh$

$$\begin{aligned} X(kh) = {} & e^{P_\alpha kh}\phi(0) + D_\alpha^{-1} 1_\alpha \left[F^*_{X_k} - e^{P_\alpha kh} F^*_{X_0}\right] \\ & + \alpha (D_\alpha^2)^{-1} P_\alpha^{-1} e^{P_\alpha (kh)} \sum_{l=0}^{k-1} \left[e^{P_\alpha (-lh)} - e^{P_\alpha (-lh-h)}\right] F^*_{X_l}, \quad k \in \mathbb{N}_0. \end{aligned}$$

Similarly, if $t \in [kh+h, kh+2h)$ for the above mentioned $k$, then

$$X(kh+h) = e^{P_\alpha(kh+h)}\phi(0) + D_\alpha^{-1}1_\alpha\left[F^*_{X_{k+1}} - e^{P_\alpha(kh+h)}F^*_{X_0}\right]$$
$$+\alpha(D_\alpha^2)^{-1}P_\alpha^{-1}e^{P_\alpha(kh)}\sum_{l=0}^{k}\left[e^{P_\alpha(-lh+h)} - e^{P_\alpha(-lh)}\right]F^*_{X_l}, \quad k \in \mathbb{N}_0.$$

By the above two equations, it yields the implicit exponential Euler difference form of CFOEs (5.1) given by

$$X_{k+1} = e^{P_\alpha h}X_k + D^{-1}D_\alpha^{-1}\left(e^{P_\alpha h} - \mathbb{I}_n\right)F^*_{X_k} + R_k^X, \quad k \in \mathbb{N}_0 \tag{5.4}$$

with initial value

$$X_s = \phi_s\ (s \in [-\tau_h, 0]_{\mathbb{Z}}, \tau_h = \max_{1\le i\le n}\sup_{t\ge 0}\lfloor \tau_i(t)/h\rfloor),$$

where $X_k = X(kh)$,

$$R_k^X = D_\alpha^{-1}1_\alpha\left(F^*_{X_{k+1}} - e^{P_\alpha h}F^*_{X_k}\right), \quad k \in \mathbb{N}_0.$$

Hereby, $R_k^X$ is called the fractional remainder term of approximate model (5.4).

**Property 5.1** *The solutions of Eq.* (5.4) *denote the numerical solutions of CFOEs* (5.1).

**Explicit Exponential Euler Difference**

If $\alpha \to \mathbb{I}_n$ in Eq. (5.4), then $1_\alpha \to 0$, $D_\alpha \to \mathbb{I}_n$ and $P_\alpha \to D$. So the fractional remainder term $R_k^X$ vanishes and Eq. (5.4) is turned into

$$X_{k+1} = e^{Dh}X_k + D^{-1}(e^{Dh} - \mathbb{I}_n)F^*_{X_k}, \quad k \in \mathbb{N}_0 \tag{5.5}$$

with initial value $X_s = \phi_s\ (s \in [-\tau_h, 0]_{\mathbb{Z}})$.

**Remark 5.1** In the old literature [19–22, 26–32], the authors studied various dynamic behaviours of discrete-time forms of many applied models by employing exponential Euler differences like Eq. (5.5). Obviously, Eq. (5.4) of this chapter is an expansion of discrete analogue in literature [19–22, 26–32].

**Remark 5.2** Equation (5.4) is implicit single step Euler form. Based on Eq. (5.4), the highly accurate forms can be discussed, e.g., Runge-Kutta form [22], linear multistep form [23], etc. So the current work lays a certain theoretical foundation for the future researches of exponential Euler methods of CFOEs.

## 5.3 Fractional PECE Algorithm

Both sides of Eq. (5.4) contain the term $X_{k+1}$, $k \in \mathbb{N}_0$. To avoid solving nonlinear differential equations, the term $X_{k+1}$ on the right-hand side of Eq. (5.4) should be replaced by a predicted value $X^p_{k+1}$, $k \in \mathbb{N}_0$. Therefore, Eq. (5.4) can be rewritten as the corrected formula shown below

$$X_{k+1} = e^{P_\alpha h} X_k + D^{-1} D_\alpha^{-1} \left(e^{P_\alpha h} - \mathbb{I}_n\right) F^*_{X_k} + R_k^{X^p}, \quad k \in \mathbb{N}_0 \tag{5.6}$$

with initial value

$$X_s = \phi_s \; (s \in [-\tau_h, 0]_{\mathbb{Z}}),$$

where

$$R_k^{X^p} = D_\alpha^{-1} 1_\alpha \left[F^*_{X^p_{k+1}} - e^{P_\alpha h} F^*_{X_k}\right], \quad F^*_{X^p_{k+1}} = F\left(kh + h, X^p_{k+1}, X^{\mathcal{T}_h, p}_{k+1}\right),$$

$$X^p_{k+1} = (x^p_{1,k+1}, \ldots, x^p_{n,k+1})^T,$$

$$X^{\mathcal{T}_h, p}_{k+1} = \left(\tilde{x}^p_1(kh + h - \lfloor \tau_1 (kh+h)/h \rfloor h), \ldots, \tilde{x}^p_n(kh + h - \lfloor \tau_n (kh+h)/h \rfloor h)\right)^T,$$

$\tilde{x}^p_i(t) = x^p_{i,k+1}$ if $t = kh + h$, otherwise $\tilde{x}^p_i(t) = x_i(t)$, $t \geq 0$, $k \in \mathbb{N}_0$, $i = 1, 2, \ldots, n$.

Let $\alpha \in [0, 1)$, $M(\alpha) = 1$ and $x \in C^1([a, b], \mathbb{R})$. Then

$${}^{CF}_{\;\;0}D^\alpha_t x(t)\Big|_{t=kh} \approx \frac{1}{1-\alpha} \sum_{l=1}^{k} \int_{lh-h}^{lh} e^{\bar{\alpha}(kh-s)} \mathrm{d}s \times \frac{x(lh+h) - x(lh)}{h}, \quad k \in \mathbb{N}_0.$$

Utilizing the above relation, it easily gains the explicit Euler difference form to Eq. (5.1) described by

$$\begin{aligned} X_{k+1} = {} & X_k - \sum_{l=1}^{k-1} e^{\bar{\alpha}(k-l)h} \left[X_{l+1} - X_l\right] \\ & + \bar{\alpha} h (1-\alpha) \left(e^{\bar{\alpha} h} - \mathbb{I}_n\right)^{-1} \left[D X_k + F^*_{X_k} - (D X_0 + F^*_{X_0}) e^{\bar{\alpha} k h}\right], \end{aligned} \tag{5.7}$$

where $k \in \mathbb{N}_0$, $\bar{\alpha} = \mathrm{diag}\left(\frac{\alpha_1}{\alpha_1 - 1}, \ldots, \frac{\alpha_n}{\alpha_n - 1}\right)$. Consequently, the predicted value $X^p_{k+1}$ can be computed by

$$\begin{aligned} X^p_{k+1} = {} & X_k - \sum_{l=1}^{k-1} e^{\bar{\alpha}(k-l)h} \left[X_{l+1} - X_l\right] \\ & + \bar{\alpha} h (1-\alpha) \left(e^{\bar{\alpha} h} - \mathbb{I}_n\right)^{-1} \left[D X_k + F^*_{X_k} - (D X_0 + F^*_{X_0}) e^{\bar{\alpha} k h}\right], \end{aligned} \tag{5.8}$$

with initial value

$$X_s = \phi_s \; (s \in [-\tau_h, 0]_{\mathbb{Z}}),$$

where $k \in \mathbb{N}_0$. Equation (5.8) is called predicted formula.

## 5.4 Global Dynamics: Exponential Synchronization

Consider Eq. (5.4) as a drive system and taking the following exponential Euler difference model as a response system

**Algorithm 2** PECE algorithm for Eq. (5.4)

(1) Predict: predict $X_{k+1}^{p}$ by Eq. (5.8).
(2) Evaluate: compute $F^*_{X^p_{k+1}}$.
(3) Correct: correct $X_{k+1}$ by Eq. (5.6).
(4) Evaluate: compute $F^*_{X_{k+1}}$ and start next iteration.

$$Y_{k+1} = e^{P_\alpha h} Y_k + D^{-1} D_\alpha^{-1} \left(e^{P_\alpha h} - \mathbb{I}_n\right) F^*_{Y_k} + R_k^Y + U_k, \quad k \in \mathbb{N}_0 \tag{5.9}$$

with initial value $Y_s = \psi_s \; (s \in [-\tau_h, 0]_{\mathbb{Z}})$, where $U_k$ is the control input and $R_k^Y$ is defined as $R_k^X$ with $X = Y$.

Let $E_k = Y_k - X_k$ and $\tilde{F}^*_{E_k} = F^*_{Y_k} - F^*_{X_k}$, $k \in \mathbb{N}_0$. Then the error system of (5.4) and (5.9) is given by

$$E_{k+1} = e^{P_\alpha h} E_k + D^{-1} D_\alpha^{-1} \left(e^{P_\alpha h} - \mathbb{I}_n\right) \tilde{F}^*_{E_k} + \left[R_k^Y - R_k^X\right] + U_k, \quad k \in \mathbb{N}_0. \tag{5.10}$$

**Theorem 5.1** *Let* $|D_\alpha|_{\det} \neq 0$, $\min_{1 \le i \le n} \inf_{t \ge 0} \lfloor \tau_i(t)/h \rfloor \ge 1$ *and the assumption below hold.*

(W) *It exists a n-order diagonal matrix* $\Lambda$ *such that*

$$A = \mathrm{diag}(a_1, \ldots, a_n) = h^{-1} \ln(\Lambda + e^{P_\alpha h}) \text{ and } A_0 = \max_{1 \le i \le n} a_i < 0.$$

*Further,*

$$\max\left\{ \|D_\alpha^{-1} 1_\alpha\|, \frac{\mathbf{P}_D^\alpha}{(1 - e^{A_0})(1 - \|D_\alpha^{-1} 1_\alpha\| \|L_F\|)} \right\} \|L_F\| < 1,$$

*where*

$$\mathbf{P}_D^{\alpha} = 2\|D^{-1}D_{\alpha}^{-1}(e^{P_{\alpha}h} - \mathbb{I}_n)\| + \|D_{\alpha}^{-1}1_{\alpha}\|.$$

*Then the drive-response systems* (5.4) *and* (5.9) *are globally exponentially synchronized under a controlling law designed by*

$$U_k = \Lambda E_k + D_{\alpha}^{-1}1_{\alpha}e^{P_{\alpha}h}\tilde{F}_{E_k}^{*}, \quad k \in \mathbb{N}_0.$$

***Proof*** By controlling law $U_k$ and the method of constant variation as Lemma 2.3 in [27], Eq. (5.10) possesses an equivalent form as described below

$$\begin{aligned} E_k = e^{Akh}E_0 + \sum_{s=0}^{k-1} e^{A(k-s-1)h} \\ \times \left[D^{-1}D_{\alpha}^{-1}\left(e^{P_{\alpha}h} - \mathbb{I}_n\right)\tilde{F}_{E_s}^{*} + D_{\alpha}^{-1}1_{\alpha}\tilde{F}_{E_{s+1}}^{*}\right], \end{aligned} \tag{5.11}$$

where $k \in \mathbb{N}_0$.

Set $\gamma_0 = \max_{s\in[-\tau_h,0]_{\mathbb{Z}}} \|E_s\| = \max_{s\in[-\tau_h,0]_{\mathbb{Z}}} \|\phi_s - \psi_s\|$. By (W), it exists $M > 1$ and $0 < \kappa \ll -A_0$ such that

$$\frac{1}{1 - \|D_{\alpha}^{-1}1_{\alpha}\|\|L_F\|}\left\{\frac{1}{M} + \frac{\mathbf{P}_D^{\alpha}\|L_F\|e^{\kappa(\tau+1)h}}{1 - e^{A_0+\kappa}}\right\} < 1.$$

Claim that $\|E_k\| \le M\gamma_0 e^{-\kappa kh}$ is satisfied for $k \in \mathbb{N}_0$. If not, then it exists $k_0 \in \mathbb{N}$ such that $\|E_{k_0}\| > M\gamma_0 e^{-\kappa k_0 h}$; $\|E_k\| \le M\gamma_0 e^{-\kappa kh}$, $\forall k \in [-\tau, k_0)_{\mathbb{Z}}$. By Eq. (5.11) and $\min_{1\le i\le n} \inf_{t\ge 0}\lfloor \tau_i(t)/h \rfloor \ge 1$, it gains

$$\begin{aligned} \|E_{k_0}\| \le \|e^{Ak_0h}\|\gamma_0 + 2\|D^{-1}D_{\alpha}^{-1}\left(e^{P_{\alpha}h} - \mathbb{I}_n\right)\|\|L_F\| \\ \times \sum_{s=0}^{k_0-1} \|e^{A(k_0-s-1)h}\| \max_{q\in[-\tau,0]_{\mathbb{Z}}} \|E_{s-q}\| \\ + \|D_{\alpha}^{-1}1_{\alpha}\|\|L_F\|\|E_{k_0}\| + \|D_{\alpha}^{-1}1_{\alpha}\|\|L_F\| \\ \times \sum_{s=0}^{k_0-1} \|e^{A(k_0-s-1)h}\| \max_{q\in[-\tau,0]_{\mathbb{Z}}} \|E_{s-q}\|, \end{aligned}$$

which induces

$$\begin{aligned}\|E_{k_0}\| &\le \frac{1}{1-\|D_\alpha^{-1}1_\alpha\|\|L_F\|}\Bigg\{e^{A_0k_0h}\gamma_0 \\ &\quad + \mathbf{P}_D^\alpha\|L_F\|\sum_{s=0}^{k_0-1} e^{(A_0+\kappa)(k_0-s-1)h}e^{\kappa(\tau+1)h}M\gamma_0 e^{-\kappa k_0 h}\Bigg\} \\ &\le M\gamma_0 e^{-\kappa k_0 h}\frac{1}{1-\|D_\alpha^{-1}1_\alpha\|\|L_F\|} \\ &\quad \times\left\{\frac{1}{M}e^{(A_0+\kappa)k_0h} + \frac{\mathbf{P}_D^\alpha\|L_F\|e^{\kappa(\tau+1)h}}{1-e^{A_0+\kappa}}\right\} \\ &\le M\gamma_0 e^{-\kappa k_0 h}.\end{aligned}$$

This is a contradiction. So $\|E_k\| \le M\gamma_0 e^{-\kappa k h}$, $\forall k \in \mathbb{N}_0$. Consequently, the error system (5.10) is globally exponentially stable. The proof is achieved. □

**Remark 5.3** It is easy to obtain global exponential stability of Eq. (5.4) by using the technique in the proof of Theorem 5.1. Similar results can be found in literature [26–32]. It should be noted that it is more complicated to discuss global stability of Eq. (5.4) owing to the existence of the fractional remainder term $R^X$. Because of this point, this chapter solves this problem from the viewpoint of control theory. By designing a controller, Eq. (5.4) can reach global exponential stability.

**Remark 5.4** Mozyrska et al. [55] discussed the exponential stability of CFOEs (5.1) with constant function $F$. For nonconstant or nonlinear function $F$, owing to the uncertainty of the fractional remainder term $R^X$ in Eq. (5.4), it is more difficulty acquiring global *(*exponential*)* stability of CFOEs (5.1) by employing the traditional ways [26–32]. By control techniques such as Theorem 5.1, it also obtains global *(*exponential*)* stability under a suitable controller. So the main aim of this chapter is to provide a meaningful guideline for the future study of global dynamics of continuous-time or discrete-time CFOEs.

## 5.5 Numerical Experiments

**Example 5.1** Considering a CFOEs with time lag described by

$$\begin{aligned}{}^{CF}_{0}D_t^\alpha x(t) &= dx(t) + 0.1|\sin(\sqrt{5}t)| \\ &\quad \times |x(t-0.1)+1| + 5\cos^2(\sqrt{7}t) + H_\alpha(t),\end{aligned} \tag{5.12}$$

where $x(s) = \psi(s)$ $(s \in [-0.1, 0])$; $H_\alpha(t) = -e^{\frac{\alpha t}{\alpha-1}}(dx(0)+5), \forall t \in (0, +\infty)$. Let $h = 0.1$. Similar to Eq. (5.4) implicit Euler form of Eq. (5.12) is depicted by

$$x_{k+1} = e^{\frac{\alpha d}{d_\alpha}h}x_k + d^{-1}d_\alpha^{-1}\left(e^{\frac{\alpha d}{d_\alpha}h}-1\right)F_{x_k}^* + R_k^x, \quad k \in \mathbb{N}_0, \tag{5.13}$$

where $x_k = x(kh)$, $x_s = \psi_s$ $(s = -1, 0)$, $d_\alpha = 1 - d(1-\alpha)$,

$$F^*_{x_k} = 0.1|\sin(\sqrt{5}kh)||x(kh - \lfloor 0.1/h \rfloor h) + 1| + 5\cos^2(\sqrt{7}kh),$$

$$R^x_k = d_\alpha^{-1}(1-\alpha)\left(F^*_{x_{k+1}} - e^{\frac{\alpha d}{d_\alpha}h} F^*_{x_k}\right), \quad k \in \mathbb{N}_0.$$

By utilizing the exponential Euler predictor-corrector method (i.e., PECE) and explicit Euler method as that in Eq. (5.7), the numerical solutions of Eq. (5.12) with different $h$, $\alpha$ and $d$ are drawn in Figs. 5.1, 5.2, 5.3, 5.4 and 5.5. Consider Eq. (5.13) as a drive model, the corresponding response model is given by

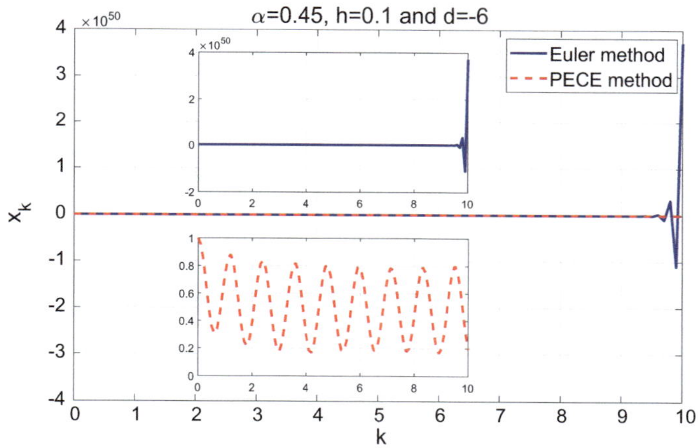


**Fig. 5.1** PECE and Euler methods for Eq. (5.12) with $\alpha = 0.45$, $h = 0.1$

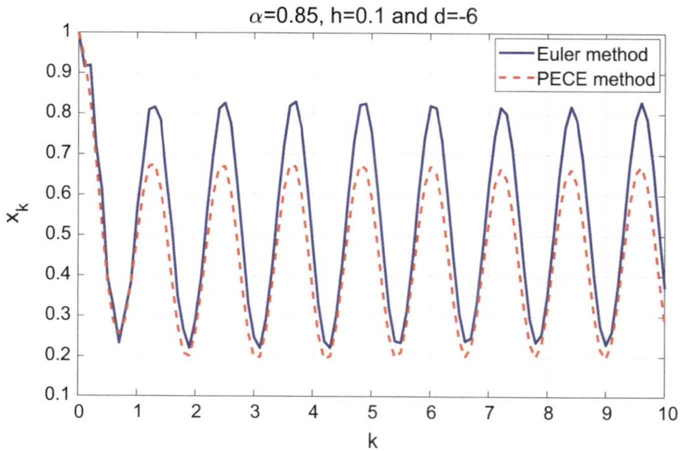


**Fig. 5.2** PECE and Euler methods for Eq. (5.12) with $\alpha = 0.85$, $h = 0.1$

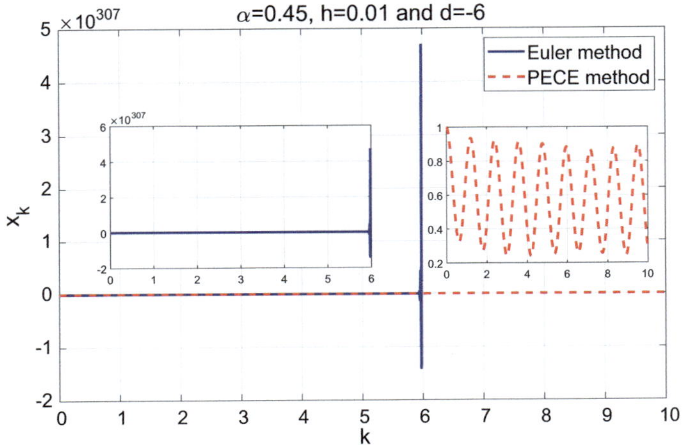


**Fig. 5.3** PECE and Euler methods for Eq. (5.12) with $\alpha = 0.45$, $h = 0.01$

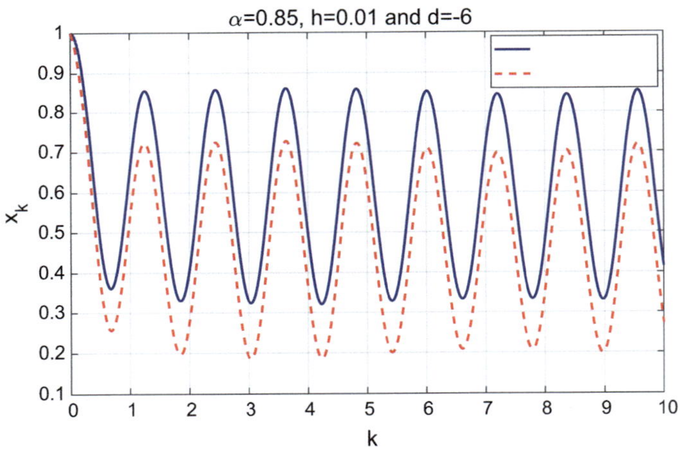


**Fig. 5.4** PECE and Euler methods for Eq. (5.12) with $\alpha = 0.85$, $h = 0.01$

$$y_{k+1} = e^{\frac{\alpha d}{d_\alpha}h} y_k + d^{-1} d_\alpha^{-1} \left( e^{\frac{\alpha d}{d_\alpha}h} - 1 \right) F_{y_k}^* + R_k^y + u_k, \quad k \in \mathbb{N}_0, \tag{5.14}$$

where $y_s = \phi_s$ $(s = -1, 0)$,

$$u_k = d_\alpha^{-1}(1-\alpha) e^{\frac{\alpha d}{d_\alpha}h} \left( F_{y_k}^* - F_{x_k}^* \right)$$

is the control input, $k \in \mathbb{N}_0$. Taking $\alpha = 0.5$ and $d = -6$ in Eq. (5.13), then it easily verifies that all conditions in Theorem 5.1 hold. That is, the drive-response models (5.13) and (5.14) are globally exponentially synchronized, see the third figure of Fig. 5.6.

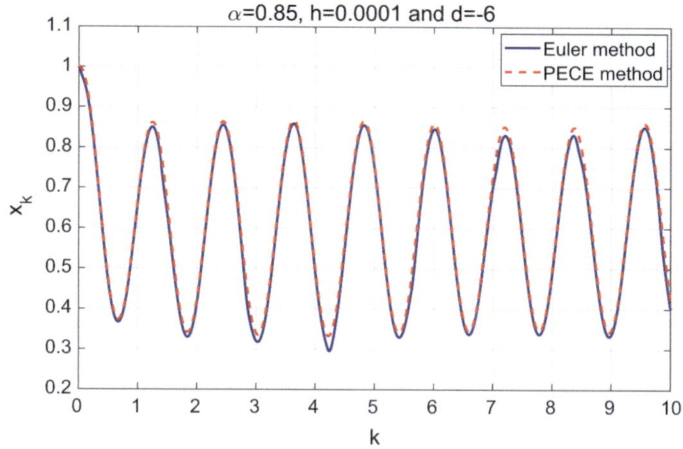


**Fig. 5.5** PECE and Euler methods for Eq. (5.12) with $\alpha = 0.85$, $h = 0.0001$

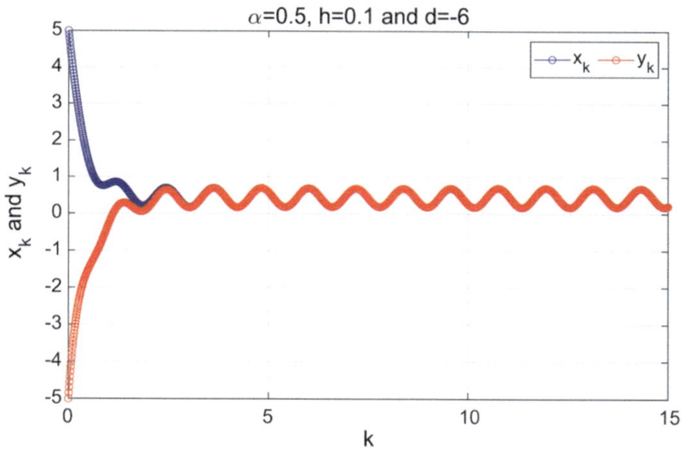


**Fig. 5.6** Global synchronization for Eqs. (5.13) and (5.14)

Figures 5.1, 5.2, 5.3, 5.4, 5.5 and 5.6 indicate that

**(a)** *The solutions produced by Euler method are divergence for smaller $\alpha$, please see* Figs. 5.1, 5.2, 5.3, 5.4 and 5.5. *Inversely, the solutions produced by PECE method always keep the almost periodic oscillations (convergence) even for smaller $\alpha$ and bigger $h$. Consequently, exponential Euler difference* (5.4) *can depict and retain the dynamical characteristics of CFOEs better than Euler difference* (5.7).

**(b)** *From the second figure in* Fig. 5.6, *the solution produced by Euler method is almost consistent with the solution produced by PECE method only for very small $h$. This is a result of the error caused by predicted formula* (5.8). *Consequently, improving the precision of predicted formula* (5.8) *can availably modify the execution efficiency of PECE and Euler methods. This work will be considered in the future.*

# Chapter 6
# A Fuzzy EED-CF-BAMNNs with Time Lags

This chapter presents the basic framework of the exponential Euler difference for a fuzzy BAM neural network with Caputo-Fabrizio operator and multiple lags. The achieved difference scheme falls within the scope of implicit Euler differences. The fractional PECE algorithm is proposed to solve it. Additionally, the chapter investigates the existence of a unique bounded asymptotically almost automorphic sequence solution, global exponential stability, and synchronization of the proposed model.

## 6.1 Introduction

Kosko [134, 135] firstly presented the BAM neural networks. This is a specific type of repetitive neural network that can store bipolar sub pairs. BAM neural networks consist of two layers of neurons, i.e., layer $X$ and layer $Y$. The neurons in a single layer are fully connected to the neurons in the next layer. In real world, BAM neural networks provide strong information processing ability in a few excellent applied domains, like information associative memory, image processing, artificial intelligence, etc. At the other hand, time lags are inevitable in electronic neural systems due to the inevitable restricted exchange rate of the enhancers. The study of BAM neural networks with time lags possesses potential interests in the design and implementation of stable neural circuits in neural networks [136–139]. Therefore, a good deal of BAM neural systems with time lags had been developed in the past. Notably, fractional-order BAM neural networks have played an exceedingly significant position in the evolution of artificial intelligence, since they provide an efficacious way to describe memory and genetic properties [140–145]. Such as, taking advantage of Lyapunov approach, linear state feedback control and partial state feedback control law, Wu et al. [145] stabilized fractional-order BAM neural networks in the form of

 
T. Zhang et al., *Mittag-Leffler Euler Difference Techniques*,
https://doi.org/10.1007/978-981-96-6228-9_6

$$
\begin{cases}
{}_0^C D_t^{\alpha} x_i(t) = -c_i x_i(t) + \sum\limits_{j=1}^{n} a_{ij} f_j(y_j(t)) + I_i, \\
{}_0^C D_t^{\alpha} y_j(t) = -d_j y_j(t) + \sum\limits_{i=1}^{m} b_{ji} g_i(x_i(t)) + P_j, \quad t > 0, \\
x_i(0) = \phi_{i0} \in \mathbb{R}, \ y_j(0) = \psi_{j0} \in \mathbb{R},
\end{cases}
$$

where ${}_0^C D_t^{\alpha}$ denotes the Caputo fractional-order derivatives from initial point 0, $\alpha \in (0, 1]$; $x_i$ and $y_j$ are the neural states; $c_i$ and $d_j$ stand for the self-inhibitions; $a_{ij}$ and $b_{ji}$ describe the synaptic connection strengths; $f_j$ and $g_i$ are the feedback functions; $I_i$ and $P_j$ denote the external inputs, $i = 1, 2, \ldots, m$, $j = 1, 2, \ldots, n$. Ali et al. in paper [140] discussed global Mittag-Leffler stability for a kind of impulsive Caputo fractional-order complex-valued BAM neural networks with time varying lags. In [141], Ali et al. also considered the equilibrium solution and its global asymptotic stability for impulsive Riemann-Liouville fractional-order fuzzy BAM neural networks with time lags.

A mass of processes have a certain regularity, however, they are not completely periodic. As a result, some more complex concepts than periodic processes exist. The initial concept of almost automorphism was introduced by Bochner in the context of differential geometry [56–58] of year 1995, and it was further elaborated by Veech [146–148] and Johnson [149]. Veech in [146] developed the concept of almost automorphism to groups and discussed some properties of these functions including the corresponding Fourier series. Johnson [149] provided an almost automorphic example, which is not almost periodic. Other important literature on the theory of almost automorphism, please see [150–153]. Almost automorphic sequence is a significant evolution of the almost periodic sequence and the concept of almost automorphic sequence was introduced in several decades ago, see the work of Veech [146]. For some relevant works in this field, please see [59, 99, 154–157]. Asymptotically almost automorphic sequence is a new space which strictly contains the sets of periodic, almost periodic and almost automorphic sequences. So, the study of asymptotically almost automorphic sequence has been a significant and interesting topic in the field of difference equations owing to the intensive evolution of the theories of difference equations and the applications in the areas of science and engineering. In recent years, many scholars had devoted themselves to the study of asymptotically periodic motions of fractional-order neural network models [100, 158–162], because fractional-order differential equations in the Caputo or Riemann-Liouville sense do not possess the periodic solutions [163]. For all the authors know, up to now, there are few papers discussing the asymptotically almost automorphic oscillations of fractional-order neural network models with Caputo or Riemann-Liouville derivative. Especially, almost no literature concentrates on the investigation of asymptotically almost automorphic sequences for discrete-time neural networks with Caputo-Fabrizio operators.

Because Caputo-Fabrizio operator is a new concept in the last few years, a number of meaningful research topics for the Caputo-Fabrizio operator equations have not been concerned by the scholars, e.g., exponential stability, chaos, bifurcation, etc. Besides, numerous research findings have made clear that the exponential Euler difference models can depict and retain the dynamics of continuous-time differential equations better than the traditional Euler differences, please see [19–22, 26–32]. On the basis of these reasons, the authors believe that the exponential Euler difference scheme for Caputo-Fabrizio BAM neural networks (CF-BAMNNs) are worthy of discussions in depth. Accordingly, by employing the variation-of-constants formula in the Caputo-Fabrizio sense, the exponential Euler difference scheme for some fuzzy CF-BAMNNs will be achieved in this chapter and some important dynamical behaviours will be discussed.

This chapter is organized as follows. In Sect. 6.2, the main discussed model are described. In Sects 6.3 and 6.4, the exponential Euler difference scheme of a fuzzy CF-BAMNNs and the corresponding fractional PECE algorithm are proposed. In Sects. 6.5–6.7, the existence of a unique bounded asymptotically almost automorphic sequence solution, global exponential stability and synchronization of discrete-time model are investigated. Some numerical experiments are employed to visually expound the current research findings in Sect. 6.8.

## 6.2 Model Description of CF-BAMNNs

Let us focus on a fuzzy CF-BAMNN with time lags

$$
\begin{cases}
{}^{CF}_{0}D_t^{\alpha_i} x_i(t) = -c_i x_i(t) + \sum\limits_{j=1}^{n} a_{ij}(t) f_j(y_j(t-\xi_{ij})) \\
\qquad + \bigwedge\limits_{j=1}^{n} \gamma_{ij}(t) f_j(y_j(t-\xi_{ij})) + \bigvee\limits_{j=1}^{n} \eta_{ij}(t) f_j(y_j(t-\xi_{ij})) \\
\qquad + \bigwedge\limits_{j=1}^{n} T_{ij}(t)\rho_j + \bigvee\limits_{j=1}^{n} Q_{ij}(t)\rho_j + I_i(t) + H_{\alpha_i}(t), \\
{}^{CF}_{0}D_t^{\beta_j} y_j(t) = -d_j y_j(t) + \sum\limits_{i=1}^{m} b_{ji}(t) g_i(x_i(t-\sigma_{ji})) \\
\qquad + \bigwedge\limits_{i=1}^{m} \varepsilon_{ji}(t) g_i(x_i(t-\sigma_{ji})) + \bigvee\limits_{i=1}^{m} \vartheta_{ji}(t) g_i(x_i(t-\sigma_{ji})) \\
\qquad + \bigwedge\limits_{i=1}^{m} W_{ji}(t)\varrho_i + \bigvee\limits_{i=1}^{m} R_{ji}(t)\varrho_i + P_j(t) + H_{\beta_j}(t),
\end{cases}
\tag{6.1}
$$

where $t \in \mathbb{R}_0$,

$$
x_i(s) = \phi_i(s), \quad y_j(s) = \psi_j(s), \quad \forall s \in [-\xi, 0], \xi = \max_{1\le i\le m, 1\le j\le n} \{\xi_{ij}, \sigma_{ji}\};
$$

${}_{0}^{CF}D_{t}^{\alpha_i}$ and ${}_{0}^{CF}D_{t}^{\beta_j}$ denote the Caputo-Fabrizio operators, $\alpha_i, \beta_j \in [0, 1)$; $\xi_{ij} \geq 0$ and $\sigma_{ji} \geq 0$ denote the transmission delays; $\bigwedge$ and $\bigvee$ denote the fuzzy AND and fuzzy OR operations, respectively; $\gamma_{ij}, \varepsilon_{ji}, \eta_{ij}, \vartheta_{ji}, T_{ij}, W_{ji}, Q_{ij}$ and $R_{ji}$ are the elements of fuzzy feedback MIN template and fuzzy feedback MAX template, respectively;

$$H_{\alpha_i}(t) = \big[c_i\phi_i(0) - \sum_{j=1}^{n} a_{ij}(0)f_j(\psi_j(-\xi_{ij})) - \bigwedge_{j=1}^{n} \gamma_{ij}(0)f_j(\psi_j(-\xi_{ij}))$$

$$- \bigvee_{j=1}^{n} \eta_{ij}(0)f_j(\psi_j(-\xi_{ij})) - \bigwedge_{j=1}^{n} T_{ij}(0)\rho_j - \bigvee_{j=1}^{n} Q_{ij}(0)\rho_j - I_i(0)\big]e^{\bar{\alpha}_i t},$$

$$H_{\beta_j}(t) = \big[d_j\psi_j(0) - \sum_{i=1}^{m} b_{ji}(0)g_i(\phi_i(-\sigma_{ji})) - \bigwedge_{i=1}^{m} \varepsilon_{ji}(0)g_i(\phi_i(-\sigma_{ji}))$$

$$- \bigvee_{i=1}^{m} \vartheta_{ji}(0)g_i(\phi_i(-\sigma_{ji})) - \bigwedge_{i=1}^{m} W_{ji}(0)\varrho_i - \bigvee_{i=1}^{m} R_{ji}(0)\varrho_i - P_j(0)\big]e^{\bar{\beta}_j t},$$

$t \in \mathbb{R}_0$, $\bar{\alpha}_i = \frac{\alpha_i}{\alpha_i - 1}$, $\bar{\beta}_j = \frac{\beta_j}{\beta_j - 1}$, $i = 1, 2, \ldots, m$, $j = 1, 2, \ldots, n$. The meanings of other parameters are defined as those in before.

If $\alpha_i \to 1^-$ and $\beta_j \to 1^-$, the remainder terms $H_{\alpha_i}$ and $H_{\beta_j}$ become zero and Eq. (6.1) is converted to

$$\begin{cases} \dfrac{\mathrm{d}x_i(t)}{\mathrm{d}t} = -c_i x_i(t) + \sum\limits_{j=1}^{n} a_{ij}(t)f_j(y_j(t-\xi_{ij})) + \bigwedge\limits_{j=1}^{n} \gamma_{ij}(t)f_j(y_j(t-\xi_{ij})) \\ \qquad + \bigvee\limits_{j=1}^{n} \eta_{ij}(t)f_j(y_j(t-\xi_{ij})) + \bigwedge\limits_{j=1}^{n} T_{ij}(t)\rho_j + \bigvee\limits_{j=1}^{n} Q_{ij}(t)\rho_j + I_i(t), \\ \dfrac{\mathrm{d}y_j(t)}{\mathrm{d}t} = -d_j y_j(t) + \sum\limits_{i=1}^{m} b_{ji}(t)g_i(x_i(t-\sigma_{ji})) + \bigwedge\limits_{i=1}^{m} \varepsilon_{ji}(t)g_i(x_i(t-\sigma_{ji})) \\ \qquad + \bigvee\limits_{i=1}^{m} \vartheta_{ji}(t)g_i(x_i(t-\sigma_{ji})) + \bigwedge\limits_{i=1}^{m} W_{ji}(t)\varrho_i + \bigvee\limits_{i=1}^{m} R_{ji}(t)\varrho_i + P_j(t), \end{cases} \tag{6.2}$$

where $t \in \mathbb{R}_0$, $i = 1, 2, \ldots, m$, $j = 1, 2, \ldots, n$.

The following assumptions are necessary for our subsequent discussions.

($A_1$) $a_{ij}, \gamma_{ij}, \eta_{ij}, T_{ij}, Q_{ij}, b_{ji}, \varepsilon_{ji}, \vartheta_{ji}, W_{ji}$ and $R_{ji}$ are continuous functions, and it exists positive numbers $a_{ij}^*, \gamma_{ij}^*, \eta_{ij}^*, T_{ij}^*, Q_{ij}^*, b_{ji}^*, \varepsilon_{ji}^*, \vartheta_{ji}^*, W_{ji}^*$ and $R_{ji}^*$ such that

$$|a_{ij}(t)| \leq a_{ij}^*, \quad |\gamma_{ij}(t)| \leq \gamma_{ij}^*, \quad |\eta_{ij}(t)| \leq \eta_{ij}^*,$$

$$|T_{ij}(t)| \leq T_{ij}^*, \quad |Q_{ij}(t)| \leq Q_{ij}^*, \quad |b_{ji}(t)| \leq b_{ji}^*,$$

$$|\varepsilon_{ji}(t)| \leq \varepsilon_{ji}^*, \quad |\vartheta_{ji}(t)| \leq \vartheta_{ji}^*, \quad |W_{ji}(t)| \leq W_{ji}^*, \quad |R_{ji}(t)| \leq R_{ji}^*,$$

where $t \in \mathbb{R}_0$, $i = 1, 2, \ldots, m$, $j = 1, 2, \ldots, n$.

($A_2$) $f_j$ and $g_i$ are continuous functions, and it exists positive numbers $L_j^f$ and $L_i^g$ such that

$$|f_j(x) - f_j(y)| \leq L_j^f |x - y|, \quad |g_i(x) - g_i(y)| \leq L_i^g |x - y|,$$

where $x, y \in \mathbb{R}$, $i = 1, 2, \ldots, m$, $j = 1, 2, \ldots, n$.

For simplify, let

$$K_{\alpha_i} = 1 + c_i(1 - \alpha_i), \quad K_{\beta_j} = 1 + d_j(1 - \beta_j), \quad p_{\alpha_i} = \frac{-c_i \alpha_i}{K_{\alpha_i}},$$

$$p_{\beta_j} = \frac{-d_j \beta_j}{K_{\beta_j}}, \quad \varpi_i^f = \sum_{j=1}^{n} (a_{ij}^* + \gamma_{ij}^* + \eta_{ij}^*) L_j^f, \quad \varpi_j^g = \sum_{j=1}^{n} (b_{ji}^* + \varepsilon_{ji}^* + \vartheta_{ji}^*) L_i^g,$$

where $i = 1, 2, \ldots, m$, $j = 1, 2, \ldots, n$.

In accordance with Lemma 1.26, the concept of weak solution to CF-BAMNNs (6.1) is introduced here.

**Definition 6.1** The function $z = (x_1, \ldots, x_m, y_1, \ldots, y_n)^T \in C(\mathbb{R}_0, \mathbb{R}^{m+n})$ is said to be a weak solution of CF-BAMNNs (6.1) if it satisfies

$$\begin{aligned}
x_i(t) = {} & e^{p_{\alpha_i} t} \phi_i(0) + \frac{1 - \alpha_i}{K_{\alpha_i}} \Bigg\{ \Bigg[ \sum_{j=1}^{n} a_{ij}(t) f_j(y_j(t - \xi_{ij})) + \bigwedge_{j=1}^{n} \gamma_{ij}(t) f_j(y_j(t - \xi_{ij})) \\
& + \bigvee_{j=1}^{n} \eta_{ij}(t) f_j(y_j(t - \xi_{ij})) + \bigwedge_{j=1}^{n} T_{ij}(t) \rho_j + \bigvee_{j=1}^{n} Q_{ij}(t) \rho_j + I_i(t) \Bigg] \\
& - e^{p_{\alpha_i} t} \Bigg[ \sum_{j=1}^{n} a_{ij}(0) f_j(y_j(-\xi_{ij})) + \bigwedge_{j=1}^{n} \gamma_{ij}(0) f_j(y_j(-\xi_{ij})) \\
& + \bigvee_{j=1}^{n} \eta_{ij}(0) f_j(y_j(-\xi_{ij})) + \bigwedge_{j=1}^{n} T_{ij}(0) \rho_j + \bigvee_{j=1}^{n} Q_{ij}(0) \rho_j + I_i(0) \Bigg] \Bigg\} \\
& + \frac{\alpha_i}{K_{\alpha_i}^2} \int_0^t e^{p_{\alpha_i}(t-s)} \Bigg[ \sum_{j=1}^{n} a_{ij}(s) f_j(y_j(s - \xi_{ij})) + \bigwedge_{j=1}^{n} \gamma_{ij}(s) f_j(y_j(s - \xi_{ij}))
\end{aligned}$$

$$+\bigvee_{j=1}^{n}\eta_{ij}(s)f_j(y_j(s-\xi_{ij}))+\bigwedge_{j=1}^{n}T_{ij}(s)\rho_j$$
$$+\bigvee_{j=1}^{n}Q_{ij}(s)\rho_j+I_i(s)\Bigg]\mathrm{d}s \tag{6.3}$$

and

$$y_j(t)=e^{p_{\beta_j}t}\psi_j(0)+\frac{1-\beta_j}{K_{\beta_j}}\Bigg\{\Bigg[\sum_{i=1}^{m}b_{ji}(t)g_i(x_i(t-\sigma_{ji}))+\bigwedge_{i=1}^{m}\varepsilon_{ji}(t)g_i(x_i(t-\sigma_{ji}))$$
$$+\bigvee_{i=1}^{m}\vartheta_{ji}(t)g_i(x_i(t-\sigma_{ji}))+\bigwedge_{i=1}^{m}W_{ji}(t)\varrho_i+\bigvee_{i=1}^{m}R_{ji}(t)\varrho_i+P_j(t)\Bigg]$$
$$-e^{p_{\beta_j}t}\Bigg[\sum_{i=1}^{m}b_{ji}(0)g_i(x_i(-\sigma_{ji}))+\bigwedge_{i=1}^{m}\varepsilon_{ji}(0)g_i(x_i(-\sigma_{ji}))$$
$$+\bigvee_{i=1}^{m}\vartheta_{ji}(0)g_i(x_i(-\sigma_{ji}))+\bigwedge_{i=1}^{m}W_{ji}(0)\varrho_i+\bigvee_{i=1}^{m}R_{ji}(0)\varrho_i+P_j(0)\Bigg]\Bigg\}$$
$$+\frac{\beta_j}{K_{\beta_j}^2}\int_0^t e^{p_{\beta_j}(t-s)}\Bigg[\sum_{i=1}^{m}b_{ji}(s)g_i(x_i(s-\sigma_{ji}))+\bigwedge_{i=1}^{m}\varepsilon_{ji}(s)g_i(x_i(s-\sigma_{ji}))$$
$$+\bigvee_{i=1}^{m}\vartheta_{ji}(s)g_i(x_i(s-\sigma_{ji}))+\bigwedge_{i=1}^{m}W_{ji}(s)\varrho_i$$
$$+\bigvee_{i=1}^{m}R_{ji}(s)\varrho_i+P_j(s)\Bigg]\mathrm{d}s, \tag{6.4}$$

where $t\in\mathbb{R}_0$, $i=1,2,\ldots,m$, $j=1,2,\ldots,n$.

**Remark 6.1** By observing Lemma 1.26, the differential of $x$ is dependent of $f$. If the right side of Eq. (6.1) only satisfies the continuity, the differential of $(x,y)^T$ is difficult to be determined owing to the uncertain differentials of the right sides of Eqs. (6.3) and (6.4). Consequently, the concept of weak solution is introduced here. If $(x,y)^T$ in Eqs. (6.3) and (6.4) is differential, then it is a classical solution (i.e., differential solution) for CF-BAMNNs (6.1).

**Lemma 6.1** ([107]) *Let* $z=(x_1,\ldots,x_m,y_1,\ldots,y_n)^T$ *and* $\tilde{z}=(\tilde{x}_1,\ldots,\tilde{x}_m,\tilde{y}_1,\ldots,\tilde{y}_n)^T$ *be two states of CF-BAMNNs* (6.1)*, then*

$$\left|\bigwedge_{j=1}^{n}\gamma_{ij}(t)f_j(y_j)-\bigwedge_{j=1}^{n}\gamma_{ij}(t)f_j(\tilde{y}_j)\right|\le\sum_{j=1}^{n}|\gamma_{ij}^{*}||f_j(y_j)-f_j(\tilde{y}_j)|,$$

$$\left|\bigvee_{j=1}^{n}\eta_{ij}(t)f_j(y_j)-\bigvee_{j=1}^{n}\eta_{ij}(t)f_j(\tilde{y}_j)\right|\le\sum_{j=1}^{n}|\eta_{ij}^{*}||f_j(y_j)-f_j(\tilde{y}_j)|,$$

$$\left|\bigwedge_{i=1}^{m}\varepsilon_{ji}(t)g_i(x_i)-\bigwedge_{i=1}^{m}\varepsilon_{ji}(t)g_i(\tilde{x}_i)\right|\le\sum_{i=1}^{m}|\varepsilon_{ji}^{*}||g_i(x_i)-g_i(\tilde{x}_i)|,$$

$$\left|\bigvee_{i=1}^{m}\vartheta_{ji}(t)g_i(x_i)-\bigvee_{i=1}^{m}\vartheta_{ji}(t)g_i(\tilde{x}_i)\right|\le\sum_{i=1}^{m}|\vartheta_{ji}^{*}||g_i(x_i)-g_i(\tilde{x}_i)|,$$

*where* $t\in\mathbb{R}_0,\ i=1,2,\ldots,m,\ j=1,2,\ldots,n.$

**Lemma 6.2** *Let* $(A_1)$ *and* $(A_2)$*,* $M(\alpha)=1$ *in Eq. (1.20),* $\min_{1\le i\le m,1\le j\le n}\{a_i,d_j\}>0$ *and the inequality below hold.*

$$\max_{1\le i\le m,1\le j\le n}\left\{\frac{1-\alpha_i}{K_{\alpha_i}}\varpi_i^f,\ \frac{1-\beta_j}{K_{\beta_j}}\varpi_j^g\right\}<1.$$

*Then CF-BAMNNs* (6.1) *has a unique weak solution* $z=(x_1,\ldots,x_m,y_1,\ldots,y_n)^T$*, which is described by Eqs.* (6.3) *and* (6.4).

***Proof*** According to the assumption, it exists a positive constant $h_\diamond$ small enough causing

$$\varsigma=\max_{1\le i\le m,1\le j\le n}\left\{K_{\alpha_i}^{-1}\left[(1-\alpha_i)+\frac{1-e^{p_{\alpha_i}h_\diamond}}{c_i}\right]\varpi_i^f,\right.$$

$$\left.K_{\beta_j}^{-1}\left[(1-\beta_j)+\frac{1-e^{p_{\beta_j}h_\diamond}}{d_j}\right]\varpi_j^g\right\}<1.$$

Define

$$\mathbb{B}_{h_\diamond}=\Big\{\varphi=(\hat{\varphi}_1,\ldots,\hat{\varphi}_m,\check{\varphi}_1,\ldots,\check{\varphi}_n)^T\in C([0,h_\diamond],\mathbb{R}^{m+n}):$$

$$\hat{\varphi}_i(s)=\phi_i(s),\check{\varphi}_j(s)=\psi_j(s),s\in[-\xi,0],i=1,2,\ldots,m,\ j=1,2,\ldots,n\Big\}$$

and

$$\mathbf{d}_{h_\diamond}(\varphi,\omega)=\max_{1\le i\le m,1\le j\le n}\sup_{t\in[-\xi,h_\diamond]}\Big\{|\hat{\varphi}_i(t)-\hat{\omega}_i(t)|,|\check{\varphi}_j(t)-\check{\omega}_j(t)|\Big\}$$

for all $\varphi=(\hat{\varphi}_1,\ldots,\hat{\varphi}_m,\check{\varphi}_1,\ldots,\check{\varphi}_n)^T$ and $\omega=(\hat{\omega}_1,\ldots,\hat{\omega}_m,\check{\omega}_1,\ldots,\check{\omega}_n)^T$ in $\mathbb{B}_{h_\diamond}$. Then $(\mathbb{B}_{h_\diamond},\mathbf{d}_{h_\diamond})$ is a complete metric space.

Based on Eqs. (6.3) and (6.4), define operator

$$\mathbf{L}:\varphi \to (x_1^{\varphi},\ldots,x_m^{\varphi},y_1^{\varphi},\ldots,y_n^{\varphi})^T,\quad \forall \varphi=(\hat{\varphi}_1,\ldots,\hat{\varphi}_m,\check{\varphi}_1,\ldots,\check{\varphi}_n)^T\in \mathbb{B}_{h_\diamond},$$

where

$$\begin{aligned}
x_i^{\varphi}(t) &= e^{p_{\alpha_i}t}\phi_i(0)+\frac{1-\alpha_i}{K_{\alpha_i}}\Bigg\{\Bigg[\sum_{j=1}^{n}a_{ij}(t)f_j(\check{\varphi}_j(t-\xi_{ij}))+\bigwedge_{j=1}^{n}\gamma_{ij}(t)f_j(\check{\varphi}_j(t-\xi_{ij}))\\
&\quad+\bigvee_{j=1}^{n}\eta_{ij}(t)f_j(\check{\varphi}_j(t-\xi_{ij}))+\bigwedge_{j=1}^{n}T_{ij}(t)\rho_j+\bigvee_{j=1}^{n}Q_{ij}(t)\rho_j+I_i(t)\Bigg]\\
&\quad-e^{p_{\alpha_i}t}\Bigg[\sum_{j=1}^{n}a_{ij}(0)f_j(\psi_j(-\xi_{ij}))+\bigwedge_{j=1}^{n}\gamma_{ij}(0)f_j(\psi_j(-\xi_{ij}))\\
&\quad+\bigvee_{j=1}^{n}\eta_{ij}(0)f_j(\psi_j(-\xi_{ij}))+\bigwedge_{j=1}^{n}T_{ij}(0)\rho_j+\bigvee_{j=1}^{n}Q_{ij}(0)\rho_j+I_i(0)\Bigg]\Bigg\}\\
&\quad+\frac{\alpha_i}{K_{\alpha_i}^2}\int_0^t e^{p_{\alpha_i}(t-s)}\Bigg[\sum_{j=1}^{n}a_{ij}(s)f_j(\check{\varphi}_j(s-\xi_{ij}))+\bigwedge_{j=1}^{n}\gamma_{ij}(s)f_j(\check{\varphi}_j(s-\xi_{ij}))\\
&\quad+\bigvee_{j=1}^{n}\eta_{ij}(s)f_j(\check{\varphi}_j(s-\xi_{ij}))+\bigwedge_{j=1}^{n}T_{ij}(s)\rho_j\\
&\quad+\bigvee_{j=1}^{n}Q_{ij}(s)\rho_j+I_i(s)\Bigg]\mathrm{d}s
\end{aligned}\tag{6.5}$$

and

$$\begin{aligned}
y_j^{\varphi}(t) &= e^{p_{\beta_j}t}\psi_j(0)+\frac{1-\beta_j}{K_{\beta_j}}\Bigg\{\Bigg[\sum_{i=1}^{m}b_{ji}(t)g_i(\hat{\varphi}_i(t-\sigma_{ji}))+\bigwedge_{i=1}^{m}\varepsilon_{ji}(t)g_i(\hat{\varphi}_i(t-\sigma_{ji}))\\
&\quad+\bigvee_{i=1}^{m}\vartheta_{ji}(t)g_i(\hat{\varphi}_i(t-\sigma_{ji}))+\bigwedge_{i=1}^{m}W_{ji}(t)\varrho_i+\bigvee_{i=1}^{m}R_{ji}(t)\varrho_i+P_j(t)\Bigg]\\
&\quad-e^{p_{\beta_j}t}\Bigg[\sum_{i=1}^{m}b_{ji}(0)g_i(\phi_i(-\sigma_{ji}))+\bigwedge_{i=1}^{m}\varepsilon_{ji}(0)g_i(\phi_i(-\sigma_{ji}))\\
&\quad+\bigvee_{i=1}^{m}\vartheta_{ji}(0)g_i(\phi_i(-\sigma_{ji}))+\bigwedge_{i=1}^{m}W_{ji}(0)\varrho_i+\bigvee_{i=1}^{m}R_{ji}(0)\varrho_i+P_j(0)\Bigg]\Bigg\}\\
&\quad+\frac{\beta_j}{K_{\beta_j}^2}\int_0^t e^{p_{\beta_j}(t-s)}\Bigg[\sum_{i=1}^{m}b_{ji}(s)g_i(\hat{\varphi}_i(s-\sigma_{ji}))+\bigwedge_{i=1}^{m}\varepsilon_{ji}(s)g_i(\hat{\varphi}_i(s-\sigma_{ji}))
\end{aligned}$$

$$+\bigvee_{i=1}^{m}\vartheta_{ji}(s)g_i(\hat{\varphi}_i(s-\sigma_{ji}))+\bigwedge_{i=1}^{m}W_{ji}(s)\varrho_i$$

$$+\bigvee_{i=1}^{m}R_{ji}(s)\varrho_i+P_j(s)\Bigg]\mathrm{d}s, \tag{6.6}$$

where $t\in\mathbb{R}_0$, $i=1,2,\ldots,m$, $j=1,2,\ldots,n$.

For any $\varphi=(\hat{\varphi}_1,\ldots,\hat{\varphi}_m,\check{\varphi}_1,\ldots,\check{\varphi}_n)^T$ and $\omega=(\hat{\omega}_1,\ldots,\hat{\omega}_m,\check{\omega}_1,\ldots,\check{\omega}_n)^T$ in $\mathbb{B}_{h_\diamond}$, it derives from Eq. (6.5), Lemma 6.1, ($A_1$) and ($A_2$) that

$$\begin{aligned}\left|x_i^{\varphi}(t)-x_i^{\omega}(t)\right| &\le \sum_{j=1}^{n}\left(a_{ij}^*+\gamma_{ij}^*+\eta_{ij}^*\right)\Bigg[\frac{1-\alpha_i}{K_{\alpha_i}}\Big|f_j(\check{\varphi}_j(t-\xi_{ij}))-f_j(\check{\omega}_j(t-\xi_{ij}))\Big|\\ &\quad+\frac{\alpha_i}{K_{\alpha_i}^2}\int_0^t e^{p_{\alpha_i}(t-s)}\Big|f_j(\check{\varphi}_j(s-\xi_{ij}))-f_j(\check{\omega}_j(s-\xi_{ij}))\Big|\mathrm{d}s\Bigg]\\ &\le K_{\alpha_i}^{-1}\left[(1-\alpha_i)+\frac{1-e^{p_{\alpha_i}t}}{c_i}\right]\varpi_i^f\,\mathbf{d}_{h_\diamond}(\varphi,\omega)\\ &\le \varsigma\mathbf{d}_{h_\diamond}(\varphi,\omega),\end{aligned}$$

where $t\in[0,h_\diamond]$, $i=1,2,\ldots,m$. Similarly, it obtains from Eq. (6.6) that

$$\left|y_j^{\varphi}(t)-y_j^{\omega}(t)\right|\le K_{\beta_j}^{-1}\left[(1-\beta_j)+\frac{1-e^{p_{\beta_j}t}}{d_j}\right]\varpi_j^g\mathbf{d}_{h_\diamond}(\varphi,\omega)\le\varsigma\mathbf{d}_{h_\diamond}(\varphi,\omega),$$

where $t\in[0,h_\diamond]$, $j=1,2,\ldots,n$. As a consequence,

$$\mathbf{d}_{h_\diamond}(\mathbf{L}\varphi,\mathbf{L}\omega)\le\varsigma\mathbf{d}_{h_\diamond}(\varphi,\omega),\quad\forall\varphi,\omega\in\mathbb{B}_{h_\diamond}.$$

This implies $\mathbf{L}$ is a contractive mapping. According to Banach contracting mapping principle, $\mathbf{L}$ has a unique fixed point $\varphi^*\in\mathbb{B}_{h_\diamond}$ and $\varphi^*$ is a unique solution of Eq. (6.1) on interval $[0,h_\diamond]$. By the extension method utilized in Theorem 1 of literature [50], it easily extends the existence interval of $\varphi^*$ to $t\in\mathbb{R}_0$. The proof is achieved. □

**Remark 6.2** By using a similar derivation as Lemma 6.2 and under some extremely strict conditions (e.g., the differentials of $f_j$, $g_i$, $I_i$, $P_j$, etc.), it can verify that CF-BAMNNs (6.1) has a unique solution $(x_1,\ldots,x_m,y_1,\ldots,y_n)^T\in C^1(\mathbb{R}_0,\mathbb{R}^{m+n})$ is depicted by Eqs. (6.3) and (6.4), respectively.

## 6.3 Exponential Euler Difference Scheme

For convenience, let $h > 0$ and

$$t_h = \lfloor t/h \rfloor = \max\{k : t/h \geq k, k \in \mathbb{N}_0\}, \quad \forall t \in \mathbb{R}_0.$$

Further, define

$$\xi_{ij}^h = \lfloor \xi_{ij}/h \rfloor, \quad \sigma_{ji}^h = \lfloor \sigma_{ji}/h \rfloor, \quad x_{i,k} = x_i(kh), \quad y_{j,k} = y_j(kh),$$

$$a_{ij,k} = a_{ij}(kh), \quad \gamma_{ij,k} = \gamma_{ij}(kh), \quad \eta_{ij,k} = \eta_{ij}(kh), \quad T_{ij,k} = T_{ij}(kh),$$

$$Q_{ij,k} = Q_{ij}(kh), \quad I_{i,k} = I_i(kh), \quad b_{ji,k} = b_{ji}(kh), \quad \varepsilon_{ji,k} = \varepsilon_{ji}(kh),$$

$$\vartheta_{ji,k} = \vartheta_{ji}(kh), \quad W_{ji,k} = W_{ji}(kh), \quad R_{ji,k} = R_{ji}(kh), \quad P_{j,k} = P_j(kh),$$

where $k \in \mathbb{N}_0$, $i = 1, 2, \ldots, m$, $j = 1, 2, \ldots, n$.

By Lemma 6.2 and Remark 6.2, CF-BAMNNs (6.1) admits a unique solution $(x, y)^T = (x_1, \ldots, x_m, y_1, \ldots, y_n)^T$ under some suitable conditions, which can be denoted by Eqs. (6.3) and (6.4). Discussing the following approximate models of Eqs. (6.3) and (6.4)

$$\begin{aligned}
x_i(t) = {} & e^{p_{\alpha_i} t} x_{i,0} + \frac{1-\alpha_i}{K_{\alpha_i}} \Bigg\{ \Bigg[ \sum_{j=1}^{n} a_{ij,t_h} f_j(y_{j,t_h - \xi_{ij}^h}) + \bigwedge_{j=1}^{n} \gamma_{ij,t_h} f_j(y_{j,t_h - \xi_{ij}^h}) \\
& + \bigvee_{j=1}^{n} \eta_{ij,t_h} f_j(y_{j,t_h - \xi_{ij}^h}) + \bigwedge_{j=1}^{n} T_{ij,t_h} \rho_j + \bigvee_{j=1}^{n} Q_{ij,t_h} \rho_j + I_{i,t_h} \Bigg] \\
& - e^{p_{\alpha_i} t_h} \Bigg[ \sum_{j=1}^{n} a_{ij,0} f_j(y_{j,-\xi_{ij}^h}) + \bigwedge_{j=1}^{n} \gamma_{ij,0} f_j(y_{j,-\xi_{ij}^h}) \\
& + \bigvee_{j=1}^{n} \eta_{ij,0} f_j(y_{j,-\xi_{ij}^h}) + \bigwedge_{j=1}^{n} T_{ij,0} \rho_j + \bigvee_{j=1}^{n} Q_{ij,0} \rho_j + I_{i,0} \Bigg] \Bigg\} \\
& + \frac{\alpha_i}{K_{\alpha_i}^2} \int_0^t e^{p_{\alpha_i}(t-s)} \Bigg[ \sum_{j=1}^{n} a_{ij,s_h} f_j(y_{j,s_h - \xi_{ij}^h}) \\
& + \bigwedge_{j=1}^{n} \gamma_{ij,s_h} f_j(y_{j,s_h - \xi_{ij}^h}) + \bigvee_{j=1}^{n} \eta_{ij,s_h} f_j(y_{j,s_h - \xi_{ij}^h}) \\
& + \bigwedge_{j=1}^{n} T_{ij,s_h} \rho_j + \bigvee_{j=1}^{n} Q_{ij,s_h} \rho_j + I_{i,s_h} \Bigg] \mathrm{d}s
\end{aligned} \tag{6.7}$$

and

$$
\begin{aligned}
y_j(t) = {} & e^{p_{\beta_j} t} y_{j,0} + \frac{1-\beta_j}{K_{\beta_j}} \Bigg\{ \Bigg[ \sum_{i=1}^{m} b_{ji,t_h} g_i(x_{i,t_h-\sigma_{ji}^h}) + \bigwedge_{i=1}^{m} \varepsilon_{ji,t_h} g_i(x_{i,t_h-\sigma_{ji}^h}) \\
& + \bigvee_{i=1}^{m} \vartheta_{ji,t_h} g_i(x_{i,t_h-\sigma_{ji}^h}) + \bigwedge_{i=1}^{m} W_{ji,t_h} \varrho_i + \bigvee_{i=1}^{m} R_{ji,t_h} \varrho_i + P_{j,t_h} \Bigg] \\
& - e^{p_{\beta_j} t_h} \Bigg[ \sum_{i=1}^{m} b_{ji,0} g_i(x_{i,-\sigma_{ji}^h}) + \bigwedge_{i=1}^{m} \varepsilon_{ji,0} g_i(x_{i,-\sigma_{ji}^h}) \\
& + \bigvee_{i=1}^{m} \vartheta_{ji,0} g_i(x_{i,-\sigma_{ji}^h}) + \bigwedge_{i=1}^{m} W_{ji,0} \varrho_i + \bigvee_{i=1}^{m} R_{ji,0} \varrho_i + P_{j,0} \Bigg] \Bigg\} \\
& + \frac{\beta_j}{K_{\beta_j}^2} \int_0^t e^{p_{\beta_j}(t-s)} \Bigg[ \sum_{i=1}^{m} b_{ji,s_h} g_i(x_{i,s_h-\sigma_{ji}^h}) \\
& + \bigwedge_{i=1}^{m} \varepsilon_{ji,s_h} g_i(x_{i,s_h-\sigma_{ji}^h}) + \bigvee_{i=1}^{m} \vartheta_{ji,s_h} g_i(x_{i,s_h-\sigma_{ji}^h}) \\
& + \bigwedge_{i=1}^{m} W_{ji,s_h} \varrho_i + \bigvee_{i=1}^{m} R_{ji,s_h} \varrho_i + P_{j,s_h} \Bigg] \mathrm{d}s,
\end{aligned} \tag{6.8}
$$

where $t \in \mathbb{R}_0$, $i = 1, 2, \ldots, m$, $j = 1, 2, \ldots, n$.

If $t \in [kh, kh+h)$ for some $k \in \mathbb{N}_0$, then Eq. (6.7) is transformed into

$$
\begin{aligned}
x_i(t) = {} & e^{p_{\alpha_i} t} x_{i,0} + \frac{1-\alpha_i}{K_{\alpha_i}} \Bigg\{ \Bigg[ \sum_{j=1}^{n} a_{ij,k} f_j(y_{j,k-\xi_{ij}^h}) + \bigwedge_{j=1}^{n} \gamma_{ij,k} f_j(y_{j,k-\xi_{ij}^h}) \\
& + \bigvee_{j=1}^{n} \eta_{ij,k} f_j(y_{j,k-\xi_{ij}^h}) + \bigwedge_{j=1}^{n} T_{ij,k} \rho_j + \bigvee_{j=1}^{n} Q_{ij,k} \rho_j + I_{i,k} \Bigg] \\
& - e^{p_{\alpha_i} kh} \Bigg[ \sum_{j=1}^{n} a_{ij,0} f_j(y_{j,-\xi_{ij}^h}) + \bigwedge_{j=1}^{n} \gamma_{ij,0} f_j(y_{j,-\xi_{ij}^h}) \\
& + \bigvee_{j=1}^{n} \eta_{ij,0} f_j(y_{j,-\xi_{ij}^h}) + \bigwedge_{j=1}^{n} T_{ij,0} \rho_j + \bigvee_{j=1}^{n} Q_{ij,0} \rho_j + I_{i,0} \Bigg] \Bigg\} \\
& + \frac{\alpha_i}{K_{\alpha_i}^2} \int_{kh}^{t} e^{p_{\alpha_i}(t-s)} \Bigg[ \sum_{j=1}^{n} a_{ij,k} f_j(y_{j,k-\xi_{ij}^h}) + \bigwedge_{j=1}^{n} \gamma_{ij,k} f_j(y_{j,k-\xi_{ij}^h})
\end{aligned}
$$

$$
\begin{aligned}
&+ \bigvee_{j=1}^{n} \eta_{ij,k} f_j(y_{j,k-\xi_{ij}^h}) + \bigwedge_{j=1}^{n} T_{ij,k}\rho_j + \bigvee_{j=1}^{n} Q_{ij,k}\rho_j + I_{i,k}\Bigg]\mathrm{d}s \\
&+ \frac{\alpha_i}{K_{\alpha_i}^2} \sum_{l=0}^{k-1} \int_{lh}^{lh+h} e^{p_{\alpha_i}(t-s)} \Bigg[\sum_{j=1}^{n} a_{ij,l} f_j(y_{j,l-\xi_{ij}^h}) \\
&+ \bigwedge_{j=1}^{n} \gamma_{ij,l} f_j(y_{j,l-\xi_{ij}^h}) + \bigvee_{j=1}^{n} \eta_{ij,l} f_j(y_{j,l-\xi_{ij}^h}) \\
&+ \bigwedge_{j=1}^{n} T_{ij,l}\rho_j + \bigvee_{j=1}^{n} Q_{ij,l}\rho_j + I_{i,l}\Bigg]\mathrm{d}s \\
= {}& e^{p_{\alpha_i}t} x_{i,0} + \frac{1-\alpha_i}{K_{\alpha_i}} \Bigg\{\Bigg[\sum_{j=1}^{n} a_{ij,k} f_j(y_{j,k-\xi_{ij}^h}) + \bigwedge_{j=1}^{n} \gamma_{ij,k} f_j(y_{j,k-\xi_{ij}^h}) \\
&+ \bigvee_{j=1}^{n} \eta_{ij,k} f_j(y_{j,k-\xi_{ij}^h}) + \bigwedge_{j=1}^{n} T_{ij,k}\rho_j + \bigvee_{j=1}^{n} Q_{ij,k}\rho_j + I_{i,k}\Bigg] \\
&- e^{p_{\alpha_i}kh} \Bigg[\sum_{j=1}^{n} a_{ij,0} f_j(y_{j,-\xi_{ij}^h}) + \bigwedge_{j=1}^{n} \gamma_{ij,0} f_j(y_{j,-\xi_{ij}^h}) \\
&+ \bigvee_{j=1}^{n} \eta_{ij,0} f_j(y_{j,-\xi_{ij}^h}) + \bigwedge_{j=1}^{n} T_{ij,0}\rho_j + \bigvee_{j=1}^{n} Q_{ij,0}\rho_j + I_{i,0}\Bigg]\Bigg\} \\
&+ \frac{1-e^{p_{\alpha_i}(t-kh)}}{c_i K_{\alpha_i}} \Bigg[\sum_{j=1}^{n} a_{ij,k} f_j(y_{j,k-\xi_{ij}^h}) + \bigwedge_{j=1}^{n} \gamma_{ij,k} f_j(y_{j,k-\xi_{ij}^h}) \\
&+ \bigvee_{j=1}^{n} \eta_{ij,k} f_j(y_{j,k-\xi_{ij}^h}) + \bigwedge_{j=1}^{n} T_{ij,k}\rho_j + \bigvee_{j=1}^{n} Q_{ij,k}\rho_j + I_{i,k}\Bigg] \\
&+ \frac{1}{c_i K_{\alpha_i}} \sum_{l=0}^{k-1} \left[e^{p_{\alpha_i}(t-lh-h)} - e^{p_{\alpha_i}(t-lh)}\right] \Bigg[\sum_{j=1}^{n} a_{ij,l} f_j(y_{j,l-\xi_{ij}^h}) \\
&+ \bigwedge_{j=1}^{n} \gamma_{ij,l} f_j(y_{j,l-\xi_{ij}^h}) + \bigvee_{j=1}^{n} \eta_{ij,l} f_j(y_{j,l-\xi_{ij}^h}) \\
&+ \bigwedge_{j=1}^{n} T_{ij,l}\rho_j + \bigvee_{j=1}^{n} Q_{ij,l}\rho_j + I_{i,l}\Bigg],
\end{aligned}
\tag{6.9}
$$

where $t \in \mathbb{R}_0$, $k \in \mathbb{N}_0$, $i = 1, 2, \ldots, m$, $j = 1, 2, \ldots, n$.

Let $t \to kh$ in Eq. (6.9), it deduces

$$x_{i,k} = e^{p_{\alpha_i}kh}x_{i,0} + \frac{1-\alpha_i}{K_{\alpha_i}}\Bigg\{\Bigg[\sum_{j=1}^{n} a_{ij,k}f_j(y_{j,k-\xi_{ij}^h}) + \bigwedge_{j=1}^{n}\gamma_{ij,k}f_j(y_{j,k-\xi_{ij}^h})$$

$$+ \bigvee_{j=1}^{n}\eta_{ij,k}f_j(y_{j,k-\xi_{ij}^h}) + \bigwedge_{j=1}^{n}T_{ij,k}\rho_j + \bigvee_{j=1}^{n}Q_{ij,k}\rho_j + I_{i,k}\Bigg]$$

$$- e^{p_{\alpha_i}kh}\Bigg[\sum_{j=1}^{n} a_{ij,0}f_j(y_{j,-\xi_{ij}^h}) + \bigwedge_{j=1}^{n}\gamma_{ij,0}f_j(y_{j,-\xi_{ij}^h}) + \bigvee_{j=1}^{n}\eta_{ij,0}f_j(y_{j,-\xi_{ij}^h})$$

$$+ \bigwedge_{j=1}^{n}T_{ij,0}\rho_j + \bigvee_{j=1}^{n}Q_{ij,0}\rho_j + I_{i,0}\Bigg]\Bigg\} + \frac{e^{p_{\alpha_i}kh}}{c_iK_{\alpha_i}}\sum_{l=0}^{k-1}\big[e^{p_{\alpha_i}(-lh-h)} - e^{p_{\alpha_i}(-lh)}\big]$$

$$\times\Bigg[\sum_{j=1}^{n} a_{ij,l}f_j(y_{j,l-\xi_{ij}^h}) + \bigwedge_{j=1}^{n}\gamma_{ij,l}f_j(y_{j,l-\xi_{ij}^h}) + \bigvee_{j=1}^{n}\eta_{ij,l}f_j(y_{j,l-\xi_{ij}^h})$$

$$+ \bigwedge_{j=1}^{n}T_{ij,l}\rho_j + \bigvee_{j=1}^{n}Q_{ij,l}\rho_j + I_{i,l}\Bigg], \tag{6.10}$$

where $k \in \mathbb{N}_0$, $i = 1, 2, \ldots, m$, $j = 1, 2, \ldots, n$.

If $t \in [kh + h, kh + 2h)$, similar to the derivation of Eq. (6.10), it gets from Eq. (6.7) that

$$x_{i,k+1} = e^{p_{\alpha_i}(kh+h)}x_{i,0} + \frac{1-\alpha_i}{K_{\alpha_i}}\Bigg\{\Bigg[\sum_{j=1}^{n} a_{ij,k+1}f_j(y_{j,k+1-\xi_{ij}^h})$$

$$+ \bigwedge_{j=1}^{n}\gamma_{ij,k+1}f_j(y_{j,k+1-\xi_{ij}^h}) + \bigvee_{j=1}^{n}\eta_{ij,k+1}f_j(y_{j,k+1-\xi_{ij}^h}) + \bigwedge_{j=1}^{n}T_{ij,k+1}\rho_j$$

$$+ \bigvee_{j=1}^{n}Q_{ij,k+1}\rho_j + I_{i,k+1}\Bigg] - e^{p_{\alpha_i}(kh+h)}\Bigg[\sum_{j=1}^{n} a_{ij,0}f_j(y_{j,-\xi_{ij}^h})$$

$$+ \bigwedge_{j=1}^{n}\gamma_{ij,0}f_j(y_{j,-\xi_{ij}^h}) + \bigvee_{j=1}^{n}\eta_{ij,0}f_j(y_{j,-\xi_{ij}^h}) + \bigwedge_{j=1}^{n}T_{ij,0}\rho_j + \bigvee_{j=1}^{n}Q_{ij,0}\rho_j$$

$$+ I_{i,0}\Bigg]\Bigg\} + \frac{e^{p_{\alpha_i}kh}}{c_iK_{\alpha_i}}\sum_{l=0}^{k}\big[e^{p_{\alpha_i}(-lh)} - e^{p_{\alpha_i}(-lh+h)}\big]\Bigg[\sum_{j=1}^{n} a_{ij,l}f_j(y_{j,l-\xi_{ij}^h})$$

$$+\bigwedge_{j=1}^{n}\gamma_{ij,l}f_j(y_{j,l-\xi_{ij}^h})+\bigvee_{j=1}^{n}\eta_{ij,l}f_j(y_{j,l-\xi_{ij}^h})$$

$$+\bigwedge_{j=1}^{n}T_{ij,l}\rho_j+\bigvee_{j=1}^{n}Q_{ij,l}\rho_j+I_{i,l}\Bigg],\tag{6.11}$$

where $k\in\mathbb{N}_0, i=1,2,\ldots,m,\ j=1,2,\ldots,n$.

Similarly, it derives from Eq. (6.8) that

$$y_{j,k}=e^{p_{\beta_j}kh}y_{j,0}+\frac{1-\beta_j}{K_{\beta_j}}\Bigg\{\Bigg[\sum_{i=1}^{m}b_{ji,k}g_i(x_{i,k-\sigma_{ji}^h})+\bigwedge_{i=1}^{m}\varepsilon_{ji,k}g_i(x_{i,k-\sigma_{ji}^h})$$

$$+\bigvee_{i=1}^{m}\vartheta_{ji,k}g_i(x_{i,k-\sigma_{ji}^h})+\bigwedge_{i=1}^{m}W_{ji,k}\varrho_i+\bigvee_{i=1}^{m}R_{ji,k}\varrho_i+P_{j,k}\Bigg]$$

$$-e^{p_{\beta_j}kh}\Bigg[\sum_{i=1}^{m}b_{ji,0}g_i(x_{i,-\sigma_{ji}^h})+\bigwedge_{i=1}^{m}\varepsilon_{ji,0}g_i(x_{i,-\sigma_{ji}^h})$$

$$+\bigvee_{i=1}^{m}\vartheta_{ji,0}g_i(x_{i,-\sigma_{ji}^h})+\bigwedge_{i=1}^{m}W_{ji,0}\varrho_i+\bigvee_{i=1}^{m}R_{ji,0}\varrho_i+P_{j,0}\Bigg]\Bigg\}$$

$$+\frac{e^{p_{\beta_j}kh}}{d_jK_{\beta_j}}\sum_{l=0}^{k-1}\Big[e^{p_{\beta_j}(-lh-h)}-e^{p_{\beta_j}(-lh)}\Big]\Bigg[\sum_{i=1}^{m}b_{ji,l}g_i(x_{i,l-\sigma_{ji}^h})$$

$$+\bigwedge_{i=1}^{m}\varepsilon_{ji,l}g_i(x_{i,l-\sigma_{ji}^h})+\bigvee_{i=1}^{m}\vartheta_{ji,l}g_i(x_{i,l-\sigma_{ji}^h})$$

$$+\bigwedge_{i=1}^{m}W_{ji,l}\varrho_i+\bigvee_{i=1}^{m}R_{ji,l}\varrho_i+P_{j,l}\Bigg]\tag{6.12}$$

and

$$y_{j,k+1}=e^{p_{\beta_j}(kh+h)}y_{j,0}+\frac{1-\beta_j}{K_{\beta_j}}\Bigg\{\Bigg[\sum_{i=1}^{m}b_{ji,k+1}g_i(x_{i,k+1-\sigma_{ji}^h})$$

$$+\bigwedge_{i=1}^{m}\varepsilon_{ji,k+1}g_i(x_{i,k+1-\sigma_{ji}^h})+\bigvee_{i=1}^{m}\vartheta_{ji,k+1}g_i(x_{i,k+1-\sigma_{ji}^h})+\bigwedge_{i=1}^{m}W_{ji,k+1}\varrho_i$$

$$+\bigvee_{i=1}^{m}R_{ji,k+1}\varrho_i+P_{j,k+1}\Bigg]-e^{p_{\beta_j}(kh+h)}\Bigg[\sum_{i=1}^{m}b_{ji,0}g_i(x_{i,-\sigma_{ji}^h})$$

$$+\bigwedge_{i=1}^{m}\varepsilon_{ji,0}g_i(x_{i,-\sigma_{ji}^h})+\bigvee_{i=1}^{m}\vartheta_{ji,0}g_i(x_{i,-\sigma_{ji}^h})+\bigwedge_{i=1}^{m}W_{ji,0}\varrho_i+\bigvee_{i=1}^{m}R_{ji,0}\varrho_i$$

$$+P_{j,0}\Big]\Big\}+\frac{e^{p_{\beta_j}kh}}{c_iK_{\beta_j}}\sum_{l=0}^{k}\left[e^{p_{\beta_j}(-lh)}-e^{p_{\beta_j}(-lh+h)}\right]$$

$$\times\Bigg[\sum_{i=1}^{m}b_{ji,l}g_i(x_{i,l-\sigma_{ji}^h})+\bigwedge_{i=1}^{m}\varepsilon_{ji,l}g_i(x_{i,l-\sigma_{ji}^h})$$

$$+\bigvee_{i=1}^{m}\vartheta_{ji,l}g_i(x_{i,l-\sigma_{ji}^h})+\bigwedge_{i=1}^{m}W_{ji,l}\varrho_i+\bigvee_{i=1}^{m}R_{ji,l}\varrho_i+P_{j,l}\Bigg], \tag{6.13}$$

where $k\in\mathbb{N}_0$, $i=1,2,\ldots,m$, $j=1,2,\ldots,n$.

By utilizing cancellation law in Eqs. (6.10)–(6.13), it obtains the exponential Euler difference model (EED-CF-BAMNNs in short) of CF-BAMNNs (6.1) denoted by

$$\begin{cases} x_{i,k+1} = e^{p_{\alpha_i}h}x_{i,k}+\dfrac{1-e^{p_{\alpha_i}h}}{c_iK_{\alpha_i}}F_{i,k}(y_{\xi_i}) \\ \qquad +\dfrac{1-\alpha_i}{K_{\alpha_i}}\left[F_{i,k+1}(y_{\xi_i})-e^{p_{\alpha_i}h}F_{i,k}(y_{\xi_i})\right], \\ y_{j,k+1} = e^{p_{\beta_j}h}y_{j,k}+\dfrac{1-e^{p_{\beta_j}h}}{d_jK_{\beta_j}}G_{j,k}(x_{\sigma_j}) \\ \qquad +\dfrac{1-\beta_j}{K_{\beta_j}}\left[G_{j,k+1}(x_{\sigma_j})-e^{p_{\beta_j}h}G_{j,k}(x_{\sigma_j})\right] \end{cases} \tag{6.14}$$

with the following initial conditions

$$x_{i,s}=\phi_{i,s},\quad y_{j,s}=\psi_{j,s},\quad s\in[-\xi_h,0]_{\mathbb{Z}},$$

$$\xi_h=\max_{1\le i\le m,1\le j\le n}\{\lfloor\xi_{ij}/h\rfloor,\lfloor\sigma_{ij}/h\rfloor\},$$

where

$$F_{i,k}(y_{\xi_i})=F_{i,k}(y_{1,k-\xi_{i1}^h},\ldots,y_{n,k-\xi_{in}^h})$$

$$=\sum_{j=1}^{n}a_{ij,k}f_j(y_{j,k-\xi_{ij}^h})+\bigwedge_{j=1}^{n}\gamma_{ij,k}f_j(y_{j,k-\xi_{ij}^h})$$

$$+\bigvee_{j=1}^{n}\eta_{ij,k}f_j(y_{j,k-\xi_{ij}^h})+\bigwedge_{j=1}^{n}T_{ij,k}\rho_j+\bigvee_{j=1}^{n}Q_{ij,k}\rho_j+I_{i,k},$$

$$G_{j,k}(x_{\sigma_j}) = G_{j,k}(x_{1,k-\sigma^h_{j1}}, \ldots, x_{m,k-\sigma^h_{jm}})$$
$$= \sum_{i=1}^{m} b_{ji,k} g_i(x_{i,k-\sigma^h_{ji}}) + \bigwedge_{i=1}^{m} \varepsilon_{ji,k} g_i(x_{i,k-\sigma^h_{ji}})$$
$$+ \bigvee_{i=1}^{m} \vartheta_{ji,k} g_i(x_{i,k-\sigma^h_{ji}}) + \bigwedge_{i=1}^{m} W_{ji,k}\varrho_i + \bigvee_{i=1}^{m} R_{ji,k}\varrho_i + P_{j,k},$$

in which $y_{\xi_i} = (y_{1,k-\xi^h_{i1}}, \ldots, y_{n,k-\xi^h_{in}})^T \in \mathbb{R}^n$, $x_{\sigma_j} = (x_{1,k-\sigma^h_{j1}}, \ldots, x_{m,k-\sigma^h_{jm}})^T \in \mathbb{R}^m$, $k \in \mathbb{N}_0$, $i = 1, 2, \ldots, m$, $j = 1, 2, \ldots, n$.

**Property 6.1** *The solutions of EED-CF-BAMNNs* (6.14) *denote the numerical solutions of CF-BAMNNs* (6.1).

Let

$$R_k^{\alpha_i} = \frac{1-\alpha_i}{K_{\alpha_i}}\Big[F_{i,k+1}(y_{\xi_i}) - e^{p_{\alpha_i}h} F_{i,k}(y_{\xi_i})\Big],$$
$$R_k^{\beta_j} = \frac{1-\beta_j}{K_{\beta_j}}\Big[G_{j,k+1}(x_{\sigma_j}) - e^{p_{\beta_j}h} G_{j,k}(x_{\sigma_j})\Big],$$

where $k \in \mathbb{N}_0$, $i = 1, 2, \ldots, m$, $j = 1, 2, \ldots, n$.

**Remark 6.3** Hereby, $R_k^{\alpha_i}$ and $R_k^{\beta_j}$ are called the fractional remainder terms of approximate model (6.14), which become zero if $\alpha_i \to 1$ and $\beta_j \to 1$, where $k \in \mathbb{N}_0$, $i = 1, 2, \ldots, m$, $j = 1, 2, \ldots, n$.

If $\alpha_i \to 1$ and $\beta_j \to 1$ in EED-CF-BAMNNs (6.14), then $K_{\alpha_i} \to 1$, $K_{\beta_j} \to 1$, $p_{\alpha_i} \to -c_i$ and $p_{\beta_j} \to -d_j$, $i = 1, 2, \ldots, m$, $j = 1, 2, \ldots, n$. As a result, EED-CF-BAMNNs (6.14) is changed to

$$\begin{cases} x_{i,k+1} = e^{-c_i h} x_{i,k} + \dfrac{1-e^{-c_i h}}{c_i}\Big[\sum_{j=1}^{n} a_{ij,k} f_j(y_{j,k-\xi^h_{ij}}) \\ \quad + \bigwedge_{j=1}^{n} \gamma_{ij,k} f_j(y_{j,k-\xi^h_{ij}}) + \bigvee_{j=1}^{n} \eta_{ij,k} f_j(y_{j,k-\xi^h_{ij}}) \\ \quad + \bigwedge_{j=1}^{n} T_{ij,k}\rho_j + \bigvee_{j=1}^{n} Q_{ij,k}\rho_j + I_{i,k}\Big], \\ y_{j,k+1} = e^{-d_j h} y_{j,k} + \dfrac{1-e^{-d_j h}}{d_j}\Big[\sum_{i=1}^{m} b_{ji,k} g_i(x_{i,k-\sigma^h_{ji}}) \\ \quad + \bigwedge_{i=1}^{m} \varepsilon_{ji,k} g_i(x_{i,k-\sigma^h_{ji}}) + \bigvee_{i=1}^{m} \vartheta_{ji,k} g_i(x_{i,k-\sigma^h_{ji}}) \\ \quad + \bigwedge_{i=1}^{m} W_{ji,k}\varrho_i + \bigvee_{i=1}^{m} R_{ji,k}\varrho_i + P_{j,k}\Big] \end{cases} \tag{6.15}$$

with the following initial conditions

$$x_{i,s} = \phi_{i,s}, \quad y_{j,s} = \psi_{j,s}, \quad s \in [-\xi_h, 0]_{\mathbb{Z}},$$

where $k \in \mathbb{N}_0$, $i = 1, 2, \ldots, m$, $j = 1, 2, \ldots, n$.

**Property 6.2** *The solutions of model* (6.15) *denote the numerical solutions of model* (6.2).

**Remark 6.4** The authors in [136] discussed the exponential Euler difference model of stochastic BAM neural networks and the form is similar to model (6.15). By using some numerical experiments, paper [136] exhibited that the exponential Euler difference models can better depict and retain the dynamics of the corresponding continuous-time models than the traditional Euler difference models. As a consequence, this chapter investigates the exponential Euler difference scheme of CF-BAMNNs, which has important guiding significance.

The variation-of-constants formula of EED-CF-BAMNNs (6.14) can be obtained directly from Lemma 2.3 in literature [27], which is described as below.

**Lemma 6.3** *If $M(\alpha) = 1$ in Eq. (1.20), then the solutions of EED-CF-BAMNNs* (6.14) *have an equivalent form of*

$$\begin{cases} x_{i,k} = e^{p_{\alpha_i} kh} x_{i,0} + \sum_{q=0}^{k-1} e^{p_{\alpha_i}(k-q-1)h} \Bigg\{ \dfrac{1 - e^{p_{\alpha_i} h}}{c_i K_{\alpha_i}} F_{i,q}(y_{\xi_i}) \\ \quad + \dfrac{1-\alpha_i}{K_{\alpha_i}} \Big[ F_{i,q+1}(y_{\xi_i}) - e^{p_{\alpha_i} h} F_{i,q}(y_{\xi_i}) \Big] \Bigg\}, \\ y_{j,k} = e^{p_{\beta_j} kh} y_{j,0} + \sum_{q=0}^{k-1} e^{p_{\beta_j}(k-q-1)h} \Bigg\{ \dfrac{1 - e^{p_{\beta_j} h}}{d_j K_{\beta_j}} G_{j,q}(x_{\sigma_j}) \\ \quad + \dfrac{1-\beta_j}{K_{\beta_j}} \Big[ G_{j,q+1}(x_{\sigma_j}) - e^{p_{\beta_j} h} G_{j,q}(x_{\sigma_j}) \Big] \Bigg\} \end{cases} \tag{6.16}$$

*with the following initial conditions*

$$x_{i,s} = \phi_{i,s}, \quad y_{j,s} = \psi_{j,s}, \quad s \in [-\xi_h, 0]_{\mathbb{Z}},$$

*where $k \in \mathbb{N}_0$, $i = 1, 2, \ldots, m$, $j = 1, 2, \ldots, n$.*

## 6.4 Fractional PECE Algorithm

By observing EED-CF-BAMNNs (6.14) carefully, it is within the scope of implicit Euler differences if $\xi_h = \max_{1\le i\le m,1\le j\le n}\{\lfloor \xi_{ij}/h\rfloor, \lfloor \sigma_{ij}/h\rfloor\} = 0$. That is, the both sides of EED-CF-BAMNNs (6.14) include $x_{i,k+1}$ or $y_{j,k+1}$, $k \in \mathbb{N}_0$, $i = 1, 2, \dots, m$, $j = 1, 2, \dots, n$. In order to avoid solving nonlinear differential equations, the terms $x_{i,k+1}$ and $y_{j,k+1}$ in the right side of EED-CF-BAMNNs (6.14) should be replaced by the predicted values $x^p_{i,k+1}$ and $y^p_{j,k+1}$, $k \in \mathbb{N}_0$, $i = 1, 2, \dots, m$, $j = 1, 2, \dots, n$. Hence, EED-CF-BAMNNs (6.14) can be rewritten to the corrected formula depicted by

$$\begin{cases} x_{i,k+1} = e^{p_{\alpha_i}h}x_{i,k} + \dfrac{1-e^{p_{\alpha_i}h}}{c_i K_{\alpha_i}}F_{i,k}(y_{\xi_i}) \\ \qquad + \dfrac{1-\alpha_i}{K_{\alpha_i}}\Big[F_{i,k+1}(y^p_\xi) - e^{p_{\alpha_i}h}F_{i,k}(y_{\xi_i})\Big], \\ y_{j,k+1} = e^{p_{\beta_j}h}y_{j,k} + \dfrac{1-e^{p_{\beta_j}h}}{d_j K_{\beta_j}}G_{j,k}(x_{\sigma_j}) \\ \qquad + \dfrac{1-\beta_j}{K_{\beta_j}}\Big[G_{j,k+1}(x^p_\sigma) - e^{p_{\beta_j}h}G_{j,k}(x_{\sigma_j})\Big] \end{cases} \tag{6.17}$$

with the following initial conditions

$$x_{i,s} = \phi_{i,s}, \quad y_{j,s} = \psi_{j,s}, \quad s \in [-\xi_h, 0]_{\mathbb{Z}},$$

where

$$\begin{cases} F_{i,k}(y^p_\xi) = \sum_{j=1}^{n} a_{ij,k}f_j(y^{*p}_{j,k-\xi^h_{ij}}) + \bigwedge_{j=1}^{n}\gamma_{ij,k}f_j(y^{*p}_{j,k-\xi^h_{ij}}) \\ \qquad + \bigvee_{j=1}^{n}\eta_{ij,k}f_j(y^{*p}_{j,k-\xi^h_{ij}}) + \bigwedge_{j=1}^{n}T_{ij,k}\rho_j \\ \qquad + \bigvee_{j=1}^{n}Q_{ij,k}\rho_j + I_{i,k}, \\ G_{j,k}(x^p_\sigma) = \sum_{i=1}^{m} b_{ji,k}g_i(x^{*p}_{i,k-\sigma^h_{ji}}) + \bigwedge_{i=1}^{m}\varepsilon_{ji,k}g_i(x^{*p}_{i,k-\sigma^h_{ji}}) \\ \qquad + \bigvee_{i=1}^{m}\vartheta_{ji,k}g_i(x^{*p}_{i,k-\sigma^h_{ji}}) + \bigwedge_{i=1}^{m}W_{ji,k}\varrho_i \\ \qquad + \bigvee_{i=1}^{m}R_{ji,k}\varrho_i + P_{j,k}, \end{cases} \tag{6.18}$$

$$x^{*p}_{i,k-\sigma^h_{ji}} = \begin{cases} x^p_{i,k}, & \text{if } \sigma^h_{ji} = 0, \\ x_{i,k-\sigma^h_{ji}}, & \text{otherwise,} \end{cases}$$

$$y^{*p}_{j,k-\xi^h_{ij}} = \begin{cases} y^p_{j,k}, & \text{if } \xi^h_{ij} = 0, \\ y_{j,k-\xi^h_{ij}}, & \text{otherwise,} \end{cases}$$

$k \in \mathbb{N}_0, i = 1, 2, \dots, m, j = 1, 2, \dots, n.$

Let $\alpha \in [0, 1)$, $M(\alpha) = 1$ and $x \in C^1([a, b], \mathbb{R})$. Then

$$\begin{aligned} {}^{CF}_{\ 0}D^{\alpha}_{t} x(t)\Big|_{t=kh} &\approx \frac{1}{1-\alpha} \sum_{l=1}^{k} \int_{lh-h}^{lh} e^{\bar{\alpha}(kh-s)} x'(lh) \mathrm{d}s \\ &\approx \frac{1}{1-\alpha} \sum_{l=1}^{k} \int_{lh-h}^{lh} e^{\bar{\alpha}(kh-s)} \mathrm{d}s \times \frac{x(lh+h) - x(lh)}{h}, \quad k \in \mathbb{N}_0, \end{aligned}$$

where $\bar{\alpha} = \frac{\alpha}{\alpha - 1}$.

Utilizing the above relation, it easily gains the explicit Euler difference form for CF-BAMNNs (6.1) described by

$$\begin{cases} x_{i,k+1} = x_{i,k} - \displaystyle\sum_{l=1}^{k-1} e^{\bar{\alpha}_i(k-l)h} \big(x_{i,l+1} - x_{i,l}\big) \\ \qquad + \dfrac{\alpha_i h}{1 - e^{\bar{\alpha}_i h}} \Big[ - c_i x_{i,k} + F_{i,k}(y_{\xi_i}) + H_{\alpha_i,k} \Big], \\ y_{j,k+1} = y_{j,k} - \displaystyle\sum_{l=1}^{k-1} e^{\bar{\beta}_j(k-l)h} \big(y_{j,l+1} - y_{j,l}\big) \\ \qquad + \dfrac{\beta_j h}{1 - e^{\bar{\beta}_j h}} \Big[ - d_j y_{j,k} + G_{j,k}(x_{\sigma_j}) + H_{\beta_j,k} \Big], \end{cases} \tag{6.19}$$

where

$$\bar{\alpha}_i = \frac{\alpha_i}{\alpha_i - 1}, \quad \bar{\beta}_j = \frac{\beta_j}{\beta_j - 1},$$

$$H_{\alpha_i,k} = \big[c_i x_{i,0} - F_{i,0}(y_{\xi_i})\big] e^{\bar{\alpha}_i kh}, \quad H_{\beta_j,k} = \big[d_j y_{j,0} - G_{j,0}(x_{\sigma_j})\big] e^{\bar{\beta}_j kh},$$

$k \in \mathbb{N}_0, i = 1, 2, \dots, m, j = 1, 2, \dots, n.$

So, the predicted values $x^p_{i,k+1}$ and $y^p_{j,k+1}$ in Eq. (6.17) can be computed by

$$
\begin{cases}
x_{i,k+1}^{p} = x_{i,k} - \sum_{l=1}^{k-1} e^{\bar{\alpha}_i(k-l)h} \left(x_{i,l+1} - x_{i,l}\right) \\
\quad + \dfrac{\alpha_i h}{1 - e^{\bar{\alpha}_i h}} \Big[ - c_i x_{i,k} + F_{i,k}(y_{\xi_i}) + H_{\alpha_i,k} \Big], \\
y_{j,k+1}^{p} = y_{j,k} - \sum_{l=1}^{k-1} e^{\bar{\beta}_j(k-l)h} \left(y_{j,l+1} - y_{j,l}\right) \\
\quad + \dfrac{\beta_j h}{1 - e^{\bar{\beta}_j h}} \Big[ - d_j y_{j,k} + G_{j,k}(x_{\sigma_j}) + H_{\beta_j,k} \Big]
\end{cases}
\tag{6.20}
$$

with the following initial conditions

$$
x_{i,s} = \phi_{i,s}, \quad y_{j,s} = \psi_{j,s}, \quad s \in [-\xi_h, 0]_{\mathbb{Z}},
$$

where $k \in \mathbb{N}_0$, $i = 1, 2, \ldots, m$, $j = 1, 2, \ldots, n$. Equation (6.20) is called predicted formula.

---

**Algorithm 3** PECE algorithm for EED-CF-BAMNNs (6.14)

---

(1) Predict: predict $x_{i,k+1}^{p}$ and $y_{j,k+1}^{p}$ by Eq. (6.20).
(2) Evaluate: compute $F_{i,k+1}$ and $G_{j,k+1}$ by Eq. (6.18).
(3) Correct: correct $x_{i,k+1}$ and $y_{j,k+1}$ by Eq. (6.17).
(4) Evaluate: compute the right sides of Eq. (6.20) and start next iteration.

---

## 6.5 Asymptotically Almost Automorphic Sequence

The concept of almost automorphism was first introduced by Bochner [56–58] in 1955. The main objective of this section is to present the definition and basic properties of almost automorphic sequence.

Let $(\mathbb{M}, \mathbf{d})$ be a metric space and $\{x_k\}_{k\in\mathbb{N}_0} \subseteq \mathbb{M}$ be a sequence.

**Definition 6.2** *([59])* A sequence $x : \mathbb{Z} \to \mathbb{R}^n$ is said to be an almost automorphic sequence in case for every sequence $\{\bar{s}_l\}_{l\in\mathbb{Z}}$, it exists a subsequence $\{s_l\}_{l\in\mathbb{Z}}$ satisfying

$$
\tilde{x}_k = \lim_{l\to\infty} x_{k+s_l}
$$

is well defined for each $k \in \mathbb{Z}$ and

$$
\lim_{l\to\infty} \tilde{x}_{k-s_l} = x_k
$$

for each $k \in \mathbb{Z}$. The collection of these functions is denoted by $\mathbb{AA}(\mathbb{Z}, \mathbb{R}^n)$.

If the convergence of Definition 6.2 is uniform for all $k \in \mathbb{Z}$, then $x$ is an almost periodic sequence. As a consequence, each almost periodic sequence is almost automorphic, but the converse is not always valid. The sequence below is an almost automorphic sequence, which is not almost periodic,

$$x_k = \operatorname{sgn}\cos(2\pi k\theta) = \begin{cases} 1 & \text{if } \cos(2\pi k\theta) > 0, \\ -1 & \text{if } \cos(2\pi k\theta) < 0, \end{cases}$$

where $\theta$ is an irrational real number.

**Lemma 6.4** ([59]) *If* $x \in \mathbb{AA}(\mathbb{Z}, \mathbb{R}^n)$ *and* $\lim_{k\to\infty} \mathbf{d}(x_k, 0) = 0$, *then* $x_k = 0$ *for all* $k \in \mathbb{Z}$.

**Lemma 6.5** ([59]) *Let* $f(k, x) : \mathbb{Z} \times \mathbb{R}^n \to \mathbb{R}^n$ *be almost automorphic in* $k \in \mathbb{Z}$ *uniformly for* $x \in B$, *where* $B \subseteq \mathbb{R}^n$ *is bounded. Further, if* $f$ *is Lipschitz in* $x \in \mathbb{R}^n$ *uniformly for* $k \in \mathbb{Z}$, *then for arbitrary sequence* $x = \{x_k\}_{k\in\mathbb{Z}} \in \mathbb{AA}(\mathbb{Z}, \mathbb{R}^n)$, $F(k) := f(k, x_k)$ *is almost automorphic.*

**Definition 6.3** *([59])* A sequence $x : \mathbb{N}_0 \to \mathbb{R}^n$ is said to be an asymptotically almost automorphic sequence in case it can be decomposed as

$$x_k = x_k^{\aleph} + x_k^0, \quad \forall k \in \mathbb{N}_0,$$

where $x^{\aleph} = \{x_k^{\aleph}\}_{k\in\mathbb{N}_0} \in \mathbb{AA}(\mathbb{Z}, \mathbb{R}^n)$ and $x^0 \in N_0(\mathbb{N}_0, \mathbb{R}^n)$,

$$N_0(\mathbb{N}_0, \mathbb{R}^n) = \{x^0 : \mathbb{N}_0 \to \mathbb{R}^n | \lim_{k\to\infty} \mathbf{d}(x_k^0, 0) = 0\}.$$

The collection of these functions is denoted by $\mathbb{AAA}(\mathbb{N}_0, \mathbb{R}^n)$.

**Lemma 6.6** ([59]) *The decomposition of an asymptotically almost automorphic sequence is unique.*

Considering EED-CF-BAMNNs (6.14) with the following initial conditions

$$x_{i,s} = \phi_{i,s}, \quad y_{j,s} = \psi_{j,s}, \quad s \in [-\xi_h, 0]_{\mathbb{Z}}.$$

Define

$$\phi_0 = \max_{1\le i\le m, 1\le j\le n} \sup_{s\in[-\xi_h,0]_{\mathbb{Z}}} \left\{|\phi_{i,s}|, |\psi_{j,s}|\right\}.$$

Let $\mathbb{N}_{\xi_h} = [-\xi_h, +\infty)_{\mathbb{Z}}$, $\mathbb{M} = \mathbb{R}^{m+n}$ and

$$\mathbb{C}(\mathbb{N}_{\xi_h}, \mathbb{M}) = \Big\{ z = (x_1, \ldots, x_m, y_1, \ldots, y_n)^T : \mathbb{N}_{\xi_h} \to \mathbb{M} | x_{i,s} = \phi_{i,s},$$

$$y_{j,s} = \psi_{j,s}, s \in [-\xi_h, 0]_{\mathbb{Z}}, i = 1, 2, \ldots, m, j = 1, 2, \ldots, n \Big\}.$$

Clearly, $\mathbb{C}(\mathbb{N}_{\xi_h}, \mathbb{M})$ is a complete metric space with the distance

$$\mathbf{d}(z, \tilde{z}) = \sup_{k\in\mathbb{N}_{\xi_h}} \|z - \tilde{z}\| = \sup_{k\in\mathbb{N}_{\xi_h}} \max_{1\le i\le m, 1\le j\le n} \{|x_{i,k} - \tilde{x}_{i,k}|, |y_{j,k} - \tilde{y}_{j,k}|\}$$

for any $z = (x_1, \ldots, x_m, y_1, \ldots, y_n)^T$ and $\tilde{z} = (\tilde{x}_1, \ldots, \tilde{x}_m, \tilde{y}_1, \ldots, \tilde{y}_n)^T$ in $\mathbb{C}(\mathbb{N}_{\xi_h}, \mathbb{M})$.

Define

$$\mathbb{AAA}_c(\mathbb{M}) = \{x \in \mathbb{AAA}(\mathbb{N}_0, \mathbb{M}) \cap \mathbb{C}(\mathbb{N}_{\xi_h}, \mathbb{M}) | \mathbf{d}(x, 0) \le c\}$$

for some $c > 0$.

For simplicity, let

$$\eta_{\alpha_i} = \frac{(1-\alpha_i)(1+e^{p_{\alpha_i} h})}{|K_{\alpha_i}|}, \quad \eta_{\beta_j} = \frac{(1-\beta_j)(1+e^{p_{\beta_j} h})}{|K_{\beta_j}|},$$

$$\mathcal{F}_i^0 = \sum_{j=1}^{n} a_{ij}^* |f_j(0)| + \bigwedge_{j=1}^{n} \gamma_{ij}^* |f_j(0)| + \bigvee_{j=1}^{n} \eta_{ij}^* |f_j(0)| + \bigwedge_{j=1}^{n} T_{ij}^* |\rho_j| + \bigvee_{j=1}^{n} Q_{ij}^* |\rho_j| + I_i^*,$$

$$\mathcal{G}_j^0 = \sum_{i=1}^{m} b_{ji}^* |g_i(0)| + \bigwedge_{i=1}^{m} \varepsilon_{ji}^* |g_i(0)| + \bigvee_{i=1}^{m} \vartheta_{ji}^* |g_i(0)| + \bigwedge_{i=1}^{m} W_{ji}^* |\varrho_i| + \bigvee_{i=1}^{m} R_{ji}^* |\varrho_i| + P_j^*,$$

where $k \in \mathbb{N}_0$, $i = 1, 2, \ldots, m$, $j = 1, 2, \ldots, n$.

**Proposition 6.1** *Let* $(A_1)$ *and* $(A_2)$ *hold. If* $z = (x_1, \ldots, x_m, y_1, \ldots, y_n)^T \in \mathbb{AAA}_c(\mathbb{M})$, *then*

$$\left|F_{i,k}(y_{\xi_i})\right| \le \varpi_i^f c + \mathcal{F}_i^0, \quad \left|G_{j,k}(x_{\sigma_j})\right| \le \varpi_j^g c + \mathcal{G}_j^0,$$

*where* $k \in \mathbb{N}_0$, $i = 1, 2, \ldots, m$, $j = 1, 2, \ldots, n$.

***Proof*** Let

$$F_{i,k}(0) = \sum_{j=1}^{n} a_{ij,k} f_j(0) + \bigwedge_{j=1}^{n} \gamma_{ij,k} f_j(0)$$

$$+ \bigvee_{j=1}^{n} \eta_{ij,k} f_j(0) + \bigwedge_{j=1}^{n} T_{ij,k}\rho_j + \bigvee_{j=1}^{n} Q_{ij,k}\rho_j + I_{i,k},$$

where $k \in \mathbb{N}_0$, $i = 1, 2, \ldots, m$. From the definitions of $F_{i,k}$ and $G_{j,k}$, $(A_1)$ and $(A_2)$ and Lemma 6.1, it has

$$\begin{aligned}\left|F_{i,k}(y_{\xi_i})\right| &\le \left|F_{i,k}(y_{\xi_i}) - F_{i,k}(0)\right| + \left|F_{i,k}(0)\right| \\ &\le \sum_{j=1}^{n}\left(a_{ij}^* + \gamma_{ij}^* + \eta_{ij}^*\right)\left|f_j(y_{j,k-\xi_{ij}^h}) - f_j(0)\right| + \mathcal{F}_i^0 \\ &\le \varpi_i^f c + \mathcal{F}_i^0,\end{aligned}$$

where $k \in \mathbb{N}_0$, $i = 1, 2, \ldots, m$. Similarly, $\left|G_{j,k}(x_{\sigma_j})\right| \le \varpi_j^g c + \mathcal{G}_j^0$, $\forall k \in \mathbb{N}_0$, $j = 1, 2, \ldots, n$. The proof is achieved. □

In the light of Proposition 6.1, it calculates

$$\frac{1-\alpha_i}{K_{\alpha_i}}\left|F_{i,k+1}(y_{\xi_i}) - e^{p_{\alpha_i}h}F_{i,k}(y_{\xi_i})\right| \le \eta_{\alpha_i}(\varpi_i^f c + \mathcal{F}_i^0) \tag{6.21}$$

and

$$\frac{1-\beta_j}{K_{\beta_j}}\left|G_{j,k+1}(x_{\sigma_j}) - e^{p_{\beta_j}h}G_{j,k}(x_{\sigma_j})\right| \le \eta_{\beta_j}(\varpi_j^g c + \mathcal{G}_j^0), \tag{6.22}$$

where $k \in \mathbb{N}_0$, $i = 1, 2, \ldots, m$, $j = 1, 2, \ldots, n$. The estimations (6.21) and (6.22) are important for the subsequent study of this chapter.

**Proposition 6.2** *Let $z = (x_1, \ldots, x_m, y_1, \ldots, y_n)^T$ and $\tilde{z} = (\tilde{x}_1, \ldots, \tilde{x}_m, \tilde{y}_1, \ldots, \tilde{y}_n)^T$ belong to $\mathbb{AAA}_c(\mathbb{M})$. If $(A_1)$ and $(A_2)$ hold, then*

$$\left|F_{i,k}(y_{\xi_i}) - F_{i,k}(\tilde{y}_{\xi_i})\right| \le \varpi_i^f \mathbf{d}(z, \tilde{z}), \quad \left|G_{j,k}(x_{\sigma_j}) - G_{j,k}(\tilde{x}_{\sigma_j})\right| \le \varpi_j^g \mathbf{d}(z, \tilde{z}),$$

*where $k \in \mathbb{N}_0$, $i = 1, 2, \ldots, m$, $j = 1, 2, \ldots, n$.*

***Proof*** From the definitions of $F_{i,k}$ and $G_{j,k}$, it results in

$$\begin{aligned}\left|F_{i,k}(y_{\xi_i}) - F_{i,k}(\tilde{y}_{\xi_i})\right| &\le \left[\sum_{j=1}^{n} a_{ij}^* + \bigwedge_{j=1}^{n}\gamma_{ij}^* + \bigvee_{j=1}^{n}\eta_{ij}^*\right]\left|f_j(y_{j,k-\xi_{ij}^h}) - f_j(\tilde{y}_{j,k-\xi_{ij}^h})\right| \\ &\le \sum_{j=1}^{n}(a_{ij}^* + \gamma_{ij}^* + \eta_{ij}^*)L_j^f\left|y_{j,k-\xi_{ij}^h} - \tilde{y}_{j,k-\xi_{ij}^h}\right| \\ &\le \varpi_i^f \mathbf{d}(z, \tilde{z})\end{aligned} \tag{6.23}$$

and $\left|G_{j,k}(x_{\sigma_j}) - G_{j,k}(\tilde{x}_{\sigma_j})\right| \le \varpi_j^g \mathbf{d}(z, \tilde{z})$, where $k \in \mathbb{N}_0$, $i = 1, 2, \ldots, m$, $j = 1, 2, \ldots, n$. The proof is achieved. □

By (6.23), it obtains

$$\frac{1-\alpha_i}{|K_{\alpha_i}|}\Big|F_{i,k+1}(y_{\xi_i})-F_{i,k+1}(\tilde{y}_{\xi_i})\Big| \\ \leq \frac{(1-\alpha_i)\varpi_i^f}{|K_{\alpha_i}|}\max_{1\leq j\leq n}\Big|y_{j,k+1-\xi_{ij}^h}-\tilde{y}_{j,k+1-\xi_{ij}^h}\Big|, \tag{6.24}$$

and similarly,

$$\frac{1-\beta_j}{K_{\beta_j}}\Big|G_{j,k+1}(x_{\sigma_j})-G_{j,k+1}(\tilde{x}_{\sigma_j})\Big| \\ \leq \frac{(1-\beta_j)\varpi_j^g}{|K_{\beta_j}|}\max_{1\leq i\leq m}\Big|x_{i,k+1-\sigma_{ji}^h}-\tilde{x}_{i,k+1-\sigma_{ji}^h}\Big|, \tag{6.25}$$

where $k\in\mathbb{N}_0,\ i=1,2,\ldots,m,\ j=1,2,\ldots,n$. (6.24) and (6.25) are crucial to verify global exponential stability and synchronization of EED-CF-BAMNNs (6.14).

The assumption below is necessary for the subsequent discussion.

(A$_3$) $a_{ij}$, $\gamma_{ij}$, $\eta_{ij}$, $T_{ij}$, $Q_{ij}$, $b_{ji}$, $\varepsilon_{ji}$, $\vartheta_{ji}$, $W_{ji}$ and $R_{ji}$ are asymptotically almost automorphic sequences, $i=1,2,\ldots,m,\ j=1,2,\ldots,n$.

**Proposition 6.3** *Let (A$_1$)–(A$_3$) hold. If $z=(x_1,\ldots,x_m,y_1,\ldots,y_n)^T\in\mathbb{AAA}_c(\mathbb{M})$, then*

$$F_i(y_{\xi_i})\in\mathbb{AAA}(\mathbb{N}_0,\mathbb{R}),\quad G_j(x_{\sigma_j})\in\mathbb{AAA}(\mathbb{N}_0,\mathbb{R}),$$

*where $k\in\mathbb{N}_0$, $i=1,2,\ldots,m$, $j=1,2,\ldots,n$.*

***Proof*** By (A$_3$), it has the decompositions as described below

$$x_{i,k}=x_{i,k}^{\aleph}+x_{i,k}^{0},\quad y_{j,k}=y_{j,k}^{\aleph}+y_{j,k}^{0},$$

$$a_{ij,k}=a_{ij,k}^{\aleph}+a_{ij,k}^{0},\quad \gamma_{ij,k}=\gamma_{ij,k}^{\aleph}+\gamma_{ij,k}^{0},\quad \eta_{ij,k}=\eta_{ij,k}^{\aleph}+\eta_{ij,k}^{0},$$

$$T_{ij,k}=T_{ij,k}^{\aleph}+T_{ij,k}^{0},\quad Q_{ij,k}=Q_{ij,k}^{\aleph}+Q_{ij,k}^{0},\quad I_{i,k}=I_{i,k}^{\aleph}+I_{i,k}^{0},$$

$$b_{ji,k}=b_{ji,k}^{\aleph}+b_{ji,k}^{0},\quad \varepsilon_{ji,k}=\varepsilon_{ji,k}^{\aleph}+\varepsilon_{ji,k}^{0},\quad \vartheta_{ji,k}=\vartheta_{ji,k}^{\aleph}+\vartheta_{ji,k}^{0},$$

$$W_{ji,k}=W_{ji,k}^{\aleph}+W_{ji,k}^{0},\quad R_{ji,k}=R_{ji,k}^{\aleph}+R_{ji,k}^{0},\quad P_{j,k}=P_{j,k}^{\aleph}+P_{j,k}^{0},$$

where $x_i^{\aleph}$, $y_j^{\aleph}$, $a_{ij}^{\aleph}$, $\gamma_{ij}^{\aleph}$, $\eta_{ij}^{\aleph}$, $T_{ij}^{\aleph}$, $Q_{ij}^{\aleph}$, $b_{ji}^{\aleph}$, $\varepsilon_{ji}^{\aleph}$, $\vartheta_{ji}^{\aleph}$, $W_{ji}^{\aleph}$, $R_{ji}^{\aleph}\in\mathbb{AA}(\mathbb{Z},\mathbb{R})$, and $x_i^0$, $y_j^0$, $a_{ij}^0$, $\gamma_{ij}^0$, $\eta_{ij}^0$, $T_{ij}^0$, $Q_{ij}^0$, $b_{ji}^0$, $\varepsilon_{ji}^0$, $\vartheta_{ji}^0$, $W_{ji}^0$, $R_{ji}^0\in N_0(\mathbb{N}_0,\mathbb{R})$, $i=1,2,\ldots,m$, $j=1,2,\ldots,n$.

In view of the definition of $F_{i,k}$, it gets

$$F_{i,k}(y_{\xi_i})=F_{i,k}^1+F_{i,k}^2+F_{i,k}^3,$$

where

$$
\begin{aligned}
F_{i,k}^1 &= \sum_{j=1}^{n} a_{ij,k}^{\aleph} f_j(y_{j,k-\xi_{ij}^h}^{\aleph}) + \bigwedge_{j=1}^{n} \gamma_{ij,k}^{\aleph} f_j(y_{j,k-\xi_{ij}^h}^{\aleph}) \\
&\quad + \bigvee_{j=1}^{n} \eta_{ij,k}^{\aleph} f_j(y_{j,k-\xi_{ij}^h}^{\aleph}) + \bigwedge_{j=1}^{n} T_{ij,k}^{\aleph} \rho_j \\
&\quad + \bigvee_{j=1}^{n} Q_{ij,k}^{\aleph} \rho_j + I_{i,k}^{\aleph}, \quad \forall k \in \mathbb{Z},
\end{aligned}
$$

$$
\begin{aligned}
F_{i,k}^2 &= \sum_{j=1}^{n} a_{ij,k}^{\aleph} \Big[ f_j(y_{j,k-\xi_{ij}^h}^{\aleph} + y_{j,k-\xi_{ij}^h}^{0}) - f_j(y_{j,k-\xi_{ij}^h}^{\aleph}) \Big] \\
&\quad + \bigwedge_{j=1}^{n} \gamma_{ij,k}^{\aleph} \Big[ f_j(y_{j,k-\xi_{ij}^h}^{\aleph} + y_{j,k-\xi_{ij}^h}^{0}) - f_j(y_{j,k-\xi_{ij}^h}^{\aleph}) \Big] \\
&\quad + \bigvee_{j=1}^{n} \eta_{ij,k}^{\aleph} \Big[ f_j(y_{j,k-\xi_{ij}^h}^{\aleph} + y_{j,k-\xi_{ij}^h}^{0}) - f_j(y_{j,k-\xi_{ij}^h}^{\aleph}) \Big] \\
&\quad + \bigwedge_{j=1}^{n} T_{ij,k}^{\aleph} \rho_j + \bigvee_{j=1}^{n} Q_{ij,k}^{\aleph} \rho_j + I_{i,k}^{\aleph}, \quad \forall k \in \mathbb{N}_0,
\end{aligned}
$$

$$
\begin{aligned}
F_{i,k}^3 &= \sum_{j=1}^{n} a_{ij,k}^{0} f_j(y_{j,k-\xi_{ij}^h}) + \bigwedge_{j=1}^{n} \gamma_{ij,k}^{0} f_j(y_{j,k-\xi_{ij}^h}) \\
&\quad + \bigvee_{j=1}^{n} \eta_{ij,k}^{0} f_j(y_{j,k-\xi_{ij}^h}) + \bigwedge_{j=1}^{n} T_{ij,k}^{0} \rho_j \\
&\quad + \bigvee_{j=1}^{n} Q_{ij,k}^{0} \rho_j + I_{i,k}^{0}, \quad \forall k \in \mathbb{N}_0,
\end{aligned}
$$

where $i = 1, 2, \ldots, m$. By Lemma 6.5, $F_i^1 \in \mathbb{AA}(\mathbb{Z}, \mathbb{R})$, $i = 1, 2, \ldots, m$. Further, it easily proves that

$$\lim_{k \to \infty} \|F_{i,k}^2\| = 0 = \lim_{k \to \infty} \|F_{i,k}^3\|$$

by employing assumption $(A_2)$, the boundedness of $f_j$ and other parameters, $i = 1, 2, \ldots, m$, $j = 1, 2, \ldots, n$. Consequently, $F_i(y_{\xi_i}) \in \mathbb{AAA}(\mathbb{N}_0, \mathbb{R})$, $i = 1, 2, \ldots, m$. Similarly, $G_j(x_{\sigma_j}) \in \mathbb{AAA}(\mathbb{N}_0, \mathbb{R})$, $j = 1, 2, \ldots, n$. The proof is achieved. □

Define $c = \dfrac{\zeta_2}{1-\zeta_1}$, where

$$\zeta_1 = \max_{1\le i\le m, 1\le j\le n}\left\{\left[\frac{1}{c_i K_{\alpha_i}} + \frac{\eta_{\alpha_i}}{1-e^{p_{\alpha_i}h}}\right]\varpi_i^f, \left[\frac{1}{d_j K_{\beta_j}} + \frac{\eta_{\beta_j}}{1-e^{p_{\beta_j}h}}\right]\varpi_j^g\right\},$$

$$\zeta_2 = \phi_0 + \max_{1\le i\le m, 1\le j\le n}\left\{\left[\frac{1}{c_i K_{\alpha_i}} + \frac{\eta_{\alpha_i}}{1-e^{p_{\alpha_i}h}}\right]\mathcal{F}_i^0, \left[\frac{1}{d_j K_{\beta_j}} + \frac{\eta_{\beta_j}}{1-e^{p_{\beta_j}h}}\right]\mathcal{G}_j^0\right\}.$$

**Theorem 6.1** *Let* $(A_1)$*–*$(A_3)$ *and the assumption below be fulfilled.*

($A_4$) $\min_{1\le i\le m, 1\le j\le n}\{a_i, d_j\} > 0$ *and* $\zeta_1 < 1$.

*Then EED-CF-BAMNNs* (6.14) *possesses a unique asymptotically almost automorphic sequence solution* $z^*$ *satisfying* $\mathbf{d}(z^*, 0) \le c$.

***Proof*** According to Eq. (6.16), for any $z = (x_1, \ldots, x_m, y_1, \ldots, y_n)^T \in \mathbb{AAA}_c(\mathbb{M})$, define the operator $\mathbf{T} : \mathbb{AAA}_c(\mathbb{M}) \to \mathbb{C}(\mathbb{N}_{\xi_0}, \mathbb{M})$ as follows

$$(\mathbf{T}z)_k = (\Phi_{1,k}^z, \ldots, \Phi_{m,k}^z, \Psi_{1,k}^z, \ldots, \Psi_{n,k}^z)^T,$$

where

$$\Phi_{i,s}^z = \phi_i(s), \quad \Psi_{j,s}^z = \psi_{j,s}, \quad \forall s \in [-\xi_h, 0]_{\mathbb{Z}}$$

and

$$\begin{cases} \Phi_{i,k}^z = e^{p_{\alpha_i}kh}x_{i,0} + \dfrac{1-e^{p_{\alpha_i}h}}{c_i K_{\alpha_i}}\displaystyle\sum_{q=0}^{k-1} e^{p_{\alpha_i}(k-q-1)h}F_{i,q}(y_{\xi_i}) \\ \qquad + \dfrac{1-\alpha_i}{K_{\alpha_i}}\displaystyle\sum_{q=0}^{k-1} e^{p_{\alpha_i}(k-q-1)h}\Big[F_{i,q+1}(y_{\xi_i}) - e^{p_{\alpha_i}h}F_{i,q}(y_{\xi_i})\Big], \\ \Psi_{j,k}^z = e^{p_{\beta_j}kh}y_{j,0} + \dfrac{1-e^{p_{\beta_j}h}}{d_j K_{\beta_j}}\displaystyle\sum_{q=0}^{k-1} e^{p_{\beta_j}(k-q-1)h}G_{j,q}(x_{\sigma_j}) \\ \qquad + \dfrac{1-\beta_j}{K_{\beta_j}}\displaystyle\sum_{q=0}^{k-1} e^{p_{\beta_j}(k-q-1)h}\Big[G_{j,q+1}(x_{\sigma_j}) - e^{p_{\beta_j}h}G_{j,q}(x_{\sigma_j})\Big] \end{cases} \tag{6.26}$$

for all $k \in \mathbb{N}_0$, $i = 1, 2, \ldots, m$, $j = 1, 2, \ldots, n$.

By the first equation of Eq. (6.26), it derives from Proposition 6.1, (6.21) and ($A_3$) that

$$\left|\Phi_{i,k}^z\right| \le \phi_0 + \left[\frac{1}{c_i K_{\alpha_i}} + \frac{\eta_{\alpha_i}}{1-e^{p_{\alpha_i}h}}\right](\varpi_i^f c + \mathcal{F}_i^0) \le \zeta_1 c + \zeta_2 = c, \tag{6.27}$$

where $k \in \mathbb{N}_0, i = 1, 2, \ldots, m$. Similar to the derivation in (6.27), it computes from the second equation of Eq. (6.26) that

$$\left|\Psi_{j,k}^{z}\right| \leq \zeta_1 c + \zeta_2 = c, \tag{6.28}$$

where $k \in \mathbb{N}_0,\ j = 1, 2, \ldots, n$. Inequalities (6.27) and (6.28) imply

$$\mathbf{d}(\mathbf{T}z, 0) \leq c, \quad \forall z \in \mathbb{AAA}_c(\mathbb{M}).$$

Next, $\mathbf{T}z \in \mathbb{AAA}(\mathbb{N}_0, \mathbb{M})$ should be verified for arbitrary $z \in \mathbb{AAA}_c(\mathbb{M})$. Let $z \in \mathbb{AAA}_c(\mathbb{M})$. According to Proposition 6.3, it has the decompositions as

$$F_i(y_{\xi_i}) = F_i^{\aleph}(y_{\xi_i}) + F_i^{0}(y_{\xi_i}), \quad G_j(x_{\sigma_j}) = G_j^{\aleph}(x_{\sigma_j}) + G_j^{0}(x_{\sigma_j}),$$

where $F_i^{\aleph}, G_j^{\aleph} \in \mathbb{AA}(\mathbb{Z}, \mathbb{R}), F_i^{0}, G_j^{0} \in N_0(\mathbb{N}_0, \mathbb{R}), i = 1, 2, \ldots, m, j = 1, 2, \ldots, n$.

By employing the first equation of Eq. (6.26), it derives

$$\Phi_{i,k}^{z} = \hat{\Phi}_{i,k}^{z} + \check{\Phi}_{i,k}^{z},$$

where

$$\begin{aligned}
\hat{\Phi}_{i,k}^{z} = {} & \frac{1 - e^{p_{\alpha_i} h}}{c_i K_{\alpha_i}} \sum_{q=-\infty}^{k-1} e^{p_{\alpha_i}(k-q-1)h} F_{i,q}^{\aleph}(y_{\xi_i}) \\
& + \frac{1 - \alpha_i}{K_{\alpha_i}} \sum_{q=-\infty}^{k-1} e^{p_{\alpha_i}(k-q-1)h} \Big[F_{i,q+1}^{\aleph}(y_{\xi_i}) - e^{p_{\alpha_i} h} F_{i,q}^{\aleph}(y_{\xi_i})\Big], \quad \forall k \in \mathbb{Z},
\end{aligned}$$

$$\begin{aligned}
\check{\Phi}_{i,k}^{z} = {} & e^{p_{\alpha_i} kh} x_{i,0} + \frac{1 - e^{p_{\alpha_i} h}}{c_i K_{\alpha_i}} \sum_{q=0}^{k-1} e^{p_{\alpha_i}(k-q-1)h} F_{i,q}^{0}(y_{\xi_i}) \\
& + \frac{1 - \alpha_i}{K_{\alpha_i}} \sum_{q=0}^{k-1} e^{p_{\alpha_i}(k-q-1)h} \Big[F_{i,q+1}^{0}(y_{\xi_i}) - e^{p_{\alpha_i} h} F_{i,q}^{0}(y_{\xi_i})\Big] \\
& - \frac{1 - e^{p_{\alpha_i} h}}{c_i K_{\alpha_i}} \sum_{q=-\infty}^{-1} e^{p_{\alpha_i}(k-q-1)h} F_{i,q}^{\aleph}(y_{\xi_i}) - \frac{1 - \alpha_i}{K_{\alpha_i}} \sum_{q=-\infty}^{-1} e^{p_{\alpha_i}(k-q-1)h} \\
& \times \Big[F_{i,q+1}^{\aleph}(y_{\xi_i}) - e^{p_{\alpha_i} h} F_{i,q}^{\aleph}(y_{\xi_i})\Big], \quad \forall k \in \mathbb{N}_0,
\end{aligned}$$

where $i = 1, 2, \ldots, m$. By the almost automorphism of $F_i^{\aleph}$, for every sequence $\{\bar{s}_l\}_{l \in \mathbb{Z}}$, it exists a subsequence $\{s_l\}_{l \in \mathbb{Z}}$ satisfying

$$\tilde{F}^{\aleph}_{i,k} = \lim_{l\to\infty} F^{\aleph}_{i,k+s_l}$$

for each $k \in \mathbb{Z}$ and

$$\lim_{l\to\infty} \tilde{F}^{\aleph}_{i,k-s_l} = F^{\aleph}_{i,k}$$

for each $k \in \mathbb{Z}$. So

$$\begin{aligned}
\hat{\Phi}^{z}_{i,k+s_l} &= \frac{1-e^{p_{\alpha_i}h}}{c_i K_{\alpha_i}} \sum_{q=-\infty}^{k+s_l-1} e^{p_{\alpha_i}(k+s_l-q-1)h} F^{\aleph}_{i,q}(y_{\xi_i}) \\
&\quad + \frac{1-\alpha_i}{K_{\alpha_i}} \sum_{q=-\infty}^{k+s_l-1} e^{p_{\alpha_i}(k+s_l-q-1)h} \Big[ F^{\aleph}_{i,q+1}(y_{\xi_i}) - e^{p_{\alpha_i}h} F^{\aleph}_{i,q}(y_{\xi_i}) \Big], \\
&= \frac{1-e^{p_{\alpha_i}h}}{c_i K_{\alpha_i}} \sum_{q=-\infty}^{k-1} e^{p_{\alpha_i}(k-q-1)h} F^{\aleph}_{i,q+s_l}(y_{\xi_i}) + \frac{1-\alpha_i}{K_{\alpha_i}} \sum_{q=-\infty}^{k-1} e^{p_{\alpha_i}(k-q-1)h} \\
&\quad \times \Big[ F^{\aleph}_{i,q+s_l+1}(y_{\xi_i}) - e^{p_{\alpha_i}h} F^{\aleph}_{i,q+s_l}(y_{\xi_i}) \Big],
\end{aligned} \tag{6.29}$$

where $l, k \in \mathbb{Z}, i = 1, 2, \ldots, m$. It should be noted that $F^{\aleph}_i(y_{\xi_i}) = F^1_i$ and $F^0_i(y_{\xi_i}) = F^2_i + F^3_i$, where $F^l_i$ is defined in the proof of Proposition 6.3, $l = 1, 2, 3$, $i = 1, 2, \ldots, m$. Similar to Proposition 6.1, it must exist a number $C_1 > 0$ large enough such that

$$\left|F^{\aleph}_{i,k}(y_{\xi_i})\right| < C_1 \text{ for } k \in \mathbb{Z}, \quad \left|F^{0}_{i,k}(y_{\xi_i})\right| < C_1 \text{ for } k \in \mathbb{N}_0,$$

where $i = 1, 2, \ldots, m$. Noting that

$$\left| \sum_{q=-\infty}^{k-1} e^{p_{\alpha_i}(k-q-1)h} F^{\aleph}_{i,q+l}(y_{\xi_i}) \right| \le \frac{C_1}{1-e^{p_{\alpha_i}h}}, \quad \forall l, k \in \mathbb{Z}.$$

By the Lebesgue dominated convergence theorem, it gets from Eq. (6.29) that

$$\begin{aligned}
\lim_{l\to\infty} \hat{\Phi}^{z}_{i,k+s_l} &= \frac{1-e^{p_{\alpha_i}h}}{c_i K_{\alpha_i}} \sum_{q=-\infty}^{k-1} e^{p_{\alpha_i}(k-q-1)h} \tilde{F}^{\aleph}_{i,q}(y_{\xi_i}) \\
&\quad + \frac{1-\alpha_i}{K_{\alpha_i}} \sum_{q=-\infty}^{k-1} e^{p_{\alpha_i}(k-q-1)h} \Big[ \tilde{F}^{\aleph}_{i,q+1}(y_{\xi_i}) \\
&\quad - e^{p_{\alpha_i}h} \tilde{F}^{\aleph}_{i,q}(y_{\xi_i}) \Big] = \boldsymbol{\Phi}^{z}_{i,k}
\end{aligned}$$

for each $k \in \mathbb{Z}$, $i = 1, 2, \ldots, m$. Similarly, it achieves $\lim_{l\to\infty} \mathbf{\Phi}^z_{i,k-s_l} = \hat{\Phi}^z_{i,k}$ for each $k \in \mathbb{Z}$, $i = 1, 2, \ldots, m$. Consequently, $\hat{\Phi}^z_i \in \mathbb{AA}(\mathbb{Z}, \mathbb{R})$, $i = 1, 2, \ldots, m$.

On the other hand, for arbitrary $\epsilon > 0$, it has a positive integer $T$ such that $|F^0_{i,k}(y_{\xi_i})| \leq \epsilon$ for $k \in [T, +\infty)_{\mathbb{Z}}$, $i = 1, 2, \ldots, m$. For $k \geq 2T$, it gains

$$
\begin{aligned}
|\check{\Phi}^z_{i,k}| \leq & e^{p_{\alpha_i}kh}\phi_0 + \frac{1-e^{p_{\alpha_i}h}}{c_iK_{\alpha_i}} \sum_{q=0}^{\lfloor k/2\rfloor-1} e^{p_{\alpha_i}(k-q-1)h}|F^0_{i,q}(y_{\xi_i})| \\
& + \frac{1-\alpha_i}{K_{\alpha_i}} \sum_{q=0}^{\lfloor k/2\rfloor-1} e^{p_{\alpha_i}(k-q-1)h}\Big|F^0_{i,q+1}(y_{\xi_i}) - e^{p_{\alpha_i}h}F^0_{i,q}(y_{\xi_i})\Big| \\
& + \frac{1-e^{p_{\alpha_i}h}}{c_iK_{\alpha_i}} \sum_{q=\lfloor k/2\rfloor}^{k-1} e^{p_{\alpha_i}(k-q-1)h}|F^0_{i,q}(y_{\xi_i})| \\
& + \frac{1-\alpha_i}{K_{\alpha_i}} \sum_{q=\lfloor k/2\rfloor}^{k-1} e^{p_{\alpha_i}(k-q-1)h}\Big|F^0_{i,q+1}(y_{\xi_i}) - e^{p_{\alpha_i}h}F^0_{i,q}(y_{\xi_i})\Big| \\
& + \left[\frac{1}{c_iK_{\alpha_i}} + \frac{\eta_{\alpha_i}}{1-e^{p_{\alpha_i}h}}\right]e^{p_{\alpha_i}kh}C_1 \\
\leq & e^{p_{\alpha_i}kh}\phi_0 + \left[\frac{1}{c_iK_{\alpha_i}} + \frac{\eta_{\alpha_i}}{1-e^{p_{\alpha_i}h}}\right]C_1 e^{\frac{1}{2}p_{\alpha_i}kh} \\
& + \left[\frac{1}{c_iK_{\alpha_i}} + \frac{\eta_{\alpha_i}}{1-e^{p_{\alpha_i}h}}\right]\epsilon + \left[\frac{1}{c_iK_{\alpha_i}} + \frac{\eta_{\alpha_i}}{1-e^{p_{\alpha_i}h}}\right]C_1 e^{p_{\alpha_i}kh},
\end{aligned}
$$

which indicates $\lim_{k\to\infty}|\check{\Phi}^z_{i,k}| = 0$, i.e., $\check{\Phi}^z_i \in N_0(\mathbb{N}_0, \mathbb{R})$, $i = 1, 2, \ldots, m$. As a result, $\Phi^z_i \in \mathbb{AAA}(\mathbb{N}_0, \mathbb{R})$, $i = 1, 2, \ldots, m$. Similarly, $\Psi^z_j \in \mathbb{AAA}(\mathbb{N}_0, \mathbb{R})$, $j = 1, 2, \ldots, n$. Combining the above analyses, $\mathbf{T}z \in \mathbb{AAA}(\mathbb{N}_0, \mathbb{M})$ if $z \in \mathbb{AAA}_c(\mathbb{M})$. Further,

$$\mathbf{T} : \mathbb{AAA}_c(\mathbb{M}) \to \mathbb{AAA}_c(\mathbb{M}).$$

In the end, we demonstrate $\mathbf{T}$ is a contractive mapping. For any

$$z = (x_1, \ldots, x_m, y_1, \ldots, y_n)^T, \tilde{z} = (\tilde{x}_1, \ldots, \tilde{x}_m, \tilde{y}_1, \ldots, \tilde{y}_n)^T \in \mathbb{AAA}_c(\mathbb{M}),$$

it acquires from the first equation of Eq. (6.26), Proposition 6.2 and (6.24) that

$$
\begin{aligned}
\left|\Phi^z_{i,k} - \Phi^{\tilde{z}}_{i,k}\right| \leq & \sum_{q=0}^{k-1} e^{p_{\alpha_i}(k-q-1)h}\bigg\{\frac{1-e^{p_{\alpha_i}h}}{c_iK_{\alpha_i}}\left|F_{i,q}(y_{\xi_i}) - F_{i,q}(\tilde{y}_{\xi_i})\right| \\
& + \frac{1-\alpha_i}{K_{\alpha_i}}\bigg|\Big[F_{i,q+1}(y_{\xi_i}) - F_{i,q+1}(\tilde{y}_{\xi_i})\Big]
\end{aligned}
$$

$$
\begin{aligned}
&\quad - e^{p_{\alpha_i} h}\Big[F_{i,q}(y_{\xi_i}) - F_{i,q}(\tilde{y}_{\xi_i})\Big]\Big|\Big\} \\
&\leq \left[\frac{1}{c_i K_{\alpha_i}} + \frac{\eta_{\alpha_i}}{1 - e^{p_{\alpha_i} h}}\right]\varpi_i^f \mathbf{d}(z, \tilde{z}) \\
&\leq \zeta_1 \mathbf{d}(z, \tilde{z}),
\end{aligned}
$$

where $k \in \mathbb{N}_0$, $i = 1, 2, \ldots, m$. Similarly, it obtains $\left|\Psi_{j,k}^{z} - \Psi_{j,k}^{\tilde{z}}\right| \leq \zeta_1 \mathbf{d}(z, \tilde{z})$, where $k \in \mathbb{N}_0$, $j = 1, 2, \ldots, n$. As a result,

$$
\mathbf{d}(\mathbf{T}z, \mathbf{T}\tilde{z}) \leq \zeta_1 \mathbf{d}(z, \tilde{z}), \quad \forall z, \tilde{z} \in \mathbb{AAA}_c(\mathbb{M}).
$$

By ($A_4$), $\mathbf{T}$ is a contractive mapping and $\mathbf{T}$ possesses a unique fixed point $z^* = \mathbf{T}z^* \in \mathbb{AAP}_c(\mathbb{M})$ and $z^*$ solves EED-CF-BAMNNs (6.14). The proof is achieved. □

Let $\alpha_i \to 1, \beta_j \to 1 (i = 1, 2, \ldots, m, j = 1, 2, \ldots, n)$ in EED-CF-BAMNNs (6.14). Then it gets

**Corollary 6.1** *Equation* (6.15) *has a unique bounded asymptotically almost automorphic sequence solution if ($A_1$)–($A_4$) hold for* $\alpha_i = \beta_j = 1$, $i = 1, 2, \ldots, m$, $j = 1, 2, \ldots, n$.

**Remark 6.5** The literature [163] revealed that Caputo or Riemann-Liouville fractional-order differential equations do not produce the periodic oscillations and they only possess asymptotically periodicity. As a result, many excellent results for asymptotically periodic oscillations of Caputo fractional-order neural network models were achieved in literature [158–162]. Remarkably, few papers discuss the asymptotically almost automorphic oscillations of fractional-order neural network models until now. Particularly for the asymptotically almost automorphic sequences of neural network models in the Caputo-Fabrizio sense, almost no literature involves. Consequently, the work of Theorem 6.1 fills this gap and lays a certain theoretical foundation for the future researches of asymptotically almost automorphic sequences of discrete-time Caputo-Fabrizio models.

**Remark 6.6** Observing Theorem 6.1, ($A_4$) is more complicated than other conditions. To ensure the validness of the conditions in Theorem 6.1, in application, the following rules should be followed.

**Rule1** The coefficients of EED-CF-BAMNNs (6.14) excluding $c_i$ and $d_j$ are best to be selected with some small constants, and $c_i$, $d_j$ should be selected with some large positive constants, $i = 1, 2, \ldots, m$, $j = 1, 2, \ldots, n$.

**Rule2** The discrete step length $h$ is best to be selected with a small positive constant.

**Rule3** Choosing the activation functions $f_j$ and $g_i$ in EED-CF-BAMNNs (6.14) with some small enough constants $L_j^f$ and $L_i^g$, $i = 1, 2, \ldots, m$, $j = 1, 2, \ldots, n$.

## 6.6 Global Dynamics I: Exponential Stability

Let

$$z = (x_1, \ldots, x_m, y_1, \ldots, y_n)^T, \quad \tilde{z} = (\tilde{x}_1, \ldots, \tilde{x}_m, \tilde{y}_1, \ldots, \tilde{y}_n)^T$$

be arbitrary two solutions of EED-CF-BAMNNs (6.14) with initial values $x_{i,s} = \phi_{i,s}$, $y_{j,s} = \psi_{j,s}$, $\tilde{x}_{i,s} = \tilde{\phi}_{i,s}$ and $\tilde{y}_{j,s} = \tilde{\psi}_{j,s}$ for $s \in [-\xi_h, 0]_{\mathbb{Z}}$, respectively, $i = 1, 2, \ldots, m$, $j = 1, 2, \ldots, n$.

**Definition 6.4** EED-CF-BAMNNs (6.14) is global exponentially stable if it exists constants $M > 0$ and $0 < \kappa < 1$ such that

$$\|z_k - \tilde{z}_k\| \leq M\gamma e^{-\kappa k}, \quad \forall k \in \mathbb{N}_0,$$

where $\gamma = \max_{1\leq i\leq m, 1\leq j\leq n} \sup_{s\in[-\xi_h,0]_{\mathbb{Z}}} \left\{|\phi_{i,s} - \psi_{i,s}|, |\tilde{\phi}_{j,s} - \tilde{\psi}_{j,s}|\right\}$.

**Theorem 6.2** *Let* $(A_1)$, $(A_2)$ *and* $(A_4)$ *hold and* $\min_{1\leq i\leq m, 1\leq j\leq n}\{\xi_{ij}^h, \sigma_{ji}^h\} \geq 1$. *Then EED-CF-BAMNNs* (6.14) *is globally exponentially stable.*

***Proof*** Define

$$W_k = \max_{1\leq i\leq m, 1\leq j\leq n} \{u_{i,k}, v_{j,k}\}, \quad u_{i,k} = |x_{i,k} - \tilde{x}_{i,k}|, \quad v_{j,k} = |y_{j,k} - \tilde{y}_{j,k}|,$$

where $k \in \mathbb{N}_0$, $i = 1, 2, \ldots, m$, $j = 1, 2, \ldots, n$.

According to the first equation of Eqs. (6.16), (6.23) and (6.24), it concludes

$$\begin{aligned}
u_{i,k} \leq\, & e^{p_{\alpha_i}kh}\gamma + \frac{(1 - e^{p_{\alpha_i}h})\varpi_i^f}{c_i K_{\alpha_i}} \sum_{q=0}^{k-1} e^{p_{\alpha_i}(k-q-1)h} \max_{1\leq j\leq n} \max_{l\in[0,\xi_h]_{\mathbb{Z}}} v_{j,q-l} \\
& + \frac{(1-\alpha_i)\varpi_i^f}{K_{\alpha_i}} \sum_{q=0}^{k-1} e^{p_{\alpha_i}(k-q-1)h} \Bigg[ \max_{1\leq j\leq n} \max_{l\in[1,\xi_h]_{\mathbb{Z}}} v_{j,q+1-l} \\
& + e^{p_{\alpha_i}h} \max_{1\leq j\leq n} \max_{l\in[0,\xi_h]_{\mathbb{Z}}} v_{j,q-l} \Bigg] \\
\leq\, & e^{p_{\alpha_i}kh}\gamma + \left[\frac{1 - e^{p_{\alpha_i}h}}{c_i K_{\alpha_i}} + \eta_{\alpha_i}\right] \varpi_i^f \sum_{q=0}^{k-1} e^{p_{\alpha_i}(k-q-1)h} \max_{l\in[0,\xi_h]_{\mathbb{Z}}} W_{q-l} \\
\leq\, & e^{p_{\alpha_i}kh}\gamma + \zeta_1(1 - e^{p_{\alpha_i}h}) \sum_{q=0}^{k-1} e^{p_{\alpha_i}(k-q-1)h} \max_{l\in[0,\xi_h]_{\mathbb{Z}}} W_{q-l},
\end{aligned} \tag{6.30}$$

where $k \in \mathbb{N}_0$, $i = 1, 2, \ldots, m$. Similarly, it gets from the second equation of Eq. (6.16) that

$$v_{j,k} \le e^{p_{\beta_j} kh}\gamma + \zeta_1(1 - e^{p_{\beta_j} h}) \sum_{q=0}^{k-1} e^{p_{\beta_j}(k-q-1)h} \max_{l\in[0,\xi_h]_{\mathbb{Z}}} W_{q-l}, \tag{6.31}$$

where $k \in \mathbb{N}_0$, $j = 1, 2, \ldots, n$.

Let $p_0 = \max_{1\le i\le m, 1\le j\le n}\{p_{\alpha_i}, p_{\beta_j}\}$. In accordance with ($A_4$), it exists $M > 1$ large enough and $0 < \kappa < -p_0$ small enough such that

$$\frac{1}{M} + \zeta_1 e^{\kappa(1+\xi_h)h} \max_{1\le i\le m, 1\le j\le n}\left\{\frac{1 - e^{p_{\alpha_i} h}}{1 - e^{(p_{\alpha_i}+\kappa)h}}, \frac{1 - e^{p_{\beta_j} h}}{1 - e^{(p_{\beta_j}+\kappa)h}}\right\} < 1.$$

Supposing that

$$W_k \le M\gamma e^{-\kappa kh}, \quad \forall k \in \mathbb{N}_0. \tag{6.32}$$

If not, then it exists $k_0 \in \mathbb{N}$ such that

$$W_{k_0} > M\gamma e^{-\kappa k_0 h}; \quad W_k \le M\gamma e^{-\kappa kh}, \quad \forall k \in [0, k_0)_{\mathbb{Z}}. \tag{6.33}$$

From (6.30) to (6.31), it derives from (6.33) that

$$\begin{aligned}
W_{k_0} &\le e^{p_0 k_0 h}\gamma + \zeta_1 M\gamma \max_{1\le i\le m, 1\le j\le n}\Bigg\{(1 - e^{p_{\alpha_i} h}) \\
&\quad \times \sum_{q=0}^{k_0-1} e^{p_{\alpha_i}(k_0-q-1)h} e^{-\kappa(q-\xi_h)h}, (1 - e^{p_{\beta_j} h}) \\
&\quad \times \sum_{q=0}^{k_0-1} e^{p_{\beta_j}(k_0-q-1)h} e^{-\kappa(q-\xi_h)h}\Bigg\} \\
&\le e^{p_0 k_0 h}\gamma + \zeta_1 M\gamma e^{-\kappa k_0 h} e^{\kappa(1+\xi_h)h} \max_{1\le i\le m, 1\le j\le n}\Bigg\{(1 - e^{p_{\alpha_i} h}) \\
&\quad \times \sum_{q=0}^{k_0-1} e^{(p_{\alpha_i}+\kappa)(k_0-q-1)h}, (1 - e^{p_{\beta_j} h}) \sum_{q=0}^{k_0-1} e^{(p_{\beta_j}+\kappa)(k_0-q-1)h}\Bigg\} \\
&\le M\gamma e^{-\kappa k_0 h}\Bigg[\frac{1}{M} e^{(p_0+\kappa)k_0 h} \\
&\quad + \zeta_1 e^{\kappa(1+\xi_h)h} \max_{1\le i\le m, 1\le j\le n}\left\{\frac{1 - e^{p_{\alpha_i} h}}{1 - e^{(p_{\alpha_i}+\kappa)h}}, \frac{1 - e^{p_{\beta_j} h}}{1 - e^{(p_{\beta_j}+\kappa)h}}\right\}\Bigg]
\end{aligned}$$

$$\le M\gamma e^{-\kappa k_0 h}\left[\frac{1}{M}+\zeta_1 e^{\kappa(1+\xi_h)h}\right.$$

$$\left.\times\max_{1\le i\le m,1\le j\le n}\left\{\frac{1-e^{p_{\alpha_i}h}}{1-e^{(p_{\alpha_i}+\kappa)h}},\frac{1-e^{p_{\beta_j}h}}{1-e^{(p_{\beta_j}+\kappa)h}}\right\}\right]$$

$$\le M\gamma e^{-\kappa k_0 h},$$

which is contradiction with the first part of (6.33). Consequently, (6.32) is valid, i.e., EED-CF-BAMNNs (6.14) is globally exponentially stable. The proof is achieved. □

Let

$$\zeta_1^*=\max_{1\le i\le m,1\le j\le n}\left\{\frac{(1-\alpha_i)\varpi_i^f}{|K_{\alpha_i}|},\frac{(1-\beta_j)\varpi_j^g}{|K_{\beta_j}|}\right\}.$$

**Theorem 6.3** *Let* $(A_1)$ *and* $(A_2)$ *hold and* $\xi_h=\max_{1\le i\le m,1\le j\le n}\{\xi_{ij}^h,\sigma_{ji}^h\}=0$. *Then EED-CF-BAMNNs* (6.14) *is globally exponentially stable if the following assumption is fulfilled.*

$(A_5)$ $\min_{1\le i\le m,1\le j\le n}\{c_i,d_j\}>0$ *and* $\zeta_1+\zeta_1^*<1$, *where* $\zeta_1$ *is defined as that in Theorem 6.2.*

***Proof*** Let $W$, $u_i$, $v_j$ and $p_0$ be defined as those in the proof of Theorem 6.2, $i=1,2,\ldots,m$, $j=1,2,\ldots,n$. If $\xi_h=0$, by the first equation of Eqs. (6.16), (6.23) and (6.24), it achieves

$$\begin{aligned}u_{i,k}\le& e^{p_{\alpha_i}kh}\gamma+\frac{1-e^{p_{\alpha_i}h}}{c_iK_{\alpha_i}}\varpi_i^f\sum_{q=0}^{k-1}e^{p_{\alpha_i}(k-q-1)h}\max_{1\le j\le n}v_{j,q}\\&+\frac{(1-\alpha_i)\varpi_i^f}{K_{\alpha_i}}\sum_{q=0}^{k-2}e^{p_{\alpha_i}(k-q-1)h}\max_{1\le j\le n}v_{j,q+1}\\&+\frac{(1-\alpha_i)e^{p_{\alpha_i}h}\varpi_i^f}{K_{\alpha_i}}\sum_{q=0}^{k-1}e^{p_{\alpha_i}(k-q-1)h}\max_{1\le j\le n}v_{j,q}+\frac{(1-\alpha_i)\varpi_i^f}{K_{\alpha_i}}v_{j,k}\\\le& e^{p_{\alpha_i}kh}\gamma+\zeta_1(1-e^{p_{\alpha_i}h})\sum_{q=0}^{k-1}e^{p_{\alpha_i}(k-q-1)h}W_q+\zeta_1^*W_k,\end{aligned}\tag{6.34}$$

where $k\in\mathbb{N}_0$, $i=1,2,\ldots,m$. Similarly,

$$v_{j,k}\le e^{p_{\beta_j}kh}\gamma+\zeta_1(1-e^{p_{\beta_j}h})\textstyle\sum_{q=0}^{k-1}e^{p_{\beta_j}(k-q-1)h}W_q+\zeta_1^*W_k,\tag{6.35}$$

where $k\in\mathbb{N}_0$, $j=1,2,\ldots,n$. Summarizing (6.34) and (6.35), it results in

$$W_k \le e^{p_0 kh}\gamma + \zeta_1 \max_{1\le i\le m, 1\le j\le n}\Bigg\{(1-e^{p_{\alpha_i}h}) \times \sum_{q=0}^{k_0-1} e^{p_{\alpha_i}(k_0-q-1)h} W_q, (1-e^{p_{\beta_j}h})\sum_{q=0}^{k_0-1} e^{p_{\beta_j}(k_0-q-1)h} W_q\Bigg\} + \zeta_1^* W_k,$$

equivalently,

$$W_k \le \frac{e^{p_0 kh}\gamma}{1-\zeta_1^*} + \frac{\zeta_1}{1-\zeta_1^*}\max_{1\le i\le m, 1\le j\le n}\Bigg\{(1-e^{p_{\alpha_i}h}) \times \sum_{q=0}^{k-1} e^{p_{\alpha_i}(k-q-1)h} W_q, (1-e^{p_{\beta_j}h})\sum_{q=0}^{k-1} e^{p_{\beta_j}(k-q-1)h} W_q\Bigg\}, \tag{6.36}$$

where $k \in \mathbb{N}_0$.

In the light of ($A_5$), it exists $M > 1$ and $0 < \kappa < -p_0$ such that

$$\frac{1}{M(1-\zeta_1^*)} + \frac{\zeta_1 e^{\kappa h}}{1-\zeta_1^*}\max_{1\le i\le m, 1\le j\le n}\left\{\frac{1-e^{p_{\alpha_i}h}}{1-e^{(p_{\alpha_i}+\kappa)h}}, \frac{1-e^{p_{\beta_j}h}}{1-e^{(p_{\beta_j}+\kappa)h}}\right\} < 1.$$

Supposing that

$$W_k \le M\gamma e^{-\kappa kh}, \quad \forall k \in \mathbb{N}_0. \tag{6.37}$$

If not, then it exists $k_0 \in \mathbb{N}$ such that

$$W_{k_0} > M\gamma e^{-\kappa k_0 h}; \quad W_k \le M\gamma e^{-\kappa kh}, \quad \forall k \in [0, k_0)_{\mathbb{Z}}. \tag{6.38}$$

From (6.36), it derives from (6.38) that

$$\begin{aligned} W_{k_0} &\le \frac{e^{p_0 k_0 h}\gamma}{1-\zeta_1^*} + \frac{\zeta_1}{1-\zeta_1^*}\max_{1\le i\le m, 1\le j\le n}\Bigg\{(1-e^{p_{\alpha_i}h})\sum_{q=0}^{k_0-1} e^{p_{\alpha_i}(k_0-q-1)h} W_q, \\ &\quad (1-e^{p_{\beta_j}h})\sum_{q=0}^{k_0-1} e^{p_{\beta_j}(k_0-q-1)h} W_q\Bigg\} M\gamma e^{-\kappa qh} \\ &\le \frac{e^{p_0 k_0 h}\gamma}{1-\zeta_1^*} + \frac{\zeta_1}{1-\zeta_1^*}\max_{1\le i\le m, 1\le j\le n}\Bigg\{(1-e^{p_{\alpha_i}h})\sum_{q=0}^{k_0-1} e^{(p_{\alpha_i}+\kappa)(k_0-q-1)h}, \\ &\quad (1-e^{p_{\beta_j}h})\sum_{q=0}^{k_0-1} e^{(p_{\beta_j}+\kappa)(k_0-q-1)h}\Bigg\} e^{\kappa h} M\gamma e^{-\kappa k_0 h} \end{aligned}$$

$$\leq M\gamma e^{-\kappa k_0 h}\left[\frac{1}{M(1-\zeta_1^*)}\right.$$

$$\left.+\frac{\zeta_1 e^{\kappa h}}{1-\zeta_1^*}\max_{1\leq i\leq m, 1\leq j\leq n}\left\{\frac{1-e^{p_{\alpha_i}h}}{1-e^{(p_{\alpha_i}+\kappa)h}}, \frac{1-e^{p_{\beta_j}h}}{1-e^{(p_{\beta_j}+\kappa)h}}\right\}\right]$$

$$\leq M\gamma e^{-\kappa k_0 h},$$

which is a contradiction with the first inequality of (6.38). Thus, (6.37) is valid, i.e., EED-CF-BAMNNs (6.14) is globally exponentially stable. The proof is achieved. □

Together with Theorems 6.1–6.3, it derives

**Theorem 6.4** *If* $(A_1)$*–*$(A_4)$ *or* $(A_1)$*–*$(A_3)$*,* $(A_5)$ *hold, then EED-CF-BAMNNs* (6.14) *admits a unique bounded globally exponentially stable asymptotically almost automorphic sequence.*

If $\alpha_i \to 1, \beta_j \to 1 (i = 1, 2, \ldots, m, j = 1, 2, \ldots, n)$ in EED-CF-BAMNNs (6.14), then it gains

**Corollary 6.2** *Equation* (6.15) *has a unique bounded globally exponentially stable asymptotically almost automorphic sequence if* $(A_1)$*–*$(A_4)$ *or* $(A_1)$*–*$(A_3)$*,* $(A_5)$ *hold for* $\alpha_i = \beta_j = 1, i = 1, 2, \ldots, m, j = 1, 2, \ldots, n.$

**Remark 6.7** Remarkably, the fractional remainder terms ($R^{\alpha_i}$ and $R^{\beta_j}$, $i = 1, 2, \ldots, m, j = 1, 2, \ldots, n$) of EED-CF-BAMNNs (6.14) increase the complexity and difficulty of calculations in the study of asymptotically almost automorphic sequence and global exponential stability to EED-CF-BAMNNs (6.14), please see the proofs of Theorems 6.1 and 6.2.

**Remark 6.8** By employing exponential Euler difference form (6.15), literature [19, 21, 22, 26–32, 136] researched the periodicity, almost periodicity, almost automorphism and stability of various discrete-time models. For instance, Abbas and Xia [31] investigated almost automorphic sequence solution of an exponential Euler discrete-time cellular neural networks with delay. Abbas [32] considered pseudo almost automorphic sequence solution of an exponential Euler discrete-time BAM neural networks. Besides, the authors [136] discussed mean almost periodic dynamics to exponential Euler discrete-time stochastic BAM neural networks. Thus, the work of this chapter supplements and extends the corresponding results in literatures [19, 21, 22, 26–32, 136].

## 6.7 Global Dynamics II: Exponential Synchronization

This section discusses global exponential synchronization of EED-CF-BAMNNs (6.14) from the viewpoint of control theory. Consider EED-CF-BAMNNs (6.14) as a master system and taking the following exponential Euler discrete-time model as a slave system

$$\begin{cases} \tilde{x}_{i,k+1} &= e^{p_{\alpha_i}h}\tilde{x}_{i,k} + \dfrac{1-e^{p_{\alpha_i}h}}{c_i K_{\alpha_i}} F_{i,k}(\tilde{y}_\xi) \\ &\quad + \frac{1-\alpha_i}{K_{\alpha_i}}\Big[F_{i,k+1}(\tilde{y}_\xi) - e^{p_{\alpha_i}h}F_{i,k}(\tilde{y}_\xi)\Big] + U_{i,k}, \\ \tilde{y}_{j,k+1} &= e^{p_{\beta_j}h}\tilde{y}_{j,k} + \dfrac{1-e^{p_{\beta_j}h}}{d_j K_{\beta_j}} G_{j,k}(\tilde{x}_\sigma) \\ &\quad + \frac{1-\beta_j}{K_{\beta_j}}\Big[G_{j,k+1}(\tilde{x}_\sigma) - e^{p_{\beta_j}h}G_{j,k}(\tilde{x}_\sigma)\Big] + V_{j,k}, \end{cases} \tag{6.39}$$

where

$$(U_1, \ldots, U_m, V_1, \ldots, V_n)^T$$

is the control input, $k \in \mathbb{N}_0$, $i = 1, 2, \ldots, m$, $j = 1, 2, \ldots, n$.

Assume that

$$z = (x_1, \ldots, x_m, y_1, \ldots, y_n)^T$$

is a solution of master system (6.14) with initial values $x_{i,s} = \phi_{i,s}$ and $y_{j,s} = \psi_{j,s}$ for $s \in [-\xi_h, 0]_{\mathbb{Z}}$, and

$$\tilde{z} = (\tilde{x}_1, \ldots, \tilde{x}_m, \tilde{y}_1, \ldots, \tilde{y}_n)^T$$

is a solution of slave system (6.39) with initial values $\tilde{x}_{i,s} = \tilde{\phi}_{i,s}$ and $\tilde{y}_{j,s} = \tilde{\psi}_{j,s}$ for $s \in [-\xi_h, 0]_{\mathbb{Z}}$, $i = 1, 2, \ldots, m$, $j = 1, 2, \ldots, n$. Define

$$e_{i,k} = \tilde{x}_{i,k} - x_{i,k}, \quad \tilde{e}_{j,k} = \tilde{y}_{j,k} - y_{j,k},$$

where $k \in \mathbb{N}_{\xi_h}$, $i = 1, 2, \ldots, m$, $j = 1, 2, \ldots, n$. Then the error system of master system (6.14) and slave system (6.39) is described by

$$\begin{cases} e_{i,k+1} &= e^{p_{\alpha_i}h}e_{i,k} + \dfrac{1-e^{p_{\alpha_i}h}}{c_i K_{\alpha_i}}\Big[F_{i,k}(\tilde{y}_\xi) - F_{i,k}(y_{\xi_i})\Big] \\ &\quad + \dfrac{1-\alpha_i}{K_{\alpha_i}}\Big\{\Big[F_{i,k+1}(\tilde{y}_\xi) - F_{i,k+1}(y_{\xi_i})\Big] \\ &\quad - e^{p_{\alpha_i}h}\Big[F_{i,k}(\tilde{y}_\xi) - F_{i,k}(y_{\xi_i})\Big]\Big\} + U_{i,k}, \\ \tilde{e}_{j,k+1} &= e^{p_{\beta_j}h}\tilde{e}_{j,k} + \dfrac{1-e^{p_{\beta_j}h}}{d_j K_{\beta_j}}\Big[G_{j,k}(\tilde{x}_\sigma) - G_{j,k}(x_{\sigma_j})\Big] \\ &\quad + \dfrac{1-\alpha_i}{K_{\alpha_i}}\Big\{\Big[G_{j,k+1}(\tilde{x}_\sigma) - G_{j,k+1}(x_{\sigma_j})\Big] \\ &\quad - e^{p_{\alpha_i}h}\Big[G_{j,k}(\tilde{x}_\sigma) - G_{j,k}(x_{\sigma_j})\Big]\Big\} + V_{j,k}, \end{cases} \tag{6.40}$$

where $k \in \mathbb{N}_0$, $i = 1, 2, \ldots, m$, $j = 1, 2, \ldots, n$.

Design a controlling law designed by

$$U_{i,k} = \theta_i e_{i,k}, \quad V_{j,k} = \tilde{\theta}_j \tilde{e}_{j,k}, \tag{6.41}$$

where $k \in \mathbb{N}_{\xi_h}$, $i = 1, 2, \ldots, m$, $j = 1, 2, \ldots, n$.

Let

$$\mu_0 = \max_{1 \le i \le m, 1 \le j \le n} \{\mu_{\alpha_i}, \mu_{\beta_j}\}, \quad \mu_{\alpha_i} = \frac{\ln(e^{p_{\alpha_i} h} + \theta_i)}{h}, \quad \mu_{\beta_j} = \frac{\ln(e^{p_{\beta_j} h} + \tilde{\theta}_j)}{h},$$

where $i = 1, 2, \ldots, m$, $j = 1, 2, \ldots, n$.

**Theorem 6.5** *The master-slave systems* (6.14) *and* (6.39) *are globally exponentially synchronized under controller* (6.41) *in case that* ($A_1$) *and* ($A_2$) *and the following assumptions are fulfilled.*

($A_6$) $-e^{p_{\alpha_i} h} < \theta_i < 1 - e^{p_{\alpha_i} h}$ *and* $-e^{p_{\beta_j} h} < \tilde{\theta}_j < 1 - e^{p_{\beta_j} h}$, $i = 1, 2, \ldots, m$, $j = 1, 2, \ldots, n$.

($A_7$) $\min_{1 \le i \le m, 1 \le j \le n} \{\xi_{ij}^h, \sigma_{ji}^h\} \ge 1$ *and* $\chi < 1$, *where*

$$\chi = \frac{1}{1 - e^{\mu_0 h}} \max_{1 \le i \le m, 1 \le j \le n} \left\{ \left[ \frac{1 - e^{p_{\alpha_i} h}}{c_i K_{\alpha_i}} + \eta_{\alpha_i} \right] \varpi_i^f, \left[ \frac{1 - e^{p_{\beta_j} h}}{d_j K_{\beta_j}} + \eta_{\beta_j} \right] \varpi_j^g \right\}.$$

***Proof*** On the basis of controlling law (6.41), the error system (6.40) is turned into

$$\begin{cases} e_{i,k+1} = e^{\mu_{\alpha_i} h} e_{i,k} + \dfrac{1 - e^{p_{\alpha_i} h}}{c_i K_{\alpha_i}} \Big[ F_{i,k}(\tilde{y}_\xi) - F_{i,k}(y_{\xi_i}) \Big] \\ \qquad + \dfrac{1 - \alpha_i}{K_{\alpha_i}} \Big\{ \Big[ F_{i,k+1}(\tilde{y}_\xi) - F_{i,k+1}(y_{\xi_i}) \Big] \\ \qquad - e^{p_{\alpha_i} h} \Big[ F_{i,k}(\tilde{y}_\xi) - F_{i,k}(y_{\xi_i}) \Big] \Big\}, \\ \tilde{e}_{j,k+1} = e^{\mu_{\beta_j} h} \tilde{e}_{j,k} + \dfrac{1 - e^{p_{\beta_j} h}}{d_j K_{\beta_j}} \Big[ G_{j,k}(\tilde{x}_\sigma) - G_{j,k}(x_{\sigma_j}) \Big] \\ \qquad + \dfrac{1 - \beta_j}{K_{\beta_j}} \Big\{ \Big[ G_{j,k+1}(\tilde{x}_\sigma) - G_{j,k+1}(x_{\sigma_j}) \Big] \\ \qquad - e^{p_{\beta_j} h} \Big[ G_{j,k}(\tilde{x}_\sigma) - G_{j,k}(x_{\sigma_j}) \Big] \Big\}, \end{cases} \tag{6.42}$$

where $k \in \mathbb{N}_0$, $i = 1, 2, \ldots, m$, $j = 1, 2, \ldots, n$.

Similar to Eq. (6.16), the solution of Eq. (6.42) can be expressed by

$$\begin{cases} e_{i,k} = e^{\mu_{\alpha_i} kh} e_{i,0} \\ \quad + \sum_{q=0}^{k-1} e^{\mu_{\alpha_i}(k-q-1)h} \Bigg\{ \frac{1-e^{p_{\alpha_i} h}}{c_i K_{\alpha_i}} \Big[ F_{i,q}(\tilde{y}_\xi) - F_{i,q}(y_{\xi_i}) \Big] \\ \quad + \frac{1-\alpha_i}{K_{\alpha_i}} \bigg( \Big[ F_{i,q+1}(\tilde{y}_\xi) - F_{i,q+1}(y_{\xi_i}) \Big] \\ \quad - e^{p_{\alpha_i} h} \Big[ F_{i,q}(\tilde{y}_\xi) - F_{i,q}(y_{\xi_i}) \Big] \bigg) \Bigg\}, \\ \tilde{e}_{j,k} = e^{\mu_{\beta_j} kh} \tilde{e}_{j,0} \\ \quad + \sum_{q=0}^{k-1} e^{\mu_{\beta_j}(k-q-1)h} \Bigg\{ \frac{1-e^{p_{\beta_j} h}}{d_j K_{\beta_j}} \Big[ G_{j,q}(\tilde{x}_\sigma) - G_{j,q}(x_{\sigma_j}) \Big] \\ \quad + \frac{1-\beta_j}{K_{\beta_j}} \bigg( \Big[ G_{j,q+1}(\tilde{x}_\sigma) - G_{j,q+1}(x_{\sigma_j}) \Big] \\ \quad - e^{p_{\alpha_i} h} \Big[ G_{j,q}(\tilde{x}_\sigma) - G_{j,q}(x_{\sigma_j}) \Big] \bigg) \Bigg\}, \end{cases} \tag{6.43}$$

where $k \in \mathbb{N}_0$, $i = 1, 2, \ldots, m$, $j = 1, 2, \ldots, n$.

Define $\gamma = \max_{1 \le i \le m, 1 \le j \le n} \sup_{s \in [-\xi_h, 0]_{\mathbb{Z}}} \big\{ |\phi_{i,s} - \psi_{i,s}|, |\tilde{\phi}_{j,s} - \tilde{\psi}_{j,s}| \big\}$ and

$$E_k = \max_{1 \le i \le m, 1 \le j \le n} \{ |e_{i,k}|, |\tilde{e}_{j,k}| \}, \quad \forall k \in \mathbb{N}_0, i = 1, 2, \ldots, m, j = 1, 2, \ldots, n.$$

Similar to the arguments as those in (6.30) and (6.31), it gets from Eq. (6.43) that

$$\begin{aligned} |e_{i,k}| &\le e^{\mu_{\alpha_i} kh} \gamma + \frac{|1-e^{p_{\alpha_i} h}| \varpi_i^f}{|c_i K_{\alpha_i}|} \sum_{q=0}^{k-1} e^{\mu_{\alpha_i}(k-q-1)h} \max_{1 \le j \le n} \max_{l \in [0,\xi_h]_{\mathbb{Z}}} |\tilde{e}_{j,q-l}| \\ &\quad + \frac{(1-\alpha_i)\varpi_i^f}{|K_{\alpha_i}|} \sum_{q=0}^{k-1} e^{\mu_{\alpha_i}(k-q-1)h} \max_{1 \le j \le n} \max_{l \in [1,\xi_h]_{\mathbb{Z}}} |\tilde{e}_{j,q+1-l}| \\ &\quad + \frac{(1-\alpha_i)\varpi_i^f}{|K_{\alpha_i}|} \sum_{q=0}^{k-1} e^{\mu_{\alpha_i}(k-q-1)h} e^{p_{\alpha_i} h} \max_{1 \le j \le n} \max_{l \in [0,\xi_h]_{\mathbb{Z}}} |\tilde{e}_{j,q-l}| \\ &\le e^{\mu_0 kh} \gamma + \left[ \frac{|1-e^{p_{\alpha_i} h}|}{|c_i K_{\alpha_i}|} + \frac{(1-\alpha_i)(1+e^{p_{\alpha_i} h})}{|K_{\alpha_i}|} \right] \\ &\quad \times \varpi_i^f \sum_{q=0}^{k-1} e^{\mu_0 (k-q-1)h} \max_{l \in [0,\xi_h]_{\mathbb{Z}}} E_{q-l} \\ &\le e^{\mu_0 kh} \gamma + \chi (1 - e^{\mu_0 h}) \sum_{q=0}^{k-1} e^{\mu_0 (k-q-1)h} \max_{l \in [0,\xi_h]_{\mathbb{Z}}} E_{q-l} \end{aligned} \tag{6.44}$$

and

$$
\begin{aligned}
|\tilde{e}_{j,k}| &\le e^{\mu_0 kh}\gamma + \left[\frac{|1-e^{p_{\beta_j}h}|}{|d_j K_{\beta_j}|} + \frac{(1-\beta_j)(1+e^{p_{\beta_j}h})}{|K_{\beta_j}|}\right] \\
&\quad \times \varpi_j^g \sum_{q=0}^{k-1} e^{\mu_0(k-q-1)h} \max_{l\in[0,\xi_h]_{\mathbb{Z}}} E_{q-l} \\
&\le e^{\mu_0 kh}\gamma + \chi(1-e^{\mu_0 h}) \sum_{q=0}^{k-1} e^{\mu_0(k-q-1)h} \max_{l\in[0,\xi_h]_{\mathbb{Z}}} E_{q-l},
\end{aligned} \tag{6.45}
$$

where $k \in \mathbb{N}_0,\ i = 1, 2, \ldots, m,\ j = 1, 2, \ldots, n$. Inequalities (6.44) and (6.45) indicate

$$E_k \le e^{\mu_0 kh}\gamma + \chi(1-e^{\mu_0 h}) \textstyle\sum_{q=0}^{k-1} e^{\mu_0(k-q-1)h} \max_{l\in[0,\xi_h]_{\mathbb{Z}}} E_{q-l}, \quad \forall k \in \mathbb{N}_0 \tag{6.46}$$

Via ($A_7$), it exists $M_1 > 1$ large enough and $0 < \kappa_1 < -\mu_0$ small enough such that

$$\frac{1}{M_1} + \frac{\chi(1-e^{\mu_0 h})e^{\kappa_1(1+\mu_0)h}}{1-e^{(\mu_0+\kappa_1)h}} < 1.$$

Supposing that

$$E_k \le M_1\gamma e^{-\kappa_1 kh}, \quad \forall k \in \mathbb{N}_0.$$

If not, then it exists $k_1 \in \mathbb{N}$ such that

$$E_{k_1} > M_1\gamma e^{-\kappa_1 k_1 h}; \quad E_k \le M_1\gamma e^{-\kappa_1 kh}, \quad \forall k \in [0, k_1)_{\mathbb{Z}}. \tag{6.47}$$

In view of (6.46), it calculates from (6.47) that

$$
\begin{aligned}
E_{k_1} &\le e^{\mu_0 k_1 h}\gamma + \chi(1-e^{\mu_0 h}) \sum_{q=0}^{k_1-1} e^{\mu_0(k_1-q-1)h} M_1\gamma e^{-\kappa_1(q-\mu_0)h} \\
&\le e^{\mu_0 k_1 h}\gamma + \chi(1-e^{\mu_0 h}) \sum_{q=0}^{k_1-1} e^{(\mu_0+\kappa_1)(k_1-q-1)h} e^{\kappa_1(1+\mu_0)h} M_1\gamma e^{-\kappa_1 k_1 h} \\
&\le M_1\gamma e^{-\kappa_1 k_1 h}\left[\frac{1}{M_1} e^{(\mu_0+\kappa_1)k_1 h} + \frac{\chi(1-e^{\mu_0 h})e^{\kappa_1(1+\mu_0)h}}{1-e^{(\mu_0+\kappa_1)h}}\right] \\
&\le M_1\gamma e^{-\kappa_1 k_1 h}\left[\frac{1}{M_1} + \frac{\chi(1-e^{\mu_0 h})e^{\kappa_1(1+\mu_0)h}}{1-e^{(\mu_0+\kappa_1)h}}\right] \\
&\le M_1\gamma e^{-\kappa_1 k_1 h},
\end{aligned}
$$

which is contradiction with the first part of (6.47). Consequently,

$$E_k \leq M_1 \gamma e^{-\kappa_1 kh}, \quad \forall k \in \mathbb{N}_0.$$

So the error system (6.40) is globally exponentially stable. The proof is achieved. □

**Theorem 6.6** *Suppose that* $\xi_h = \max_{1\leq i\leq m, 1\leq j\leq n}\{\xi_{ij}^h, \sigma_{ji}^h\} = 0$, *(*$A_1$*), (*$A_2$*) and (*$A_6$*) and the following assumption hold.*

($A_8$) $\chi + \zeta_1^* < 1$, *where* $\zeta_1^*$ *is defined as that in Theorem 6.3.*

*Then the master-slave systems* (6.14) *and* (6.39) *are exponentially synchronized under controller* (6.41).

***Proof*** Let $\mu_0$ and $E$ be defined as those in the proof of Theorem 6.5. Similar to the derivations as those in (6.44) and (6.45), it infers from (6.23) and (6.24) that

$$\begin{aligned} e_{i,k} \leq& e^{\mu_0 kh}\gamma + \frac{|1-e^{p_{\alpha_i}h}|}{|c_i K_{\alpha_i}|}\varpi_i^f \sum_{q=0}^{k-1} e^{\mu_{\alpha_i}(k-q-1)h} E_q \\ &+ \frac{(1-\alpha_i)\varpi_i^f}{|K_{\alpha_i}|}\sum_{q=0}^{k-2} e^{\mu_{\alpha_i}(k-q-1)h} E_{q+1} + \frac{(1-\alpha_i)e^{p_{\alpha_i}h}\varpi_i^f}{|K_{\alpha_i}|} \\ &\times \sum_{q=0}^{k-1} e^{\mu_{\alpha_i}(k-q-1)h} E_q + \frac{(1-\alpha_i)\varpi_i^f}{|K_{\alpha_i}|} E_k \\ \leq& e^{\mu_0 kh}\gamma + \chi(1-e^{\mu_0 h})\sum_{q=0}^{k-1} e^{\mu_0(k-q-1)h} E_q + \zeta_1^* E_k, \end{aligned} \tag{6.48}$$

where $k \in \mathbb{N}_0$, $i = 1, 2, \ldots, m$. Similarly,

$$\tilde{e}_{j,k} \leq e^{\mu_0 kh}\gamma + \chi(1-e^{\mu_0 h})\sum_{q=0}^{k-1} e^{\mu_0(k-q-1)h} E_q + \zeta_1^* E_k, \tag{6.49}$$

where $k \in \mathbb{N}_0$, $j = 1, 2, \ldots, n$. Summarizing (6.48) and (6.49), it results in

$$E_k \leq e^{\mu_0 kh}\gamma + \chi(1-e^{\mu_0 h})\sum_{q=0}^{k-1} e^{\mu_0(k-q-1)h} E_q + \zeta_1^* E_k,$$

equivalently,

$$E_k \leq \frac{e^{\mu_0 kh}\gamma}{1-\zeta_1^*} + \frac{\chi(1-e^{\mu_0 h})}{1-\zeta_1^*}\sum_{q=0}^{k-1} e^{\mu_0(k-q-1)h} E_q, \quad \forall k \in \mathbb{N}_0. \tag{6.50}$$

The surplus proof of this theorem is the same as the proof of Theorem 6.3, so it is omitted. The proof is achieved. □

If $\alpha_i \to 1$ and $\beta_j \to 1$, then the slave system (6.39) is shifted to

$$\begin{cases} \tilde{x}_{i,k+1} &= e^{-c_i h}\tilde{x}_{i,k} + \dfrac{1-e^{-c_i h}}{c_i}F_{i,k}(\tilde{y}_\xi) + U_{i,k}, \\ \tilde{y}_{j,k+1} &= e^{-d_j h}\tilde{y}_{j,k} + \dfrac{1-e^{-d_j h}}{d_j}G_{j,k}(\tilde{x}_\sigma) + V_{j,k}, \end{cases} \tag{6.51}$$

where $k \in \mathbb{N}_0$, $i = 1, 2, \ldots, m$, $j = 1, 2, \ldots, n$.

In accordance with Theorems 6.5 and 6.6, it gains

**Corollary 6.3** *The master-slave systems* (6.15) *and* (6.51) *are globally exponentially synchronized under controller* (6.41) *if* $(A_1)$ *and* $(A_2)$*,* $(A_6)$ *and* $(A_7)$ *or* $(A_1)$ *and* $(A_2)$*,* $(A_6)$ *and* $(A_8)$ *hold for* $\alpha_i = \beta_j = 1$, $i = 1, 2, \ldots, m$, $j = 1, 2, \ldots, n$.

**Remark 6.9** In Sects. 6.5 and 6.6, $c_i > 0$ and $d_j > 0$ are necessary, $i = 1, 2, \ldots, m$, $j = 1, 2, \ldots, n$. However, these conditions are not indispensable in the results of this section.

**Remark 6.10** Noticeably, EED-CF-BAMNNs (6.14) is within the scope of implicit Euler differences if $\xi_h = 0$, i.e., $x_{i,k+1}$ and $y_{j,k+1}$ appear in the right side of EED-CF-BAMNNs (6.14), $k \in \mathbb{N}_0$, $i = 1, 2, \ldots, m$, $j = 1, 2, \ldots, n$. This feature extremely increases the complexity and difficulty of discussions of global exponential stability and synchronization to EED-CF-BAMNNs (6.14), please see the proofs of Theorems 6.3 and 6.6 for details.

**Remark 6.11** In allusion to the classical continuous-time Caputo or Riemann-Liouville fractional-order models with time lags, it is difficulty to discuss global exponential stability and synchronization. As a result, numerous old literature focused on global Mittag-Leffler stability (synchronization) or global asymptotic stability (synchronization), see [158–162, 164–166]. Noticeably, CFOEs is a new concept recently, so a number of meaningful research topics for the models with CFOEs have not been concerned by the scholars, e.g., discrete-time analogue, almost automorphism, stability, chaos, bifurcation, synchronization, etc. Apparently, the current works fill some of these gaps and have important guiding significance.

## 6.8 Numerical Experiments

**Example 6.1** Considering a fuzzy CF-BAMNN with time lags in the form of

$$\begin{cases} {}^{CF}_{0}D^{\alpha}_{t} x_1(t) = -c_1 x_1(t) + 0.1\sin(\sqrt{2}t)\sum_{j=1}^{2} f_j(y_j(t-\xi)) \\ \qquad +0.1\cos(\sqrt{5}t)\bigwedge_{j=1}^{2} f_j(y_j(t-\xi)) + 0.1\cos(\sqrt{5}t) + H_{\alpha_1}(t), \\ {}^{CF}_{0}D^{\alpha}_{t} x_2(t) = -c_2 x_2(t) + 0.2\sin^2(\sqrt{7}t)\sum_{j=1}^{2} f_j(y_j(t-\xi)) \\ \qquad +0.2\bigvee_{j=1}^{2} Q_{2j}(t) + 0.1|\sin(\sqrt{7}t)| + H_{\alpha_2}(t), \\ {}^{CF}_{0}D^{\beta}_{t} y_1(t) = -d_1 y_1(t) + 0.1\cos^2(\sqrt{2}t)\sum_{i=1}^{2} g_i(x_i(t-\sigma)) \\ \qquad +0.2\cos t\bigvee_{i=1}^{2} g_i(x_i(t-\sigma)) + 0.1\sin(\sqrt{3}t) + H_{\beta_1}(t), \\ {}^{CF}_{0}D^{\beta}_{t} y_2(t) = -d_2 y_2(t) + 0.15\sin^2(\sqrt{2}t)\sum_{i=1}^{2} g_i(x_i(t-\sigma)) \\ \qquad +0.1\bigwedge_{i=1}^{2} W_{2i}(t) + 0.1\cos^2 t + H_{\beta_2}(t), \end{cases} \tag{6.52}$$

where $t \in \mathbb{R}_0$, $x_i(s) = \phi_i(s)$ and $y_j(s) = \psi_j(s)$ for $s \in [-\max\{\xi, \sigma\}, 0]$ with $\xi, \sigma \geq 0$; $f_j(u) = 0.1|u|$, $g_i(u) = 0.1\cos u$, $i, j = 1, 2$;

$$\begin{pmatrix} Q_{21}(t) \\ Q_{22}(t) \end{pmatrix} = 0.1\begin{pmatrix} \operatorname{sgn}\cos(2\sqrt{3}\pi t) \\ \cos^2 t \end{pmatrix}, \quad \begin{pmatrix} W_{21}(t) \\ W_{22}(t) \end{pmatrix} = 0.1\begin{pmatrix} \cos(\sqrt{5}t) \\ |\sin(\sqrt{7}t)| \end{pmatrix},$$

$$H_{\alpha_1}(t) = e^{\bar{\alpha}t}\left[c_1\phi_1(0) - 0.1\bigwedge_{j=1}^{2} f_j(\psi_j(-\xi)) - 0.1\right],$$

$$H_{\alpha_2}(t) = e^{\bar{\alpha}t}\left[c_2\phi_2(0) - 0.2\bigvee_{j=1}^{2} Q_{2j}(0)\right],$$

$$H_{\beta_1}(t) = e^{\bar{\beta}t}\left[d_1\psi_1(0) - 0.1\sum_{i=1}^{2} g_i(\phi_i(-\sigma)) - 0.2\bigvee_{i=1}^{2} g_i(\phi_i(-\sigma))\right],$$

$$H_{\beta_2}(t) = e^{\bar{\beta}t}\left[d_2\psi_2(0) - 0.1\bigwedge_{i=1}^{2} W_{2i}(0) - 0.1\right],$$

where $t \in \mathbb{R}_0$, $\bar{\alpha} = \frac{\alpha}{\alpha-1}$, $\bar{\beta} = \frac{\beta}{\beta-1}$.

Let

$$F_{1,k}^x = 0.1\sin(\sqrt{2}kh)\sum_{j=1}^{2} f_j(y_{j,k}) + 0.1\cos(\sqrt{5}kh)\bigwedge_{j=1}^{2} f_j(y_{j,k}) + 0.1\cos(\sqrt{5}kh),$$

$$F_{2,k}^x = 0.2\sin^2(\sqrt{7}kh)\sum_{j=1}^{2} f_j(y_{j,k}) + 0.2\bigvee_{j=1}^{2} Q_{2j}(kh) + 0.1|\sin(\sqrt{7}kh)|,$$

$$F_{1,k}^y = 0.1\cos^2(\sqrt{2}kh)\sum_{i=1}^{2} g_i(x_{i,k-1}) + 0.2\cos(kh)\bigvee_{i=1}^{2} g_i(x_{i,k-1}) + 0.1\sin(\sqrt{3}kh),$$

$$F_{2,k}^y = 0.15\sin^2(\sqrt{2}kh)\sum_{i=1}^{2} g_i(x_{i,k-1}) + 0.1\bigwedge_{i=1}^{2} W_{2i}(kh) + 0.1\cos^2(kh), \quad \forall k \in \mathbb{N}_0.$$

Taking $h = \sigma = 0.1$, $\xi = 0$, $c_1 = c_2 = 8$ and $d_1 = d_2 = 9$. By using the exponential Euler difference scheme in EED-CF-BAMNNs (6.14), the exponential Euler discrete-time analogue of Eq. (6.52) is depicted by

$$\begin{cases} x_{i,k+1} = e^{p_\alpha h}x_{i,k} + \dfrac{1-e^{p_\alpha h}}{c_i K_\alpha}F_{i,k}^x + \dfrac{1-\alpha}{K_\alpha}\Big[F_{i,k+1}^x - e^{p_\alpha h}F_{i,k}^x\Big], \\ y_{j,k+1} = e^{p_\beta h}y_{j,k} + \dfrac{1-e^{p_\beta h}}{d_j K_\beta}F_{j,k}^y + \dfrac{1-\beta}{K_\beta}\Big[F_{j,k+1}^y - e^{p_\beta h}F_{j,k}^y\Big], \end{cases} \tag{6.53}$$

where $k \in \mathbb{N}_0$, $x_{i,s} = \phi_{i,s}$ and $y_{j,s} = \psi_{j,s}$ for $s \in [-\max\{\lfloor\xi/h\rfloor, \lfloor\sigma/h\rfloor\}, 0]_{\mathbb{Z}}$, $i, j = 1, 2$, $K_\alpha = 1 + 8(1-\alpha)$, $K_\beta = 1 + 9(1-\beta)$, $p_\alpha = \frac{-8\alpha}{K_\alpha}$ and $p_\beta = \frac{-9\beta}{K_\beta}$.

Using the explicit Euler scheme in Eq. (6.19), it gets

$$\begin{cases} x_{i,k+1} = x_{i,k} - \sum_{l=1}^{k-1} e^{\bar{\alpha}(k-l)h} \left[x_{i,l+1} - x_{i,l}\right] \\ \qquad + \frac{\alpha h}{1 - e^{\bar{\alpha}h}} \Big[ - 8x_{i,k} + F_{i,k}^{x} + \big(8x_{i,0} - F_{i,0}^{x}\big)e^{\bar{\alpha}kh} \Big], \\ y_{j,k+1} = y_{j,k} - \sum_{l=1}^{k-1} e^{\bar{\beta}(k-l)h} \left[y_{j,l+1} - y_{j,l}\right] \\ \qquad + \frac{\beta h}{1 - e^{\bar{\beta}h}} \Big[ - 9y_{ij,k} + F_{j,k}^{y} + \big(9y_{j,0} - F_{j,0}^{y}\big)e^{\bar{\beta}kh} \Big], \end{cases} \tag{6.54}$$

where $k \in \mathbb{N}_0$, $x_{i,s} = \phi_{i,s}$ and $y_{j,s} = \psi_{j,s}$ for $s \in [-\max\{\lfloor \xi/h \rfloor, \lfloor \sigma/h \rfloor\}, 0]_{\mathbb{Z}}$, $i, j = 1, 2$.

By utilizing the explicit Euler method in Eq. (6.54) and the exponential Euler predictor-corrector method, the numerical solutions of Eq. (6.52) with different orders $\alpha, \beta \in \{0.5, 0.8, 0.9, 0.95\}$ are drawn in Figs. 6.1, 6.2, 6.3, 6.4, 6.5, 6.6, 6.7 and 6.8.

**Remark 6.12** Figures 6.1, 6.2, 6.3, 6.4, 6.5, 6.6, 6.7 and 6.8 indicate that the solution produced by Euler scheme is divergence for smaller $\alpha, \beta \in \{0.5, 0.8\}$, please see Figs. 6.1, 6.2, 6.3 and 6.4. Inversely, the solutions produced by PECE

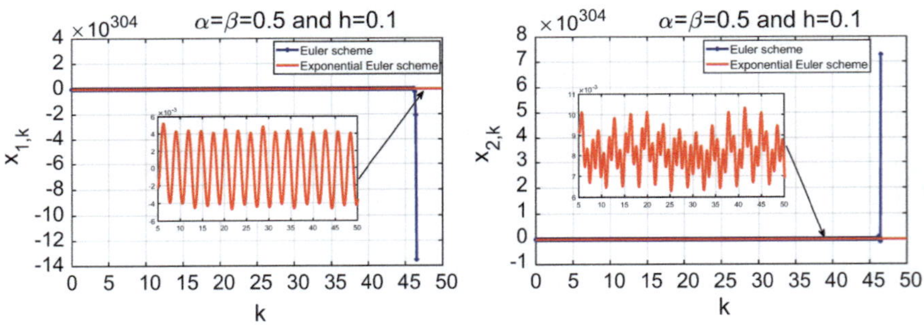


**Fig. 6.1** $x_1$ and $x_2$ of Eqs. (6.53) and (6.54) with $\alpha = \beta = 0.5$

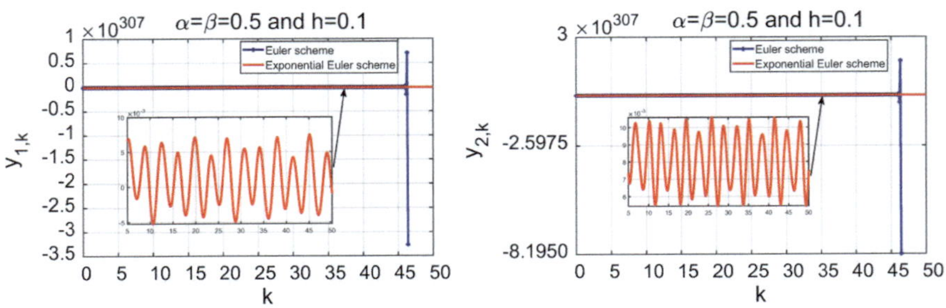


**Fig. 6.2** $y_1$ and $y_2$ of Eqs. (6.53) and (6.54) with $\alpha = \beta = 0.5$

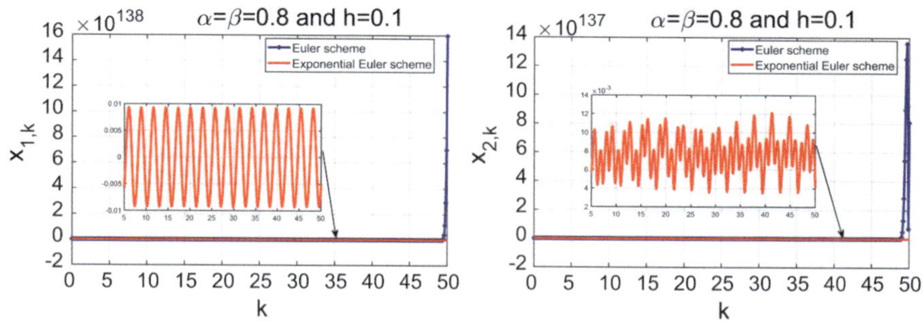


**Fig. 6.3** $x_1$ and $x_2$ of Eqs. (6.53) and (6.54) with $\alpha = \beta = 0.8$

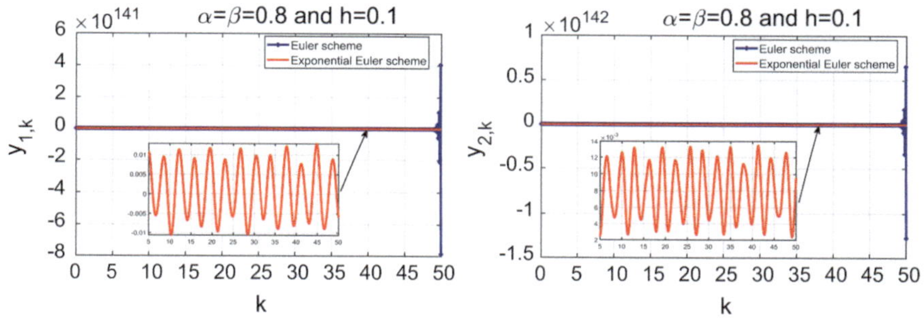


**Fig. 6.4** $y_1$ and $y_2$ of Eqs. (6.53) and (6.54) with $\alpha = \beta = 0.8$

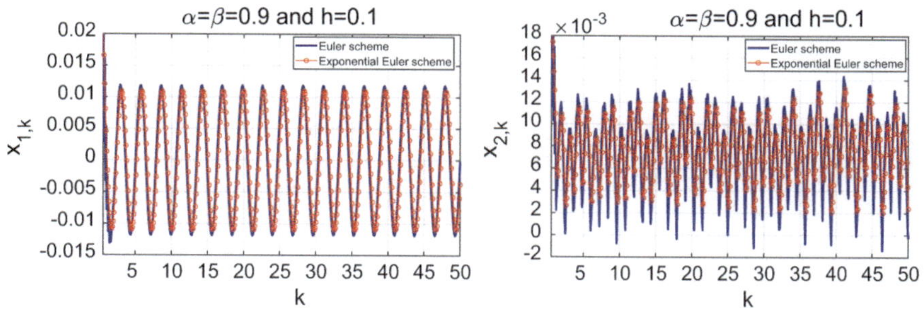


**Fig. 6.5** $x_1$ and $x_2$ of Eqs. (6.53) and (6.54) with $\alpha = \beta = 0.9$

method (i.e., exponential Euler scheme) always keep the almost automorphic oscillations (convergence) for arbitrary $\alpha, \beta \in \{0.5, 0.8, 0.9, 0.95\}$. Consequently, exponential Euler scheme can depict and retain the dynamics of the continuous-time models better than Euler scheme.

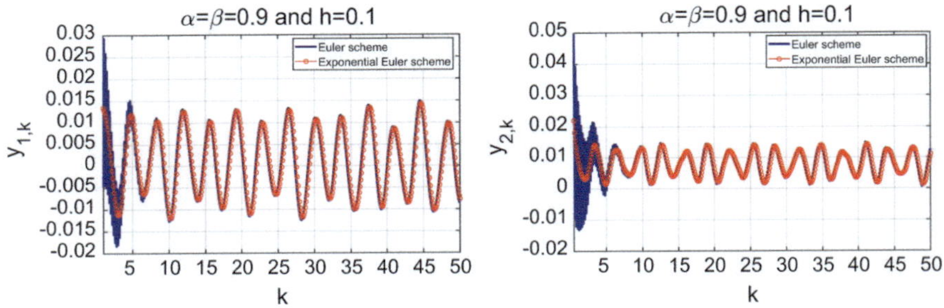


**Fig. 6.6** $y_1$ and $y_2$ of Eqs. (6.53) and (6.54) with $\alpha = \beta = 0.9$

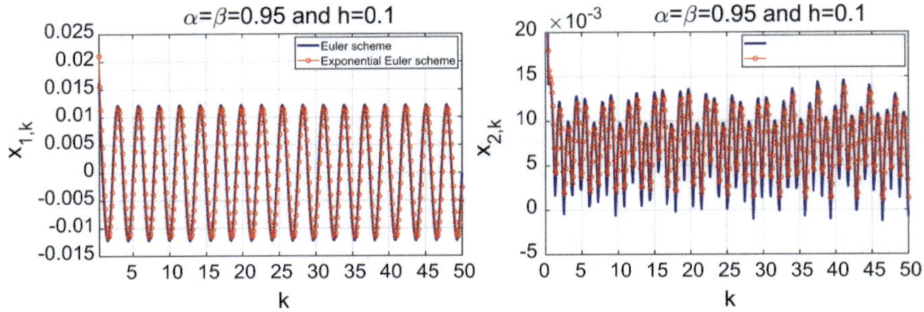


**Fig. 6.7** $x_1$ and $x_2$ of Eqs. (6.53) and (6.54) with $\alpha = \beta = 0.95$

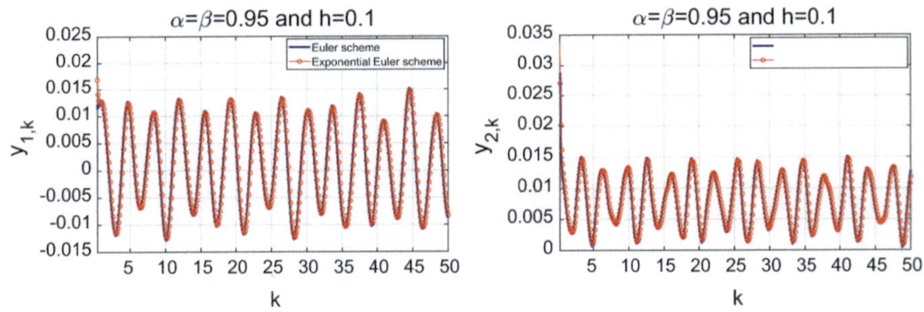


**Fig. 6.8** $y_1$ and $y_2$ of Eqs. (6.53) and (6.54) with $\alpha = \beta = 0.95$

**Example 6.2** Taking $\alpha = 0.5$ and $\beta = 0.6$ in Eq. (6.53), it computes

$$K_\alpha = 5, \quad K_\beta = 4.6, \quad p_\alpha = -0.8, \quad p_\beta = -1.1739,$$

$$\eta_\alpha = 0.1923, \quad \eta_\beta = 0.2512, \quad \zeta_1 = \max\{0.1515, 0.1604\} = 0.1604,$$

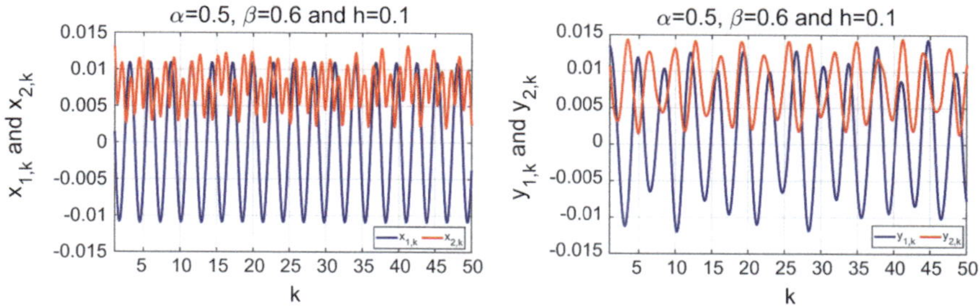


**Fig. 6.9** Asymptotically almost automorphic sequences in Eq. (6.53)

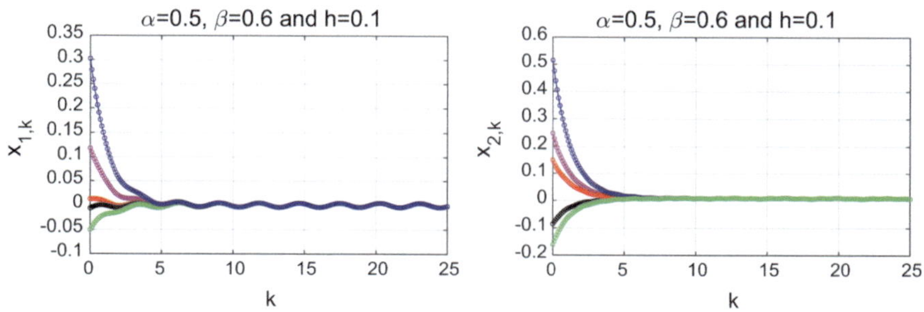


**Fig. 6.10** Global exponential stability of state variables $x_1$ and $x_2$ in Eq. (6.53)

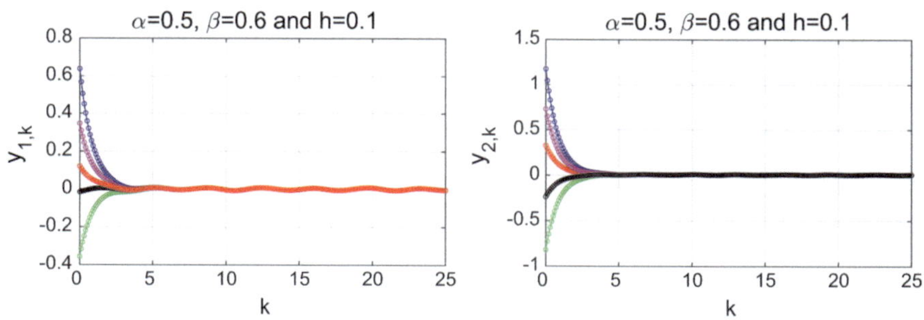


**Fig. 6.11** Global exponential stability of state variables $y_1$ and $y_2$ in Eq. (6.53)

which indicate all conditions in Theorems 6.1 and 6.2 hold. That is, Eq. (6.53) admits a unique bounded asymptotically almost automorphic sequence solution, which is globally exponentially stable. Figures 6.9, 6.10 and 6.11 describe the trajectories and exponential convergence of states $x_{11}$, $x_{12}$, $y_{11}$ and $y_{12}$ in Eq. (6.53), respectively.

**Example 6.3** Considering Eq. (6.53) as a master system and taking the following exponential Euler discrete-time model as a slave system

$$\begin{cases} x^*_{i,k+1} &= e^{p_{\alpha_i}h} x^*_{i,k} + \dfrac{1-e^{p_{\alpha_i}h}}{c_i K_{\alpha_i}} F^{x^*}_{i,k} \\ & \quad + \dfrac{1-\alpha_i}{K_{\alpha_i}} \Big[ F^{x^*}_{i,k+1} - e^{p_{\alpha_i}h} F^{x^*}_{i,k} \Big] + u_{i,k}, \\ y^*_{j,k+1} &= e^{p_{\beta_j}h} y^*_{j,k} + \dfrac{1-e^{p_{\beta_j}h}}{d_j K_{\beta_j}} F^{y^*}_{j,k} \\ & \quad + \dfrac{1-\beta_j}{K_{\beta_j}} \Big[ F^{y^*}_{j,k+1} - e^{p_{\beta_j}h} F^{y^*}_{j,k} \Big] + v_{j,k}, \end{cases} \tag{6.55}$$

where $x^*_{i,s} = \phi^*_{i,s}$ and $y^*_{j,s} = \psi^*_{j,s}$ for $s \in [-\max\{\lfloor \xi/h \rfloor, \lfloor \sigma/h \rfloor\}, 0]_{\mathbb{Z}}$, $u_{i,k}$, $v_{j,k}$ are the control inputs, $i, j = 1, 2$,

$$F^{x^*}_{1,k} = 0.1 \sin(\sqrt{2}kh) \sum_{j=1}^{2} f_j(y^*_{j,k}) + 0.1 \cos(\sqrt{5}kh) \bigwedge_{j=1}^{2} f_j(y^*_{j,k}) + 0.1 \cos(\sqrt{5}kh),$$

$$F^{x^*}_{2,k} = 0.2 \sin^2(\sqrt{7}kh) \sum_{j=1}^{2} f_j(y^*_{j,k}) + 0.2 \bigvee_{j=1}^{2} Q_{2j}(kh) + 0.1 |\sin(\sqrt{7}kh)|,$$

$$F^{y^*}_{1,k} = 0.1 \cos^2(\sqrt{2}kh) \sum_{i=1}^{2} g_i(x^*_{i,k-1}) + 0.2 \cos(kh) \bigvee_{i=1}^{2} g_i(x^*_{i,k-1}) + 0.1 \sin(\sqrt{3}kh),$$

$$F^{y^*}_{2,k} = 0.15 \sin^2(\sqrt{2}kh) \sum_{i=1}^{2} g_i(x^*_{i,k-1}) + 0.1 \bigwedge_{i=1}^{2} W_{2i}(kh) + 0.1 \cos^2(kh), \quad \forall k \in \mathbb{N}_0.$$

Taking $h = 0.1, \xi = \sigma = 0.1, \alpha = 0.95, \beta = 0.98, c_1 = c_2 = -2$ and $d_1 = d_2 = -1$. Corresponding to EED-CF-BAMNNs (6.14), it calculates

$$a^*_{ij} = 0.2, \quad \gamma^*_{ij} = 0.1, \quad \eta^*_{ij} = 0, \quad b^*_{ji} = 0.15,$$

$$\varepsilon^*_{ji} = 0, \quad \vartheta^*_{ji} = 0.2, \quad L^f_j = 0.1, \quad L^g_j = 0.1,$$

$$p_\alpha = 2.1111, \quad p_\beta = 1, \quad e^{p_\alpha h} = 1.2337, \quad e^{p_\beta h} = 1.1052,$$

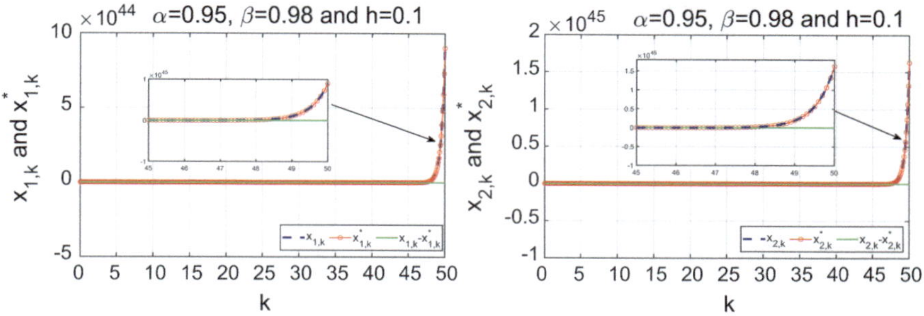


**Fig. 6.12** Global exponential synchronization of $x_1$ and $x_2$ in systems (6.53) and (6.55)

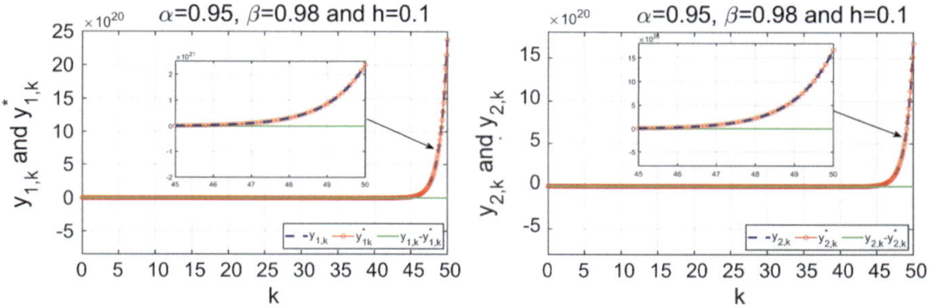


**Fig. 6.13** Global exponential synchronization of $y_1$ and $y_2$ in systems (6.53) and (6.55)

where $i, j = 1, 2$. Design a controller by

$$\begin{cases} u_{1,k} = -0.7337(x_{1,k}^* - x_{1,k}), \\ u_{2,k} = -0.7337(x_{2,k}^* - x_{2,k}), \\ v_{1,k} = -0.6052(y_{1,k}^* - y_{1,k}), \\ v_{2,k} = -0.6052(y_{2,k}^* - y_{2,k}), \end{cases} \tag{6.56}$$

where $k \in \mathbb{N}_0$. Corresponding to Theorem 6.5, it calculates

$$\mu_0 = -6.9315, \quad \chi_1 = 0.0305,$$

which indicate all conditions of Theorem 6.5 hold. Then the master-slave systems (6.53) and (6.55) are globally exponentially synchronized via controller (6.56), see Figs. 6.12 and 6.13.

# Chapter 7
# Summaries and Perspectives

The book proposes several types of Mittag-Leffler and exponential Euler differences for different types of differential equations, such as FODEs in the Caputo sense and CFOEs, by using the variation-of-constants formulas in nonlocal calculus. From the point of view of the theory of differential equations, some basic theories are set up for the further study of the differential equations generated by the Mittag-Leffler and exponential Euler differences. In addition, some important dynamic properties are studied in this book, such as the well-posed problems, global exponential stability, global exponential synchronization, almost automorphism, and so on. It lays certain theoretical foundations for future research on the Mittag-Leffler and exponential Euler difference models and opens up a new research area in the theories of numerical computations and difference equations. Based on the work presented in this book, there are many issues that warrant further discussion in the future. For example,

(1) The investigation of the case of $\mathbf{0}_n \not< \alpha \not\leq \mathbf{1}_n$ in Eq. (2.4) is worthwhile.
(2) For other types of nonlocal differential equations, such as the Riemann-Liouville type, the Mittag-Leffler Euler differences should be studied.
(3) The effects of time delay, impulse, and randomness are worth discussing in future research.
(4) From the viewpoint of numerical theory, the Mittag-Leffler Euler differences can be refined into more precise numerical methods, such as the Mittag-Leffler Runge-Kutta, Mittag-Leffler multistep and Mittag-Leffler Rosenbrock type differences.
(5) More dynamical behaviours ought to be discussed, e.g., periodicity, almost periodicity, bifurcation, chaos, control, optimization, etc.
(6) Other models that are governed by the Mittag-Leffler or exponential Euler differences are also relevant for research. For example, population models in mathematical biology, network models in information engineering, and heat equations in physics, etc.



T. Zhang et al., *Mittag-Leffler Euler Difference Techniques*,
https://doi.org/10.1007/978-981-96-6228-9_7

# References


1. Robinson, J.C.: An Introduction to Ordinary Differential Equations. Cambridge University Press, United Kingdom (2004)
2. Tatari, F., Modares, H.: Deterministic and stochastic fixed-time stability of discrete-time autonomous systems. IEEE/CAA J. Auto. Sin. **10**(4), 945–956 (2023)
3. Priyanka, K.S.R., Soundararajan, G., Kashkynbayev, A., Nagamani, G.: Exponential $H_\infty$ synchronization and anti-synchronization of delayed discrete-time complex-valued neural networks with uncertainties. Math. Comput. Simulat. **207**, 301–321 (2023)
4. Haddad, W.M., Lee, J., Bhat, S.P.: Asymptotic and finite-time semistability for nonlinear discrete-time systems with application to network consensus. IEEE Tran. Auto. Cont. **68**(2), 766–781 (2023)
5. Zhang, S., Zhao, J.: Dwell-time-dependent $H_\infty$ bumpless transfer control for discrete-time switched interval type-2 fuzzy systems. IEEE Tran. Fuzzy Syst. **30**(7), 2426–2437 (2022)
6. Zhang, H., Hu, J., Liu, G.P., Yu, X.: Event-triggered secure control of discrete systems under cyber-attacks using an observer-based sliding mode strategy. Inform. Sci. **587**, 587–606 (2022)
7. Wang, J., Zhang, Y., Su, L., Park, J.H., Shen, H.: Model-based fuzzy $l_2$-$l_\infty$ filtering for discrete-time semi-Markov jump nonlinear systems using semi-Markov kernel. IEEE Tran. Fuzzy Syst. **30**(7), 2289–2299 (2022)
8. Oliveira, P.M., Palma, J.M., Lacerda, M.J.: $H_2$ state-feedback control for discrete-time cyber-physical uncertain systems under DoS attacks. Appl. Math. Comput. **425**, 127091 (2022)
9. Naik, P.A., Eskandari, Z., Avazzadeh, Z., Zu, J.: Multiple bifurcations of a discrete-time prey-predator model with mixed functional response. Inter. J. Bifurcat. Chaos **32**(04), 2250050 (2022)
10. Li, F., Wu, Z., Yang, C., Shi, Y., Huang, T., Gui, W.: A novel learning-based asynchronous sliding mode control for discrete-time semi-Markov jump systems. Automatica **143**, 110428 (2022)
11. Lu, H., Xie, X., Peng, C.: Relaxed conditions of observer design of discrete-time Takagi-Sugeno fuzzy systems via a new multi-instant gain-scheduling scheme. IEEE Tran. Fuzzy Syst. **30**(7), 2759–2768 (2022)
12. Li, X., Mehran, K., Bao, Z.: Membership function, time delay-dependent $\eta$-exponential stabilization of the positive discrete-time polynomial fuzzy model control system. IEEE Tran. Fuzzy Syst. **30**(7), 2197–2209 (2022)
13. Arias, C.F., Blé, G., Falconi, M.: Dynamics of a discrete-time predator-prey system with Holling II functional response. Qual. Theor. Dyn. Syst. **21**, 31 (2022)
14. Chen, Q., Teng, Z., Wang, F.: Fold-flip and strong resonance bifurcations of a discrete-time mosquito model. Chaos Solitons Fract. **144**, 110704 (2021)
15. Butcher, J.C.: Numerical Methods for Ordinary Differential Equations. Wiley, United Kingdom (2016)




T. Zhang et al., *Mittag-Leffler Euler Difference Techniques*,
https://doi.org/10.1007/978-981-96-6228-9

16. Zhang, T.W., Qu, H.Z., Zhou, J.W.: Asymptotically almost periodic synchronization in fuzzy competitive neural networks with Caputo-Fabrizio operator. Fuzzy Set. Syst. **471**, 108676 (2023)
17. Zhang, T.W., Qu, H.Z., Liu, Y.T., Zhou, J.W.: Weighted pseudo $\theta$-almost periodic sequence solution and guaranteed cost control for discrete-time and discrete-space stochastic inertial neural networks. Chaos Soliton. Fract. **173**, 113658 (2023)
18. Zhang, T.W., Li, Z.H.: Switching clusters' synchronization for discrete space-time complex dynamical networks via boundary feedback controls. Pattern Recogn. **143**, 109763 (2023)
19. Mohamad, S., Gopalsamy, K.: Dynamics of a class of discrete-time neural networks and their continuous-time counterparts. Math. Comput. Simul. **53**(1–2), 1–39 (2000)
20. Bossy, M., Jabir, J.F., Martinez, K.: On the weak convergence rate of an exponential Euler scheme for SDEs governed by coefficients with superlinear growth. Bernoulli **27**(1), 312–347 (2021)
21. Hu, P., Huang, C.M.: Delay dependent asymptotic mean square stability analysis of the stochastic exponential Euler method. J. Comput. Appl. Math. **382**, 113068 (2021)
22. Hochbruck, M., Ostermann, A.: Exponential Runge-Kutta methods for parabolic problems. Appl. Numer. Math. **53**(2–4), 323–339 (2005)
23. Calvo, M.P., Palencia, C.: A class of explicit multistep exponential integrators for semilinear problems. Numer. Math. **102**(3), 367–381 (2006)
24. Ostermann, A., Thalhammer, M., Wright, W.M.: A class of exponential general linear methods. BIT Numer. Math. **46**(2), 409–431 (2006)
25. Hochbruck, M., Ostermann, A., Schweitzer, J.: Exponential Rosenbrock-type methods. SIAM J. Numer. Anal. **47**(1), 786–803 (2009)
26. Huang, Z.K., Mohamad, S., Gao, F.: Multi-almost periodicity in semi-discretizations of a general class of neural networks. Math. Comput. Simul. **101**, 43–60 (2014)
27. Zhang, T.W., Han, S.F., Zhou, J.W.: Dynamic behaviours for semi-discrete stochastic Cohen-Grossberg neural networks with time delays. J. Franklin Instit. **357**(17), 13006–13040 (2020)
28. Hu, J., Wang, J.: Global exponential periodicity and stability of discrete-time complex-valued recurrent neural networks with time-delays. Neural Netw. **66**, 119–130 (2015)
29. Huang, Z.K., Wang, Y.H., Gao, F.: The existence and global attractivity of almost periodic sequence solution of discrete-time neural networks. Phys. Lett. A **350**(3–4), 182–191 (2006)
30. Han, S.F., Liu, G.X., Zhang, T.W.: Mean almost periodicity and moment exponential stability of semi-discrete random cellular neural networks with fuzzy operations. Plos One **14**(8), e0220861 (2019)
31. Abbas, S., Xia, Y.H.: Existence and attractivity of $k$-almost automorphic sequence solution of a model of cellular neural networks with delay. Acta Math. Sci. **33**(1), 290–302 (2013)
32. Abbas, S.: Existence and attractivity of $k$-pseudo almost automorphic sequence solution of a model of bidirectional neural networks. Acta Appl. Math. **119**(1), 57–74 (2011)
33. Naik, P.A., Eskandari, Z., Yavuz, Z., Zu, J.: Complex dynamics of a discrete-time Bazykin-Berezovskaya prey-predator model with a strong Allee effect. J. Comput. Appl. Math. **413**, 114401 (2022)
34. AlSharawi, Z., Pal, N., Chattopadhyay, J.: The role of vigilance on a discrete-time predator-prey model. Discrete Cont. Dyn. Syst. B **27**(11), 6723–6744 (2022)
35. Eskandari, Z., Alidousti, J., Avazzadeh, Z., Tenreiro Machado, J.A.: Dynamics and bifurcations of a discrete-time prey-predator model with Allee effect on the prey population. Ecol. Complex. **48**, 100962 (2021)
36. Chou, Y.H., Chow, Y., Hu, X., Jang, S.R.J.: A Ricker-type predator-prey system with hunting cooperation in discrete time. Math. Comput. Simul. **190**, 570–586 (2021)
37. Guo, B.L., Pu, X.K., Huang, F.H.: Frational Patial Differential Equations and Their Numerical Solutions. Science Press, Beijing (2011)
38. Diethelm, K., Freed, A.D.: On the solution of nonlinear fractional differential equations used in the modeling of viscoplasticity. In: Keil, F., Mackens, W., Voß, H., Werther, J. (eds.) Scientific Computing in Chemical Engineering II. Springer, Berlin, Heidelberg (1999)

39. Diethelm, K., Ford, N.J., Freed, A.D.: Detailed error analysis for a fractional Adams method. Numer. Algorithms **36**(1), 31–52 (2004)
40. Diethelm, K., Luchko, Y.: Numerical solution of linear multi-term initial value problems of fractional order. J. Comput. Anal. Appl. **6**, 243–263 (2004)
41. Ford, N.J., Connolly, J.A.: Comparison of numerical methods for fractional differential equations. Commun. Pur. Appl. Anal. **5**(2), 289–307 (2006)
42. Kovács, M., Larsson, S., Saedpanah, F.: Mittag-Leffler Euler integrator for a stochastic fractional order equation with additive noise. SIAM J. Numer. Anal. **58**(1), 66–85 (2020)
43. Doan, T.S., Huong, P.T., Kloeden, P.E., Vu, A.M.: Euler-Maruyama scheme for Caputo stochastic fractional differential equations. J. Comput. Appl. Math. **380**, 112989 (2020)
44. Garrappa, R., Popolizio, M.: Generalized exponential time differencing methods for fractional order problems. Comput. Math. Appl. **62**(3), 876–890 (2011)
45. Garrappa, R., Popolizio, M.: On accurate product integration rules for linear fractional differential equations. J. Comput. Appl. Math. **235**, 1085–1097 (2011)
46. Kilbas, A.A., Srivastava, H.M., Trujillo, J.J.: Theory and Applications of Fractional Differential Equations. Elsevier, Boston (2006)
47. Jin, B.T.: Fractional Differential Equations: An Approach via Fractional Derivatives. Springer, Switzerland (2021)
48. Wang, J.R., Fečkan, M., Zhou, Y.: Presentation of solutions of impulsive fractional Langevin equations and existence results. Eur. Phys. J. Spec. Top. **222**(8), 1857–1874 (2013)
49. Zhang, T.W., Xiong, L.L.: Periodic motion for impulsive fractional functional differential equations with piecewise Caputo derivative. Appl. Math. Lett. **101**, 106072 (2020)
50. Zhang, T.W., Zhou, J.W., Liao, Y.Z.: Exponentially stable periodic oscillation and Mittag-Leffler stabilization for fractional-order impulsive control neural networks with piecewise Caputo derivatives. IEEE Tran. Cybern. **52**(9), 9670–9683 (2022)
51. Caputo, M., Fabrizio, M.: A new definition of fractional derivative without singular kernel. Progr. Fract. Differ. Appl. **1**(2), 73–85 (2015)
52. Ortigueira, M.D., Tenreiro Machado, J.: A critical analysis of the Caputo-Fabrizio operator. Commun. Nonlinear Sci. Numer. Simulat. **59**, 608–611 (2018)
53. Yépez-Martínez, H., Gómez-Aguilar, J.F.: A new modified definition of Caputo-Fabrizio fractional-order derivative and their applications to the Multi Step Homotopy Analysis Method (MHAM). J. Comput. Appl. Math. **346**, 247–260 (2019)
54. Losada, J., Nieto, J.J.: Properties of a new fractional derivative without singular kernel. Progr. Fract. Differ. Appl. **1**(2), 87–92 (2015)
55. Mozyrska, D., Torres, D.F.M., Wyrwas, M.: Solutions of systems with the Caputo-Fabrizio fractional delta derivative on time scales. Nonlinear Anal. Hyb. Syst. **32**, 168–176 (2019)
56. Bochner, S.: Curvature and Betti numbers in real and complex vector bundles, Univ. e Politec. Torino Rend. Sem. Mat. **15**, 225–253 (1955–1956)
57. Bochner, S.: A new approach to almost periodicity. Proc. Nat. Acad. Sci. USA **48**, 2039–2043 (1962)
58. Bochner, S.: Continuous mappings of almost automorphic and almost periodic functions. Proc. Nat. Acad. Sci. USA **52**, 907–910 (1964)
59. Diagana, T.: Almost Automorphic Type and Almost Periodic Type Functions in Abstract Spaces. Springer, New York (2013)
60. Affili, E., Valdinoci, E.: Decay estimates for evolution equations with classical and fractional time-derivatives. J. Differ. Eqn. **266**(7), 4027–4060 (2019)
61. Baeumer, B., Kovács, M., Sankaranarayanan, H.: Fractional partial differential equations with boundary conditions. J. Differ. Eqn. **264**(2), 1377–1410 (2018)
62. Feckan, M., Marynets, K., Wang, J.R.: Periodic boundary value problems for higher-order fractional differential systems. Math. Meth. Appl. Sci. **42**(10), 3616–3632 (2019)
63. Samko, S.G., Kilbas, A.A., Marichev, O.I.: Fractional Integrals and Derivatives: Theory and Applications. Gordon and Breach Science Publishers, Yverdon (1993)
64. Alessandretti, A., Pequito, S., Pappas, G.J., Aguiar, A.P.: Finite-dimensional control of linear discrete-time fractional-order systems. Automatica **115**, 108512 (2020)

65. Shahamatkhah, E., Tabatabaei, M.: Leader-following group consensus of discrete-time fractional-order double-integrator multi-agent systems. ISA Tran. **106**, 262–270 (2020)
66. Shahamatkhah, E., Tabatabaei, M.: Containment control of linear discrete-time fractional-order multi-agent systems with time-delays. Neurocomputing **385**, 42–47 (2020)
67. Mozyrska, D., Oziablo, P., Wyrwas, M.: Stability of fractional variable order difference systems. Fract. Calc. Appl. Anal. **22**(3), 807–824 (2019)
68. Chen, J.F.: Theory of Fractional-Order Difference Equations. Xiamen University Press, Xiaman (2011)
69. Abdeljawad, T.: On Riemann and Caputo fractional differences. Comput. Math. Appl. **62**(3), 1602–1611 (2011)
70. Baleanu, D., Wu, G.C.: Some further results of the Laplace transform for variable-order fractional difference equations. Fract. Calc. Appl. Anal. **22**(6), 1641–1654 (2019)
71. Goodrich, C., Peterson, A.: Discrete Fractional Calculus. Springer, New York (2015)
72. Wei, Y.D., Wei, Y.H., Chen, Y.Q., Wang, Y.: Mittag-Leffler stability of Nabla discrete fractional-order dynamic systems. Nonlinear Dyn. **101**(1), 407–417 (2020)
73. You, X.X., Song, Q.K., Zhao, Z.J.: Global Mittag-Leffler stability and synchronization of discrete-time fractional-order complex-valued neural networks with time delay. Neural Netw. **122**, 382–394 (2020)
74. Franco-Pérez, L., Fernández-Anaya, G., Quezada-Téllez, L.A.: On stability of nonlinear nonautonomous discrete fractional Caputo systems. J. Math. Anal. Appl. **487**(2), 124021 (2020)
75. Ouannas, A., Khennaoui, A.A., Momani, S., Pham, V.T., El-Khazali, R.: Hidden attractors in a new fractional-order discrete system: chaos, complexity, entropy, and control. Chin. Phys. B **29**(5), 050504 (2020)
76. Huang, L.L., Park, J.H., Wu, G.C., Mo, Z.W.: Variable-order fractional discrete-time recurrent neural networks. J. Comput. Appl. Math. **370**, 112633 (2020)
77. Xiang, M.Q., Zhang, B.L.: Homoclinic solutions for fractional discrete Laplacian equations. Nonlinear Anal. **198**, 111886 (2020)
78. Ciaurri, Ó., Roncal, L., Stinga, P.R., Torrea, J.L., Varona, J.L.: Nonlocal discrete diffusion equations and the fractional discrete Laplacian, regularity and applications. Adv. Math. **330**, 688–738 (2018)
79. Ortigueira, M.D., Tenreiro Machado, J.A.: New discrete-time fractional derivatives based on the bilinear transformation: definitions and properties. J. Adv. Res. **25**, 1–10 (2020)
80. Suwan, I., Owies, S., Abdeljawad, T.: Fractional $h$-differences with exponential kernels and their monotonicity properties. Math. Meth. Appl. Sci. **44**(10), 8432–8446 (2021)
81. He, J.W., Lizama, C., Zhou, Y.: The Cauchy problem for discrete time fractional evolution equations. J. Comput. Appl. Math. **370**, 112683 (2020)
82. Du, F.F., Jia, B.G.: Finite-time stability of a class of nonlinear fractional delay difference systems. Appl. Math. Lett. **98**, 233–239 (2019)
83. Allouch, N., Graef, J.R., Hamani, S.: Boundary value problem for fractional $q$-difference equations with integral conditions in Banach spaces. Fractal Fract. **6**(5), 237 (2022)
84. Wang, M., Du, F.F., Chen, C.R., Jia, B.G.: Asymptotic stability of $(q, h)$-fractional difference equations. Appl. Math. Comput. **349**, 158–167 (2019)
85. Mozyrska, D., Girejko, E.: Overview of fractional $h$-difference operators. In: Almeida, A., Castro, L., Speck, FO. (eds) Advances in Harmonic Analysis and Operator Theory. Operator Theory: Advances and Applications, vol. 229. Birkhäuser, Basel (2013)
86. Wang, X.M., Dong, X.G.: A Concise Tutorial on Computational Methods. Science Press, Beijing (2010)
87. Wu, G.C., Zeng, D.Q., Baleanu, D.: Fractional impulsive differential equations: exact solutions, integral equations and short memory case. Fract. Calc. Appl. Anal. **22**(1), 180–192 (2019)
88. Wu, G.C., Deng, Z.G., Baleanu, D., Zeng, D.Q.: New variable-order fractional chaotic systems for fast image encryption. Chaos **29**(8), 083103 (2019)

89. Liu, P., Zeng, Z.G., Wang, J.: Global synchronization of coupled fractional-order recurrent neural networks. IEEE Tran. Neural Netw. Lear. Syst. **30**(8), 2358–2368 (2019)
90. Fěckan, M., Wang, J.R.: Periodic impulsive fractional differential equations. Adv. Nonlinear Anal. **8**(1), 482–496 (2019)
91. Dai, X.J., Xiao, A.G.: A note on Euler method for the overdamped generalized Langevin equation with fractional noise. Appl. Math. Lett. **111**, 106669 (2021)
92. Yang, Z.W., Zheng, X.C., Zhang, Z.Q., Wang, H.: Strong convergence of a Euler-Maruyama scheme to a variable-order fractional stochastic differential equation driven by a multiplicative white noise. Chaos Solitons Fract. **142**, 110392 (2021)
93. Sharma, R.K., Xiong, X., Kamal, S., Ghosh, S.: Discrete-time super-twisting fractional-order differentiator with implicit Euler method. IEEE Tran. Circuit. Syst. II **68**(4), 1238–1242 (2021)
94. Malinowska, A.B., Mozyrska, D.: Sajewski, Ł: Advances in Non-integer Order Calculus and Its Applications. Springer, Switzerland (2020)
95. Bohner, M., Peterson, A.: Dynamic Equations on Time Scales: An Introduction with Applications. Birkhäuser, Boston (2001)
96. Cheban, D., Liu, Z.X.: Periodic, quasi-periodic, almost periodic, almost automorphic, Birkhoff recurrent and Poisson stable solutions for stochastic differential equations. J. Differ. Eqn. **269**(4), 3652–3685 (2020)
97. Li, Y.K., Wang, Y.L., Li, B.: Existence and finite-time stability of a unique almost periodic positive solution for fractional-order Lasota-Wazewska red blood cell models. Inter. J. Biomath. **13**(8), 2050013 (2020)
98. Liu, Z.X., Sun, K.: Almost automorphic solutions for stochastic differential equations driven by Lévy noise. J. Funct. Anal. **266**(3), 1115–1149 (2014)
99. Li, Y.K., Shen, S.P.: Almost automorphic solution of quaternion-valued BAM neural networks with time-varying delays on time scales. J. Intell. Fuzzy Syst. **37**(1), 669–685 (2019)
100. Zhang T.W., Li Y.K.: Global exponential stability of discrete-time almost automorphic Caputo-Fabrizio BAM fuzzy neural networks via exponential Euler technique. Knowl.-Based Syst. **246**, 108675 (2022)
101. Cohen, M.A., Grossberg, S.: Absolute stability and global pattern formation and parallel memory storage by competitive neural networks. IEEE Tran. Syst. Man Cybern.: Syst. **13**(5), 815–826 (1983)
102. Hu, L.Y., Ren, Y., Yang, H.J.: Exponential synchronization of stochastic Cohen-Grossberg neural networks driven by $G$-Brownian motion. Neurocomputing **350**, 13–19 (2019)
103. Li, R.X., Cao, J.D., Alsaedi, A., Ahmad, B., Alsaadi, F.E., Hayat, T.: Nonlinear measure approach for the robust exponential stability analysis of interval inertial Cohen-Grossberg neural networks. Complexity **21**(S2), 459–469 (2016)
104. Li, Y.K., Zhang, T.W.: Global exponential stability and existence of anti-periodic solutions to impulsive Cohen-Grossberg neural networks on time scales. Topol. Meth. Nonlinear Anal. **45**(2), 363–384 (2015)
105. Rao S.B., Zhang T.W., Xu L.J.: Exponential stability and synchronisation of fuzzy Mittag-Leffler discrete-time Cohen-Grossberg neural networks with time delays. Inter. J. Syst. Sci. **53**(11), 2318–2340 (2022)
106. Zadeh, L.A.: Fuzzy Sets. Inf. Control **8**(3), 338–353 (1965)
107. Yang, T., Yang, L.B.: The global stability of fuzzy cellular neural networks. IEEE Tran. Circuit. Syst. I **43**(10), 880–883 (1996)
108. Abdelaziz, M., Chérif, F.: Piecewise asymptotic almost periodic solutions for impulsive fuzzy Cohen-Grossberg neural networks. Chaos Solitons Fract. **132**, 109575 (2020)
109. Zhang, Q.H., Yang, L.H., Liu, J.Z.: Existence and stability of anti-periodic solutions for impulsive fuzzy Cohen-Grossberg neural networks on time scales. Math. Slovaca **64**(1), 119–138 (2014)
110. Pecora, L.M., Carroll, T.L.: Synchronization in chaotic system. Phys. Rev. Lett. **64**(8), 821–824 (1990)

111. Chen, L., Yin, H., Huang, T., Yuan, L., Zheng, S., Yin, L.: Chaos in fractional-order discrete neural networks with application to image encryption. Neural Netw. **125**, 174–184 (2020)
112. Lakshmanan, S., Prakash, M., Lim, C.P., Rakkiyappan, R., Balasubramaniam, P., Nahavandi, S.: Synchronization of an inertial neural network with time-varying delays and its application to secure communication. IEEE Tran. Neural Netw. Lear. Syst. **29**(1), 195–207 (2018)
113. Cheng, S., Ji, J.C., Zhou, J.: Fast synchronization of directionally coupled chaotic systems. Appl. Math. Model. **37**(1–2), 127–136 (2013)
114. Zheng, M.W., Li, L.X., Peng, H.P., Xiao, J.H., Yang, Y.X., Zhang, Y.P., Zhao, H.: Finite-time stability and synchronization of memristor-based fractional-order fuzzy cellular neural networks. Commun. Nonlinear Sci. Numer. Simulat. **59**, 272–291 (2018)
115. Kong, F., Rajan, R.: Finite-time and fixed-time synchronization control of discontinuous fuzzy Cohen-Grossberg neural networks with uncertain external perturbations and mixed time delays. Fuzzy Sets Syst. **411**, 105–135 (2021)
116. Shi, Y.C., Cao, J.D.: Finite-time synchronization of memristive Cohen-Grossberg neural networks with time delays. Neurocomputing **377**, 159–167 (2020)
117. Peng, D.X., Li, J.X., Xu, W., Li, X.D.: Finite-time synchronization of coupled Cohen-Grossberg neural networks with mixed time delays. J. Franklin Instit. **357**(16), 11349–11367 (2020)
118. Aouiti, C., Assali, E.A., El Foutayeni, Y.: Finite-time and fixed-time synchronization of inertial Cohen-Grossberg-type neural networks with time varying delays. Neural Process. Lett. **50**(3), 2407–2436 (2019)
119. Chen, Y.X., Song, Q.K., Alsaadi, F.E.: Global Mittag-Leffler stability for fractional-order quaternion-valued neural networks with piecewise constant arguments and impulses. Inter. J. Syst. Sci. **53**(8), 1756–1768 (2022)
120. Zhang, F.H., Zeng, Z.G.: Multiple Mittag-Leffler stability of delayed fractional-order Cohen-Grossberg neural networks via mixed monotone operator pair. IEEE Tran. Cybern. **51**(12), 6333–6344 (2021)
121. Agarwal, R., Hristova, S., O'Regan, D.: Mittag-Leffler stability for non-instantaneous impulsive Caputo fractional differential equations with delays. Math. Slovaca **69**(3), 583–598 (2019)
122. Gu, C.Y., Zheng, F.X., Shiri, B.: Mittag-Leffler stability analysis of tempered fractional neural networks with short memory and variable-order. Fractals **29**(08), 2140029 (2021)
123. Zhang, T.W., Li, Y.K.: $S$-asymptotically periodic fractional functional differential equations with off-diagonal matrix Mittag-Leffler function kernels. Math. Comput. Simulat. **193**, 331–347 (2022)
124. Stamov, T., Stamova, I.: Design of impulsive controllers and impulsive control strategy for the Mittag-Leffler stability behavior of fractional gene regulatory networks. Neurocomputing **424**, 54–62 (2021)
125. Huang, W.Q., Song, Q.K., Zhao, Z.J., Liu, Y.R., Alsaadi, F.E.: Robust stability for a class of fractional-order complex-valued projective neural networks with neutral-type delays and uncertain parameters. Neurocomputing **450**, 399–410 (2021)
126. Ali, M.S., Narayanan, G., Shekher, V., Alsulami, H., Saeed, T.: Dynamic stability analysis of stochastic fractional-order memristor fuzzy BAM neural networks with delay and leakage terms. Appl. Math. Comput. **369**, 124896 (2020)
127. Zhang, T.W., Li, Y.K., Zhou, J.W.: Mittag-Leffler Euler $\nabla$-differences for Caputo fractional-order systems. Results Phys. **37**, 105482 (2022)
128. Bohner, M., Peterson, A.: Advances in Dynamic Equations on Time Scales. Birkhäuser, Boston (2003)
129. Atanacković, T.M., Pilipović, S., Zorica, D.: Properties of the Caputo-Fabrizio fractional derivative and its distributional settings. Fract. Calc. Appl. Anal. **21**(1), 29–44 (2018)
130. Abbas, M.I.: On the initial value problems for the Caputo-Fabrizio impulsive fractional differential equations. Asian-Eur. J. Math. **14**(05), 2150073 (2020)
131. Ndairou, F., Area, I., Nieto, J.J., Silva, C.J., Torres, D.F.M.: Fractional model of COVID-19 applied to Galicia. Spain and Portugal. Chaos Solitons Fract. **144**, 110652 (2021)

132. Rahman, M.U., Ahmad, S., Matoog, R.T., Alshehri, N.A., Khan, T.: Study on the mathematical modelling of COVID-19 with Caputo-Fabrizio operator. Chaos Solitons Fract. **150**, 111121 (2021)
133. Zhang, T.W., Li, Y.K.: Exponential Euler scheme of multi-delay Caputo-Fabrizio fractional-order differential equations. Appl. Math. Lett. **124**, 107709 (2022)
134. Kosko, B.: Bidirectional associative memories. IEEE Tran. Syst. Man Cybern. **18**(1), 49–60 (1988)
135. Kosko, B.: Adaptive bidirectional associative memories. Appl. Opt. **26**(23), 4947–4960 (1987)
136. Han, S.F., Zhang, T.W., Liu, G.X.: Stochastic dynamics of discrete-time fuzzy random BAM neural networks with time delays. Math. Prob. Eng. **2019**, 9416234 (2019)
137. Wu, W.Q., Yang, L.: Impulsive stochastic BAM neural networks on an invariant under a translation time scale. Acta Appl. Math. **169**(1), 647–665 (2020)
138. Arbi, A., Guo, Y.X., Cao, J.D.: Convergence analysis on time scales for HOBAM neural networks in the Stepanov-like weighted pseudo almost automorphic space. Neural Comput. Appl. **33**(8), 3567–3581 (2020)
139. Sowmiya, C., Raja, R., Zhu, Q., Rajchakit, G.: Further mean-square asymptotic stability of impulsive discrete-time stochastic BAM neural networks with Markovian jumping and multiple time-varying delays. J. Franklin Instit. **356**(1), 561–591 (2019)
140. Ali, M.S., Narayanan, G., Shekher, V., Alsaedi, A., Ahmad, B.: Global Mittag-Leffler stability analysis of impulsive fractional-order complex-valued BAM neural networks with time varying delays. Commun. Nonlinear Sci. Numer. Simulat. **83**, 105088 (2020)
141. Ali, M.S., Narayanan, G., Sevgen, S., Shekher, V., Arik, S.: Global stability analysis of fractional-order fuzzy BAM neural networks with time delay and impulsive effects. Commun. Nonlinear Sci. Numer. Simulat. **78**, 104853 (2019)
142. Stamova, I., Stamov, G., Simeonov, S., Ivanov, A.: Mittag-Leffler stability of impulsive fractional-order bi-directional associative memory neural networks with time-varying delays. Tran. Instit. Meas. Control **40**(10), 3068–3077 (2018)
143. Pratap, A., Raja, R., Cao, J.D., Rihan, F.A., Seadawy, A.R.: Quasi-pinning synchronization and stabilization of fractional order BAM neural networks with delays and discontinuous neuron activations. Chaos Solitons Fract. **131**, 109491 (2020)
144. Wang, F., Yang, Y.Q., Xu, X.Y., Li, L.: Global asymptotic stability of impulsive fractional-order BAM neural networks with time delay. Neural Comput. Appl. **28**(2), 345–352 (2017)
145. Wu, A.L., Zeng, Z.G., Song, X.G.: Global Mittag-Leffler stabilization of fractional-order bidirectional associative memory neural networks. Neurocomputing **177**, 489–496 (2016)
146. Veech, W.A.: Almost automorphic functions on groups. Am. J. Math. **87**(3), 719–751 (1965)
147. Veech, W.A.: Almost automorphic functions. Proc. Nat. Acad. Sci. USA **49**(4), 462–464 (1963)
148. Veech, W.A.: On a theorem of Bochner. Ann. Math. **86**(1), 117–137 (1967)
149. Johnson, R.A.: A linear, almost periodic equation with an almost automorphic solution. Proc. Am. Math. Soc. **82**(2), 199–205 (1981)
150. Zaki, M.: Almost automorphic solutions of certain abstract differential equations. Ann. Mat. Pur. Appl. **101**, 91–114 (1974)
151. Zaki, M.: Almost automorphic integrals of almost automorphic functions. Can. Math. Bull. **15**(3), 433–436 (1972)
152. Shen, W.X., Yi, Y.F.: Almost Automorphic and Almost Periodic Dynamics in Skew-product Semiflows. Memoirs of the American Mathematical Society, Providence (1998)
153. Li, Y.K., Wang, X.H., Huo, N.N.: Besicovitch almost automorphic stochastic processes in distribution and an application to Clifford-valued stochastic neural networks. Commun. Nonlinear Sci. Numer. Simulat. **111**, 106488 (2022)
154. Diagana, T.: Existence of globally attracting almost automorphic solutions to some nonautonomous higher-order difference equations. Appl. Math. Comput. **219**(12), 6510–6519 (2013)
155. Lizama, C., Mesquita, J.G.: Almost automorphic solutions of non-autonomous difference equations. J. Math. Anal. Appl. **407**(2), 339–349 (2013)

156. Caraballo, T., Cheban, D.: Almost periodic and almost automorphic solutions of linear differential/difference equations without Favard's separation condition. I. J. Differ. Eqn. **246**(1), 108–128 (2009)
157. Caraballo, T., Cheban, D.: Almost periodic and almost automorphic solutions of linear differential/difference equations without Favard's separation condition. II. J. Differ. Eqn. **246**(3), 1164–1186 (2009)
158. Li, H., Kao, Y.G., Stamova, I., Shao, C.T.: Global asymptotic stability and $S$-asymptotic $\omega$-periodicity of impulsive non-autonomous fractional-order neural networks. Appl. Math. Comput. **410**, 126459 (2021)
159. Qu, H.Z., Zhang, T.W., Zhou, J.W.: Global stability analysis of $S$-asymptotically $\omega$-periodic oscillation in fractional-order cellular neural networks with time variable delays. Neurocomputing **399**, 390–398 (2020)
160. Wu, A.L., Zeng, Z.G.: Boundedness, Mittag-Leffler stability and asymptotical $\omega$-periodicity of fractional-order fuzzy neural networks. Neural Netw. **74**, 73–84 (2016)
161. Rakkiyappan, R., Sivaranjani, R., Velmurugan, G., Cao, J.D.: Analysis of global $O(t^{-\alpha})$ stability and global asymptotical periodicity for a class of fractional-order complex-valued neural networks with time varying delays. Neural Netw. **77**, 51–69 (2016)
162. Chen, B.S., Chen, J.J.: Global asymptotical $\omega$-periodicity of a fractional-order non-autonomous neural networks. Neural Netw. **68**, 78–88 (2015)
163. Kaslik E., Sivasundaram S.: Non-existence of periodic solutions in fractional-order dynamical systems and a remarkable difference between integer and fractional-order derivatives of periodic functions. Nonlinear Anal.: RWA **13**(3), 1489–1497 (2012)
164. Wan, P., Sun, D.H., Zhao, M., Zhao, H.: Monostability and multistability for almost-periodic solutions of fractional-order neural networks with unsaturating piecewise linear activation functions. IEEE Tran. Neural Netw. Lear. Syst. **31**(12), 5138–5152 (2020)
165. Zhang S.L., Tang M.L., Liu X.G.: Synchronization of a Riemann-Liouville fractional time-delayed neural network with two inertial terms. Circuit. Syst. Signal Process. **40**, 5280–5308 (2021)
166. Pratap, A., Raja, R., Agarwal, R.P., Cao, J.D.: Stability analysis and robust synchronization of fractional-order competitive neural networks with different time scales and impulsive perturbations. Inter. J. Adapt. Control Signal Process. **33**(11), 1635–1660 (2019)

# Index





T. Zhang et al., *Mittag-Leffler Euler Difference Techniques*,
https://doi.org/10.1007/978-981-96-6228-9

**P**



**R**



**S**



**V**

FSC
www.fsc.org

MIX
Papier aus verantwortungsvollen Quellen
Paper from responsible sources
FSC® C105338

If you have any concerns about our products,
you can contact us on
**ProductSafety@springernature.com**

In case Publisher is established outside the EU,
the EU authorized representative is:
**Springer Nature Customer Service Center GmbH**
**Europaplatz 3, 69115 Heidelberg, Germany**

Printed by Libri Plureos GmbH
in Hamburg, Germany